# Coral Reefs of the World

Volume 13

**Series editors**
Bernhard Riegl, Dania Beach, FL, USA
Richard E. Dodge, Dania Beach, FL, USA

More information about this series at http://www.springer.com/series/7539

Akira Iguchi • Chuki Hongo
Editors

# Coral Reef Studies of Japan

Springer

*Editors*
Akira Iguchi
Department of Bioresources Engineering
National Institute of Technology, Okinawa College
Nago, Okinawa, Japan

Chuki Hongo
Department of Chemistry, Biology, and Marine Science
University of the Ryukyus
Nishihara, Okinawa, Japan

ISSN 2213-719X ISSN 2213-7203 (electronic)
Coral Reefs of the World
ISBN 978-981-10-6471-5 ISBN 978-981-10-6473-9 (eBook)
https://doi.org/10.1007/978-981-10-6473-9

Library of Congress Control Number: 2017960429



Cover illustration: Coral communities of the Yamakawa Reef, Motobu, Okinawa, Japan (photo by Frederic Sinniger) and fringing reefs around Sesoko Island, Okinawa, Japan (photo by Chuki Hongo).

Printed on acid-free paper

This Springer imprint is published by Springer Nature
The registered company is Springer Nature Singapore Pte Ltd.
The registered company address is: 152 Beach Road, #21-01/04 Gateway East, Singapore 189721, Singapore

# Preface

The purpose of this book is to introduce recent important studies on coral reefs in Japan from various research fields (biology, ecology, chemistry, earth science, and conservation studies). To fulfill this objective, we have sought the contributions from young researchers and established experts who are actively working in these fields. Our intention is for this book to contribute to the systematic understanding of coral reef studies of Japan and provide motivation for the conservation and restoration of coral communities.

Akira Iguchi would like to give many thanks to Zoe Richards, David J. Miller, Takashi Nakamura, Kazuhiko Sakai, and Atsushi Suzuki for the advice and fruitful discussions during his coral reef studies. Akira Iguchi also gratefully acknowledges his family and his wife Ai Iguchi for their continued encouragement. Chuki Hongo acknowledges the help given early in his research career by Hajime Kayanne and is grateful to the following colleagues who each have contributed to his growth as a scientist: to Kaoru Sugihara and Marc Humblet for their expertise in the taxonomy of reef-building corals; to Hiroya Yamano, Saki Harii, Yoichi Ide, and Mitsunori Ishihara for collaborating on a study of the current status of coral reefs in Ryukyu Islands; to Kazuhiko Fujita, Ryuji Asami, and Hironobu Kan for collaboration with the study of reef formation history in Ryukyu Islands; to Kazuhisa Goto, Hideki Kawamata, Masashi Watanabe, and Fumihiko Imamura for collaboration with the study of the impact of tropical cyclones and tsunamis to coral reefs in the Ryukyu Islands; to Lucien F. Montaggioni for collaboration with the study of reef formation history in the Western Indian Ocean; and to Denis Wirrmann for collaboration with the study of reef formation history in New Caledonia. Field and laboratory works in Japan and other reef regions have been aided by the support of many others.

Finally, we acknowledge expert reviewers for each chapter: Dr. Zoe Richards, Dr. Shigeki Wada, Dr. Stephen Levas, Dr. Hiroyuki Fujimura, Dr. Sylvain Agostini, Dr. Tsuyoshi Watanabe, Dr. Ryuji Asami, Dr. Jens Zinke, Dr. Yoko Nozawa, Prof. Kazuhiko Sakai, Dr. Ken Okaji, Dr. Marc A. Humblet, Prof. Lucien F. Montaggioni, Dr. Kazuhisa Goto, and other anonymous advisors. We would particularly like to thank editors of Springer Japan for their help and encouragement.

Okinawa, Japan

Akira Iguchi
Chuki Hongo

# Contents

# Introduction for This Book: General Aspects of the Coral Reefs of Japan

1

Akira Iguchi and Chuki Hongo

**Abstract**

Japanese coral reef ecosystems are characterized by high biodiversity. Their inherent beauty attracts many tourists, and they provide various kinds of ecological services such as sustenance, educational opportunities, and coastal protection; however, they are easily impacted by human activities. Japanese reefs range from subtropical reefs at the southern tip of the Ryukyu Archipelago to temperate reefs at Tsushima Island in the far north which marks the limit of coral reef distribution in the Pacific Ocean. Reef development in the region is extensive and diverse, and productive fringing reefs, submerged platforms, and mesophotic reefs have accumulated. A long history research on the coral reef ecosystems in Japan has provided important knowledge on basic sciences and conservation including fundamental aspects of the biology of coral reef organisms but also has broadened our understanding of the functioning and survival of coral reef ecosystems and those occurring at environmental extremities. In this chapter, as an introduction for this book, we synthesize some of the latest multidisciplinary information that is available about the coral reef ecosystems of Japan.

**Keywords**

Japan • Kuroshio Current • Subtropical reef • Temperate reef • Fringing reef

## 1.1 Features of Coral Reefs of Japan

Coral reef ecosystems of Japan are located in the northern hemisphere at the periphery of coral reef growth in the Pacific Ocean. In the Ryukyu Archipelago, coral reefs are present from latitudes 24°N to 31°N and in the Ogasawara Islands from latitudes 26°N to 27°N. Okinotorishima Island (20°N) is a small table reef. The northern limits of coral reef development are found at Iki Island (33°48′N, 129°40′E) and Tsushima Island (34°25′N, 129°16′E) near the mainland of Japan (Yamano et al. 2001, 2012). Coral reefs are formed by calcium carbonate which calcifying organisms produce (such as scleractinian corals, coralline algae, molluscs, foraminifera, and calcareous algae). Zooxanthellate corals (Cnidaria, Anthozoa, Scleractinia) are key ecological engineers of the reef structure. Around coral reef ecosystems of Japan, over 400 coral species have been reported (Veron et al. 2009). This high diversity of coral species is maintained by the warm currents (Kuroshio Current and Tsushima Warm Current; Fig. 1.1) which enables coral species to distribute in higher latitude area and form coral reefs (around 34°N), while southern limits of coral reefs around Australia is located at Lord Howe Island (around 31°S).

The species composition of corals changes between Japan mainland and the Ryukyu Archipelago-Ogasawara Islands. Furthermore, the species composition of corals along the Ryukyu Archipelago also changes from north to south. The composition of present-day coral communities at Ryukyu

A. Iguchi (✉)
Department of Bioresources Engineering, National Institute of Technology, Okinawa College, Nago, Okinawa, Japan
e-mail: iguchi.a0218@gmail.com

C. Hongo
Department of Chemistry, Biology, and Marine Science, University of the Ryukyus, Nishihara, Okinawa, Japan
e-mail: g123001@sci.u-ryukyu.ac.jp


A. Iguchi, C. Hongo (eds.), *Coral Reef Studies of Japan*, Coral Reefs of the World 13,
https://doi.org/10.1007/978-981-10-6473-9_1

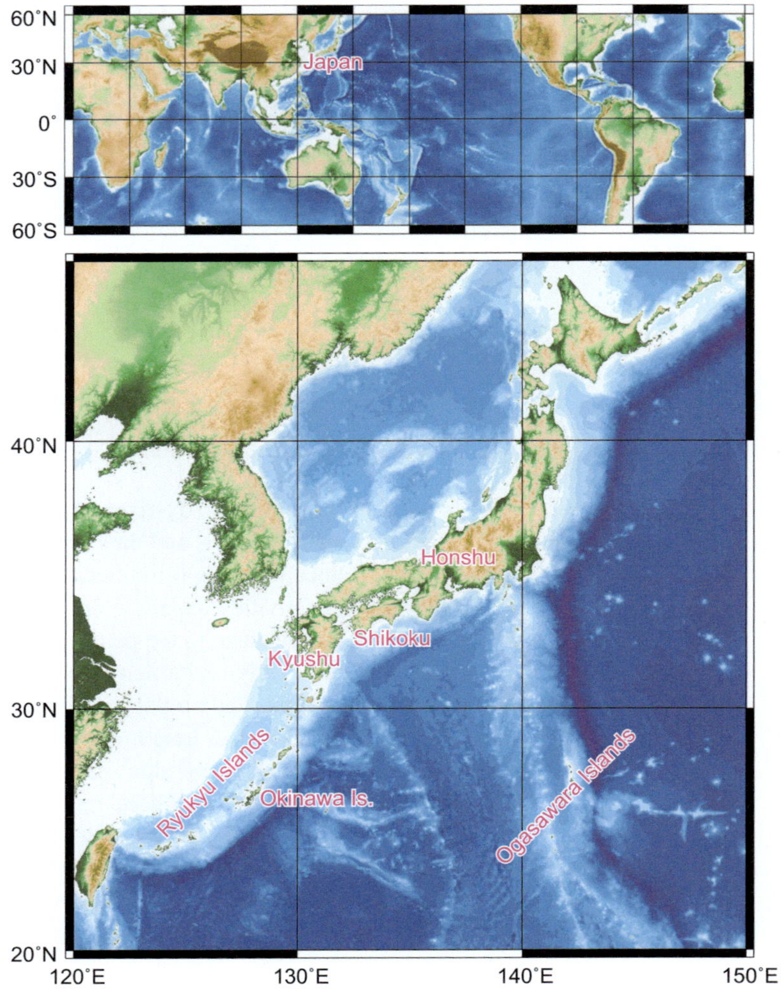


**Fig. 1.1** Map of Japan coastal area along which corals and coral reefs are distributed

Archipelago-Ogasawara Islands is dominated by species from genera *Acropora*, *Montipora*, *Pocillopora*, and *Porites* (The Japanese Coral Reef Society and Ministry of the Environment 2004). However on the temperate reefs of Iki and Tsushima Islands, the coral community is dominated by *Dipsastraea*, *Echinophyllia*, and *Caulastrea* (Sugihara et al. 2009). The occurrence of well-developed coral reefs in temperate areas which is facilitated by the warm Kuroshio and Tsushima Currents is one of the most intriguing features of the coral reefs in Japan.

As may be expected, the number of coral species is lower in the temperate areas than it is in the subtropical locations. In the subtropical locations, typical coral species dominant in the Western Pacific are frequently encountered; however, these species are not encountered in the temperate areas (Fig. 1.2). The converse pattern also occurs, for example, in the case of genus *Acropora* which is most diversified coral taxa and main reef-building corals; *A. solitaryensis* is frequently observed in temperate area (Suzuki and Fukami 2012) but not along the Ryukyu Archipelago. Hence, the seascape around subtropical area is quite different from that around temperate area (Fig. 1.3).

The main region where coral reefs are well developed is along the Ryukyu Archipelago formed by small and stepping stone islands. Fringing reefs are main type of coral reefs along the Ryukyu Archipelago. Thus, coral reef ecosystems along the Ryukyu Archipelago are easily influenced by human impacts compared to barrier reefs. In particular, Okinawajima has high human density, and the land use practices are also high. Thus, coral reef ecosystems around Okinawajima have been relatively devastated by human impacts. In addition, the Ryukyu Archipelago is a subtropical area located between temperate and tropical areas. Thus, seasonal variation of sea surface temperature is high, and the sea surface temperature in winter season is quite low for corals. Therefore, coral species living around the Ryukyu Archipelago are exposed by large environmental changes compared to those in tropical area (Fig. 1.4).

A plethora of knowledge exists about the formation of coral reefs in Japan. In the Ryukyu Archipelago, reef growth

**Fig. 1.2** Some typical *Acropora* species in each subtropical (**a**–**c**) and temperate area (**d**–**f**)

**Fig. 1.3** Seascapes of subtropical and temperate areas. (**a**) Iriomote Island, Okinawa, Japan. (**b**) Goto Islands, Nagasaki, Japan

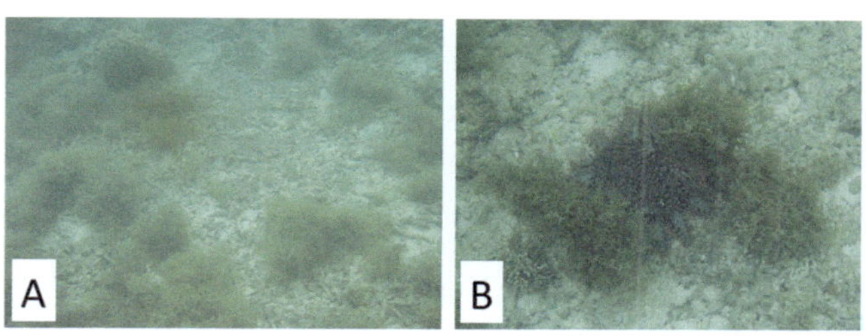


**Fig. 1.4** Seascape of coral reef affected by nutrient enrichment. (**a**) Near Sesoko Island, Okinawa, Japan. (**b**) Crown-of-thorns starfish near Sesoko Island, Okinawa, Japan

started in earliest Quaternary time (1.45–1.65 Ma, 1 Ma=1 million years ago), and that extensive reef formation dates back to −0.8 Ma (Yamamoto et al. 2006). Pleistocene reefs are exposed widely in the Ryukyu Archipelago (e.g., southern Okinawajima, Miyako Islands, and Yaeyama Islands). Holocene reefs are also widely distributed. The Holocene reefs began growing about 10,000 years ago at which time the sea surface temperature in the region had not reached the critical average temperature of 18 °C (Kayanne et al. 2004). From 10,000 to 6,000 years ago, the reefs mainly developed via the accumulation of corals and other calcareous organisms (e.g., calcareous algae) in response to a rapidly sea-level rise (Kan and Kawana 2006; Hongo and Kayanne 2009; Kan 2011). From 6000 to 4000 years ago, the upward reef growth stopped as sea level stabilized (Kan and Kawana 2006; Hongo and Kayanne 2009). After this time period, the reef expanded laterally to form flat shallow zones. Fragments of their skeletons were transported landward to shallow lagoons, and the shallow lagoons were gradually buried by sand and gravel.

The Ryukyu Archipelago is a tectonically active region. The Holocene uplifted reefs have been exposed around some islands such as Kikaijima, Kodakarajima, Okinawajima, and Kumejima Islands (Nakata et al. 1978; Koba et al. 1979, 1982; Takahashi et al. 1988; Webster et al. 1998; Sasaki et al. 1998; Sugihara et al. 2003). In Okinotorishima Island, the coral reef has been maintained throughout the past 7600 years, composed of *Pocillopora*, *Acropora*, *Porites*, *Montastraea*, and the other corals (Kayanne et al. 2012). At the northern limit of coral reefs (Iki and Tsushima Islands), the Holocene reefs consisted of *Cyphastrea*, *Dipsastraea*, *Hydnophora*, *Caulastrea*, *Echinophyllia*, and others (Yamano et al. 2001, 2012) which overlaps with the modern-day community. Hence, the coral reefs of Japan provide unique opportunities to understand geographic reef changes from the past to present and future.

## 1.2 Current and Future Coral Reef Studies of Japan

Coral reefs around Japan are characterized by unique geographical and geological features and have so far provided important knowledge on basic sciences and conservation studies regarding coral reef ecosystems. Coral reef studies are inherent multidisciplinary nature of various research fields. This book brings together all the different disciplines to encourage collaboration and lateral thinking.

Considering the base of high biodiversity in coral reef ecosystems, the study on cycles of matter is essential for understanding the ecological aspects of coral reefs. In Chap. 2, Drs. Yasuaki Tanaka (Universiti Brunei Darussalam) and Ryota Nakajima (Scripps Institution of Oceanography) provide this related topic based on their research experiences. They have so far been involved in the studies on dissolved organic matter (DOM) in coral reefs of Japan. They focus on cycles of matter in coral reefs especially for DOM production removal processes. They also introduce some candidates involved in the DOM production removal processes such as corals, benthic algae, phytoplankton, and bacteria and their roles in the processes. In particular, their main field, Shiraho Reef, Ishigaki Island, Okinawa, Japan, is a fringing reef which is common reef type in Okinawa Islands. Thus, the case study in Shiraho reef would be applicable to other reefs in tropical and subtropical islands which are widely distributed around the Coral Triangle with highest biodiversity in coral reef ecosystems.

After ecological studies above, the studies on the relationship between corals and their responses to changing environments have been introduced. Corals are main important reef builders in coral reef ecosystems and well known to be very sensitive to various environmental chances such as high seawater temperature which has been often dealt in the context of global warming. Dr. Tomihiko Higuchi (University of Tokyo) has so far carried out the studies on stress responses of corals from physiological and chemical aspects. He has published several papers on this topic by focusing on not only high seawater temperature but also other stresses (e.g., addition of $H_2O_2$). In Chap. 3, he introduces the summary of local stressors on corals around Japan. Then, he also introduces studies on stress responses of corals performed in Japan mainly focusing on laboratory experiments. He also reports studies on coral bleaching in Japan which is related not only to high seawater temperature but also to low one assumed at high latitude area in Japan and discusses the physiological aspect of bleaching mechanism.

One of the threats on corals is coral disease of which the number is reported to be increasing. Coral diseases are also known in corals around Japan coastal area, and some researchers are also tackling this issue. Dr. Naohisa Wada (University of Miyazaki), Aki Ohdera (Penn State University), and Dr. Nobuhiro Mano (Nihon University) are now conducting comprehensive approaches targeting coral diseases in Japan. In Chap. 4, they introduce historical perspective of coral diseases (mainly black band disease (BBD), White syndrome (WS), and growth anomalies (GAs)) and current situation on the related studies in Japan. They also provide the information about how coral diseases can be examined both in epidemiological and pathological approaches by including their studies. Based on the information, they provide the future direction of coral disease research in Japan.

Another threat on corals is ocean acidification which has been caused by increasing $CO_2$ from human activities since the industrial revolution. The number of studies on the effects of ocean acidification on corals is rapidly increasing. Some Japanese researchers are also reporting this topic not only from laboratory approaches but also field-based surveys. In Chap. 5, Dr. Shoji Yamamoto who is an expert in conduction ocean acidification provides an introduction to ocean acidification research in Japan. He explains basic concept of ocean acidification from the aspect of carbon chemistry and the impacts on calcifiers and marine ecosystems by citing important studies performed in Japan. He also discusses future direction of ocean acidification research in Japan and proposes what is needed for facilitating this research field.

When we consider the coral responses to environmental changes, we also need to focus on how corals have responded to past environmental changes. In this aspect, earth science approaches are very informative in coral reef studies. In fact, many studies have been carried out to understand past environmental conditions in coral reefs using coral skeletal cores. In Chap. 6, Drs. Sowa Kohki and Tanaka Kentaro provide the content related to coral paleoenvironmental studies. In their part, at first, they introduce the merit of coral skeleton for retrospective archive (coral skeletal growth as a parameter for coral physiological condition, chemical composition of coral skeleton as proxies of past seawater conditions such as sea surface temperature, sea surface salinity, etc.). After explaining the basic logic and the merit of the approach, they introduce previous coral paleoenvironmental studies. They also introduce the case studies around Japan by focusing on the merit for reconstructing past environmental conditions at both global and local scales in this area. Finally, they discuss future direction for coral paleoenvironmental studies in Japan.

We also provide biological studies on corals in Japan. One of the important life histories is reproduction which is essential for the maintenance of organisms. In Chap. 7, Drs. Naoko Isomura and Hironobu Fukami introduce the studies on coral reproductive studies in Japan. One of the famous phenomena of coral reproduction is the synchrony of coral spawning (mainly *Acropora* species). Although *Acropora* species show high synchrony of spawning, it is also reported that there are variations in the spawning patterns of *Acropora* species around high latitude area. In addition, some other coral species also show several geographic variations of spawning patterns around Japan. Based on the past coral reproductive studies including Japanese literatures, they discuss comprehensive patterns of coral spawning events around Japan, which is very informative to understand the evolutionary processes of coral reproduction. They also report the studies on fertilization and hybridization of corals in Japan and their significances. Based on previous and their studies, they provide future perspectives for Japanese coral reproduction studies.

Understanding how coral populations are maintained is one of the important topics to understand future responses of coral reef ecosystems. Thus, the information on how coral populations are connected to each other is essential for understanding the resilience of coral reef ecosystems, and many approaches have been tested in order to establish marine protected areas (MPAs). In the approaches, population genetic analyses using DNA markers have provided important information on the maintenance of coral populations. In Chap. 8, Dr. Yuichi Nakajima, who has been involved in studies on coral population genetics, at first, introduces the merit and the history of DNA markers for population genetic analyses of corals. He also explains some research examples regarding the relationship between reproductive modes of corals and the patterns of gene flow of coral populations. When we use DNA markers, we can obtain the information regarding not only connective patterns among populations but also genetic diversity of each population. The information on genetic diversity is also useful for evaluating recovery potential of coral populations, which has been discussed by citing some examples of coral species including the Japanese corals. Based on the information above, future direction for population genetics of corals in Japan is discussed from some aspects including methodological approaches such as seascape genetics and development of novel molecular markers.

Several threats on corals have been known as described above, and one of them is predation by crown-of-thorns starfish (COTS). In Japan, predation of corals by COTS has been known in most of the coral reef areas, and many trials to remove COTS for protecting corals have been historically performed. In Chap. 9, Dr. Nina Yasuda, who has been involved in COTS studies (mainly population genetic approaches), provides a review of historical COTS distribution and outbreaks in Japan. At first, presumed five species in genus *Acanthaster* and their characteristics are introduced. Then, the studies on reproduction and early life ecology of COTS are described. She also introduces population genetic analysis of COTS in Japan based on her studies including unpublished data and discusses patterns of COTS outbreak. Regarding the outbreak of COTS in Japan, the information for the past outbreak patterns (including the sources in Japanese) and the works by local people and governments to cope with COTS problems are also introduced in detail. In addition, she also proposes the possibility of poleward range expansion of COTS possibly due to the progress of global warming.

There have been already many studies on corals around shallow waters. In recent years, some researchers are focusing on corals around deeper waters (called as mesophotic coral ecosystems (MCEs)). In Chap. 10, Drs. Frederic Sinniger and Saki Harii (the University of the Ryukyus) are conducting comprehensive surveys of mesophotic coral ecosystems around Japan. In the chapter, they introduce a concept of MCEs and the environmental conditions (e.g., light, temperature) including their original measurements. They also explain current situation of studies on mesophotic coral reef ecosystems from some areas along the Ryukyu Archipelago. Not only coral composition but also other fauna in MCEs from these areas are presented in detail. In addition, they provide detailed information on some typical coral species in MCEs and their biological features (e.g., reproduction), which would be intriguing topics in comparison with corals in shallower waters. They also discuss the importance of mesophotic corals for recovery of coral communities and threats on mesophotic coral ecosystems in Japan. They also provide future perspectives for mesophotic research in Japan and the future conservation efforts.

One of important ecological services provided by coral reef ecosystems is the role of natural breakwater of coral reefs, affected by tropical cyclones. In Chap. 11, Dr. Chuki Hongo (the University of the Ryukyus) describes impacts of tropical cyclones to mechanical destruction of corals and change in reef topography. Many researches of the impact of tropical cyclones to coral reefs focus on related wind speeds. However, corals and coral reefs are directly influenced by hydrodynamic impacts. In Japan, an observation network of wave height and wave period has been maintained since 1970s because Japan is especially prone to tropical cyclones in the world. The increase of intensity of tropical cyclone and human impacts will likely cause severe impact of coral reefs at the near future. Based on these aspects, Dr. Hongo proposes future perspective of hydrodynamic impacts of tropical cyclones to coral reefs under the attacking of intensified tropical cyclones and the human impacts for the future. Additionally, Dr. Hongo prospects the strategies of keeping and/or recovery for natural breakwater for the near future.

## References


Hongo C, Kayanne H (2009) Holocene coral reef development under windward and leeward locations at Ishigaki Island, Ryukyu Islands, Japan. Sediment Geol 214:62–73

Kan H (2011) Ryukyu Islands. In: Hopley D (ed) Encyclopedia of modern coral reefs. Springer, Netherlands, pp 940–945

Kan H, Kawana T (2006) 'Catch-up' of a high-latitude barrier reef by back-reef growth during post-glacial sea-level rise, Southern Ryukyus, Japan. In: Proceedings of 10th International Coral Reef Symposium 1: 494–503

Kayanne H, Hongo C, Yamano H (2004) Coral reef landforms in Japan In: The Japanese Coral Reef Society and Ministry of the Environment (eds) Coral reefs of Japan. Ministry of the Environment, Tokyo, pp 14–19

Kayanne H, Hongo C, Okaji K, Ide Y, Hayashibara T, Yamamoto H, Mikami N, Onodera K, Ootsubo T, Takano H, Tonegawa M, Maruyama S (2012) Low species diversity of hermatypic corals on an isolated reef Okinotorishima, in the northwestern Pacific. Galaxea, J Coral Reef Stud 14:73–95

Koba M, Nakata T, Watabe S (1979) Late quaternary reef caps of Takara and Kodakara Islands, Ryukyu archipelago, and sea-level changes of late Holocene. Earth Sci (Chikyu Kagaku) 33:173–191. (In Japanese)
Koba M, Nakata T, Takahashi T (1982) Late Holocene eustatic sea-level changes deduced from geomorphological features and their $^{14}C$ dates in the Ryukyu Islands, Japan. Palaeogeogr Palaeoclimatol Palaeoecol 39:231–260
Nakata T, Takahashi T, Koba M (1978) Holocene-emerged coral reefs and sea-level changes in the Ryukyu Islands. Geogr Rev Japan 51–2:87–108
Sasaki K, Omura A, Ota Y, Murase T, Azuma T, Kobayashi M, Ikura K (1998) Holocene regressive coral reef terraces at the northern Shidooke coast of Kikai Island, central Ryukyus. Quat Res 37:349–360. (In Japanese)
Sugihara K, Nakamori T, Iryu Y, Sasaki K, Blanchon P (2003) Holocene sea-level change and tectonic uplift deduced from raised reef terraces, Kikai-jima, Ryukyu Islands, Japan. Sediment Geol 159:5–25
Sugihara K, Sonoda N, Imafuku T, Nagata S, Ibusuki T, Yamano H (2009) Latitudinal changes in hermatypic coral communities from west Kyushu to Oki Islands in Japan. J Jpn Coral Reef Soc 11:51–67
Suzuki G, Fukami H (2012) Evidence of genetic and reproductive isolation between two morphs of subtropical-dominant coral *Acropora solitaryensis* in the non-reef region of Japan. Zool Sci 29:134–140
Takahashi T, Koba M, Kan H (1988) Relationship between reef growth and sea level on the Northwest of Kume Island, the Ryukyus: data from drill holes on the Holocene coral reef. In: Proceedings of 10th international coral reef symposium 3: 491–496
The Japanese Coral Reef Society and Ministry of the Environment (2004) Coral reefs of Japan. Ministry of the Environment, Tokyo
Veron JEN, Devantier LM, Turak E, Green AL, Kininmonth S, Stafford-Smith M, Peterson N (2009) Delineating the coral triangle. Galaxea, J Coral Reef Stud 11:91–100
Webster JM, Davies PJ, Konishi K (1998) Model of fringing reef development in response to progressive sea level fall over the last 7000 years – (Kikai-jima, Ryukyu Islands, Japan). Coral Reefs 17:289–308
Yamamoto K, Iryu Y, Sato T, Chiyonobu S, Sagae K, Abe E (2006) Responses of coral reefs to increased amplitude of sea-level changes at the mid-Pleistocene climate transition. Palaeogeogr Palaeoclimatol Palaeoecol 241:160–175
Yamano H, Hori K, Yamauchi M, Yamagawa O, Ohmura A (2001) Highest-latitude coral reef at Iki Island, Japan. Coral Reefs 20:9–12
Yamano H, Sugihara K, Watanabe T, Shimamura M, Hyeong K (2012) Coral reefs at 34 N, Japan: exploring the end of environmental gradients. Geology 40:835–838

# Dissolved Organic Matter in Coral Reefs: Distribution, Production, and Bacterial Consumption

2

Yasuaki Tanaka and Ryota Nakajima

**Abstract**

Dissolved organic matter (DOM) constitutes the largest organic matter pool in coral reef waters and is released and utilized by various coral reef organisms. In this chapter, we review the distribution and fluctuation of DOM concentrations in coral reefs around the world, with a special focus on Shiraho Reef, Ishigaki Island, Okinawa, Japan, where DOM fluxes have been studied most intensively since the late 1990s. Then, we review the DOM production rates from specific reef organisms and DOM consumption rates by bacteria. Previous studies have shown that both dissolved organic carbon and nitrogen (DOC and DON, respectively) generally have a higher concentration in most coral reefs than in the surrounding ocean. At Shiraho Reef, the average ratio of the net DOC production to the net primary production on the reef flat was 18%, and the C:N ratio of DOM that was produced on the reef flat was estimated to be 9.3. The abundance of heterotrophic bacteria was also higher in most coral reefs than offshore, which indicates that bacterial growth was enhanced by reef-derived DOM. Some of the DOC that was produced in coral reefs was persistent to bacterial decomposition in the long term, which suggests that coral reef ecosystems export some reef-derived DOM to the ambient ocean, irrespective of the water residence time on the reef.

**Keywords**

DOM • C:N ratio • Biogeochemical cycles • Nutrients • Coral • Benthic algae • Phytoplankton • Primary production • Decomposition • Refractory organic matter

## 2.1 Introduction

Dissolved organic matter (DOM) in the oceans is one of the largest pools of organic matter on the Earth's surface (Hedges 1992). The amount of carbon (C) in oceanic DOM is estimated to be 700 Gt (Ogawa and Tanoue 2003), which is equivalent to that of atmospheric carbon dioxide ($CO_2$). Although most oceanic DOM is refractory organic matter with a turnover time of 1000–6000 years (Williams and Druffel 1987; Bauer et al. 1992), fresh DOM is continuously produced from various biological and chemical processes (Carlson 2002; Mostofa et al. 2013). In particular, photic zones have a higher DOM concentration in the vertical profile (Ogawa et al. 1999; Carlson 2002), which indicates that primary producers such as phytoplankton are generally major DOM producers in marine ecosystems (Meyers-Schulte and Hedges 1986; Opsahl and Benner 1997).

Marine DOM is usually defined as the organic matter that passes through glass fiber filters (Whatman GF/F, a nominal

Yasuaki Tanaka and Ryota Nakajima contributed equally to this work.

Y. Tanaka (✉)
Faculty of Science, Universiti Brunei Darussalam,
Jalan Tungku Link, Gadong BE1410, Brunei Darussalam
e-mail: yasuaki.tanaka@ubd.edu.bn

R. Nakajima
Scripps Institution of Oceanography, La Jolla, CA, USA


A. Iguchi, C. Hongo (eds.), *Coral Reef Studies of Japan*, Coral Reefs of the World 13,
https://doi.org/10.1007/978-981-10-6473-9_2

pore size 0.7 μm), which are the most widely used for filtration (Ogawa and Tanoue 2003). The organic matter that is collected on these filters is referred to as particulate organic matter (POM). The definitions of DOM and POM are based on its practical size; therefore, DOM does not exactly mean the chemically dissolved phases of organic matter. For example, 22–38% of marine bacteria are not retained on GF/F filters and are counted as DOM (Lee et al. 1995). POM includes a complex mixture of living and nonliving organic matter with a broad size range (Volkman and Tanoue 2002). For example, phytoplankton, zooplankton, large detritus, and some bacteria are included in POM (Blanchot et al. 1989).

In general oceanic surface waters, the concentration of dissolved organic carbon (DOC) ranges from 50 to 100 μmol $L^{-1}$ (μM) (Carlson 2002; Ogawa and Tanoue 2003), which is one order of magnitude higher than that of particulate organic carbon (POC). Approximately 40 μM of DOC consists of refractory compounds, which are ubiquitously distributed in the global oceans and turn over on time scales of centuries to millennia (Williams and Druffel 1987; Bauer et al. 1992), so the remaining 10–60 μM is considered to be a relatively degradable fraction. This degradable fraction can be further categorized into labile and semi-labile fractions, which turn over on time scales of minutes to days and months to years, respectively (Carlson 2002). Labile and semi-labile DOM can be utilized by heterotrophic bacteria (hereafter bacteria), and the energy is transferred to higher trophic levels, such as heterotrophic protists, through microbial food webs (Pomeroy 1974; Azam et al. 1983; Sherr and Sherr 1987). The energy transfer through microbial food webs is a particularly important process in understanding material recycling in aquatic ecosystems (Cho and Azam 1988; Fuhrman et al. 1989; Pavés and González 2008; Dinsdale and Rohwer 2011).

Coral reef ecosystems develop in shallow sea areas of tropical and subtropical climate zones. The surface oceanic seawater in tropical and subtropical zones is generally deficient in nutrients, and this oligotrophic seawater continuously flows in and out of coral reefs. Thus, the growth of phytoplankton in coral reef waters is usually limited by the availability of nutrients (Furnas et al. 2005; Rochelle-Newall et al. 2008). On the other hand, the primary production rates per unit surface area of coral reefs (e.g., 190–640 mmol C $m^{-2}$ $day^{-1}$; Atkinson 2011) have been reported to be the highest among marine ecosystems (Duarte and Cebrián 1996), although most of the primary production of coral reefs is consumed by respiration in the ecosystem (Crossland et al. 1991; Gattuso et al. 1996, 1998; Atkinson and Falter 2003). This high rate of gross primary production in coral reefs is largely attributed to benthic primary producers, the biomass of which is usually much higher than that of the planktonic primary producers in shallow coral reefs (Atkinson and Falter 2003; Atkinson 2011). Therefore, fluctuations in seawater parameters such as DOM are strongly related to benthic biological activities. For example, reef-building corals (Tanaka et al. 2009; Nakajima et al. 2010; Levas et al. 2015) and benthic algae (Haas et al. 2010a, b; Mueller et al. 2014) release DOM into the ambient seawater; conversely, sponges absorb DOM from seawater (Yahel et al. 2003; de Goeij et al. 2008, 2013; Rix et al. 2016). The produced DOM is consumed by bacteria in the reef water and sediment (Ferrier-Pagès et al. 2000; Wild et al. 2004a, b; Tanaka et al. 2011a; Nakajima et al. 2015). Because of the oligotrophic environments of most coral reefs, dissolved organic nitrogen (DON) can be an essential nitrogen (N) source for microorganisms (Seitzinger and Sanders 1999). The fluxes of DOC and DON reflect these benthic and planktonic processes and thus can be used as an important parameter to understand biogeochemical cycles in coral reefs (Wild et al. 2004a; Mari et al. 2007; Miyajima et al. 2007a; Tanaka et al. 2011a, b).

Measuring precise DOC and DON concentrations in seawater had been a challenging task for a long time and has been improved since the late 1990s (Sharp et al. 2002a, b, 2004). However, determining accurate DOC and DON concentrations is still difficult; therefore, detecting small changes in their concentrations is particularly difficult in oligotrophic seawaters, such as those near coral reefs. Among the three major geographical structures of coral reefs (fringing reefs, barrier reefs, and atolls), fringing reefs develop along shorelines and often have a shallow backreef zone called a reef flat (Iryu et al. 1995; Kennedy and Woodroffe 2002). Because the DOM concentration fluctuates to a relatively large extent on a shallow reef flat, fringing reefs are considered to be most suitable for detecting changes in DOM concentrations and studying their fluxes in the reef ecosystem (Hata et al. 2002; Miyajima et al. 2007a; Tanaka et al. 2011a, b; Wyatt et al. 2012; Thibodeau et al. 2013). Other seawater parameters such as dissolved inorganic carbon (DIC) have also been studied to evaluate their fluxes in fringing coral reefs (Kayanne et al. 1995, 2005 Gattuso et al. 1996; Wyatt et al. 2010, 2013; Watanabe et al. 2013).

### 2.1.1 Shiraho Reef

Fringing reefs are commonly seen in the Ryukyu Islands, which are located in the southwestern part of Japan (Iryu et al. 1995). Shiraho Reef, Ishigaki Island, Okinawa, Japan, is a typical fringing reef in which the biogeochemical cycles of DOM have been intensively studied (Hata et al. 2002; Miyajima et al. 2007a; Tanaka et al. 2011a, b). This reef has a well-developed reef flat that extends several hundred meters offshore from the shoreline (Fig. 2.1). On the reef flat of Shiraho Reef, seagrass beds extend from the shoreline to 200–400 m offshore. Many reef-building coral communities are found from approximately 500 to 800 m, and the dominant coral species are *Acropora* spp., *Porites* spp., and

**Fig. 2.1** Aerial photo of Shiraho Reef (Obtained from Tanaka et al. 2011b, with permission from ©Springer). The *symbols* indicate the sampling sites in Tanaka et al. (2011b). CR1, CR2, S1, and S2: coral-dominated areas. SG1 and SG2: seagrass-dominated areas. B: offshore site

*Montipora* spp. (Nakamura and Nakamori 2009). The gap zone between seagrass and corals is mainly covered with bare carbonate sands. The depth of the Shiraho reef flat is approximately 1–3 m and becomes shallower on the reef crest, which forms almost parallel to the shoreline (Fig. 2.1). The reef crest emerges above the sea surface during low tide and thus separates the seawater on the reef flat from the offshore oceanic water. During this low-tide period, the reef flat becomes a semi-closed system, although some seawater is exchanged with the offshore region through channels on the reef crest. Measuring the changes in seawater parameters on the reef flat during this low-tide period can provide information on the biological activities of benthic and planktonic organisms in the coral reef.

Land-derived groundwater with high nitrate ($NO_3^-$) concentrations is continuously discharging into the reef flat at Shiraho Reef (Kawahata et al. 2000; Umezawa et al. 2002a, b). Although most of the seawater on the reef flat contains low $NO_3^-$ concentrations (<1 μM) (Miyajima et al. 2007a, b; Tanaka et al. 2011a, b), the effect of groundwater $NO_3^-$ appeared to accumulate in benthic algae several hundred meters offshore from the shoreline, and the amount of $NO_3^-$ that was discharged from the backland depended on the land use (Umezawa et al. 2002a, b). Many other fringing reefs around the world are also susceptible to adjacent terrestrial ecosystems through groundwater seepage or river discharge (Lapointe et al. 2010; Tedetti et al. 2011).

This chapter reviews DOM studies in coral reefs around the world and provides the present understanding of biogeochemical cycles of DOM. First, the distribution and fluctuation of DOM concentrations in coral reefs are reviewed, with a special focus on Shiraho Reef. Then, DOC production rates from individual organisms, especially corals and benthic algae, are compared among reef organisms to obtain a general range of DOC production rates. Finally, the abundance and growth rate of bacteria in coral reef waters are reviewed to estimate the general DOM consumption rate by pelagic bacteria.

## 2.2 Distribution and Fluctuation of DOM

### 2.2.1 DOM Distribution

The DOM concentrations at Shiraho Reef have been measured since the late 1990s, with the observed concentrations constantly ranging between 49 and 87 μM and between 3.8 and 8.0 μM for DOC and DON, respectively (Hata et al. 2002; Miyajima et al. 2007a; Tanaka et al. 2011a, b) (Table 2.1). These observations suggest that the production and consumption processes of DOM on this coral reef have not changed drastically over the past 10 years. Compared to other coral reefs, Paopao Bay (French Polynesia), Noumea (New Caledonia), and La Saline (Reunion) exhibited similar DOC concentrations on the reef flat or in the lagoon (Table 2.1). On the other hand, much higher DOC concentrations have been observed at Tuamotu Atolls (French Polynesia), Great Astrolabe Reef (Fiji), and Curaçao (Netherlands Antilles) (Table 2.1). DON concentrations were measured at Tuamotu Atolls and Ningaloo Reef (Western Australia), which were higher than those at Shiraho Reef (Table 2.1). There are some possibilities to explain the difference in the DOC and DON concentrations among these study sites: (1) the local environmental conditions (e.g., groundwater seepage, artificial wastewater, and nutrient discharge) could affect organic matter and nutrient concentrations around the reef (Kawahata et al. 2000; Umezawa et al. 2002a, b; Lapointe et al. 2010; Tedetti et al. 2011); (2) the oceanic DOM concentrations outside the reef might affect the baseline concentration in the reef water because the oceanic seawater continuously circulates in and out of the reef (Tanaka et al. 2011b; Watanabe et al. 2013); and (3) the analytical accuracy of the DOC and DON concentrations in the seawater samples might differ between laboratories (Sharp et al. 2002a, b, 2004).

**Table 2.1** Summary of the DOC and DON concentrations in coral reefs

| Site | References | Specific sampling site | DOC (μM) | DON (μM) |
|---|---|---|---|---|
| Japan | | | | |
| Shiraho Reef, Ishigaki Island | Hata et al. (2002) | Reef flat | 60–87 | – |
| | | Offshore | 68 | – |
| | Miyajima et al. (2007a) | Reef flat | 53–76 | 4.0–7.7 |
| | | Offshore | 49–64 | 4.0–8.0 |
| | Tanaka et al. (2011a) | Reef flat | 66–75 | 4.8–5.7 |
| | | Offshore | 57–58 | 3.8 |
| | Tanaka et al. (2011b) | Reef flat | 57–77 | 3.8–5.7 |
| | | Offshore | 57 | 4.0 |
| Fukido Reef, Ishigaki Island | Thibodeau et al. (2013) | Reef flat | – | 4.5–5.5 |
| | | Offshore | – | 3.6–4.5 |
| Bora Bay, Miyako Island | Suzuki et al. (2000) | Reef flat | 64–129 | 5.6–11 |
| | | Offshore | 69–83 | 5.4–6.2 |
| Sesoko, Okinawa Island | Tanaka et al. (2011c) | Reef flat | 62 | 4.0 |
| Pacific Ocean | | | | |
| Tuamotu Atolls, French Polynesia[a] | Torréton et al. (1997) | Lagoon | 105 | – |
| | | Offshore | 87 | |
| | Pagès et al. (1997) | Lagoon | 62–159 | – |
| | | Offshore | 82–86 | – |
| | Torréton et al. (2000) | Lagoon | – | 3.9–15.3 |
| | | Offshore | – | 5.5 |
| | Bouvy et al. (2012) | Lagoon | 73–100 | – |
| | | Offshore | 80–83 | – |
| Paopao Bay, French Polynesia | Nelson et al. (2011) | Reef flat | 63–73 | – |
| | | Offshore | 75–85 | – |
| Northern Line Islands[a] | Dinsdale et al. (2008) | Lagoon | 29–69 | |
| Great Astrolabe Reef, Fiji | Torréton et al. (1997) | Lagoon | 114 | |
| | | Offshore | 102 | |
| Noumea, New Caledonia | Mari et al. (2007) | Lagoon | 62–74 | – |
| | | Offshore | 63 | – |
| | Rochelle-Newall et al. (2008) | Lagoon | 55–99 | – |
| Indian Ocean | | | | |
| Ningaloo Reef, Western Australia | Wyatt et al. (2012) | Reef flat | – | 8–13 |
| | | Offshore | – | 9–11 |
| La Saline, Reunion | Tedetti et al. (2011) | Reef flat | 63–74 | – |
| | | Offshore | 75 | – |
| Atlantic Ocean | | | | |
| Curaçao, Netherlands Antilles | van Duyl and Gast (2001) | Reef flat | 125–200 | – |
| | | Offshore | 125–225 | – |

The concentrations are separated between the interior of the reef and the offshore seawater. The interior seawater is described as either a reef flat (fringing reef) or lagoon (atoll). Some values were visually interpreted from figures

[a]Details from the Tuamotu Atolls and Northern Line Islands are shown in Table 2.4. Some additional data are also found in Table 2.4

The DOC and DON concentrations on the Shiraho reef flat were always higher than those in the offshore waters. Most of the other coral reefs around the world also exhibited higher DOM concentrations on the reef flat or in the lagoon than offshore (Table 2.1). These results indicate that most of the coral reefs that have been studied so far are functioning as a net source of DOM to the ambient open ocean. Contrarily, the DOC concentration at Paopao Bay was lower on the reef flat than offshore (Nelson et al. 2011), which indicates that the reef flat was a net sink of DOC. Whether a coral reef acts as a net sink or source of DOM could be influenced by the baseline concentration of DOM (oceanic DOM outside the reef) and the composition of the organisms in the coral reef (Haas et al. 2016). The baseline concentration of DOM indicates the degradability of DOM in the seawater: higher concentrations of DOM indicate that the seawater has more degradable DOM. When this degradable fraction of oceanic DOM is decomposed faster than the DOM production rate in the reef, the coral reef appears to be a net sink of DOM, as is the case for Paopao Bay (Nelson et al. 2011). When oceanic DOM is mainly composed of refractory organic matter and the decomposition rate is slower than the DOM production rate in the reef, the reef ecosystem appears to be a net source of DOM, as is the case for most other reefs. The DOM production rate depends on the composition of the reef organisms, as shown below.

### 2.2.2 DOM Fluctuation

The diel fluctuation of DOC and DON concentrations in coral reefs has only been observed at Shiraho Reef (Hata et al. 2002; Tanaka et al. 2011b). These studies showed that the concentrations of DOM decreased to the lowest level, which was close to oceanic DOM concentrations, during flood-tide periods (Fig. 2.2). This fluctuation in the DOM concentrations suggests that DOM was not produced in the water column but primarily from benthic communities because the contribution of benthic organisms to the bulk DOM concentration in seawater changes with the water depth (Fig. 2.3). This result was not unexpected because the benthic biomass in shallow coral reefs is generally much higher than that in the water column (Atkinson 2011). For example, if we suppose that the concentration of POC in the water column on the Shiraho reef flat is 5–10 μM (Hata et al. 2002; Miyajima et al. 2007a; Tanaka et al. 2011a, b), the POC per unit surface area corresponds to 10–20 mmol $m^{-2}$ when the water depth on the reef flat is assumed to be 2 m. POC consists of both C in microbes, such as phytoplankton, and C in detritus, so the contribution of planktonic organisms to the bulk POC must be smaller than the estimate (Nakajima et al. 2011). On the other hand, corals, which are a representative benthic organism, have a C biomass of approximately 100–400 μmol $cm^{-2}$ per unit coral skeletal surface area (Muller-Parker et al. 1994; Tanaka et al. 2007, 2009). By simply extending this C biomass to the horizontal squared area, the coral C biomass is calculated to be 1000–4000 mmol $m^{-2}$. This estimate is conservative because corals and other benthic organisms form three-dimensional structures on the sea bottom. The benthic biomass would be at least two orders of magnitude higher than the C biomass of planktonic organisms on the Shiraho reef flat.

Another possible source of DOM from the benthic environment may be groundwater seepage from the adjacent land, which is often reported at fringing reefs worldwide (Kawahata et al. 2000; Umezawa et al. 2002a, b; Lapointe et al. 2010; Tedetti et al. 2011; Cawley et al. 2012). However, if groundwater seepage and/or river water discharge are considerably responsible for DOM fluctuations, a negative correlation between the DOM concentration and salinity should be detected on the reef flat. Such a correlation was not observed at Shiraho Reef (Tanaka et al. 2011b), which indicates that the DOM fluctuation was caused by autochthonous sources.

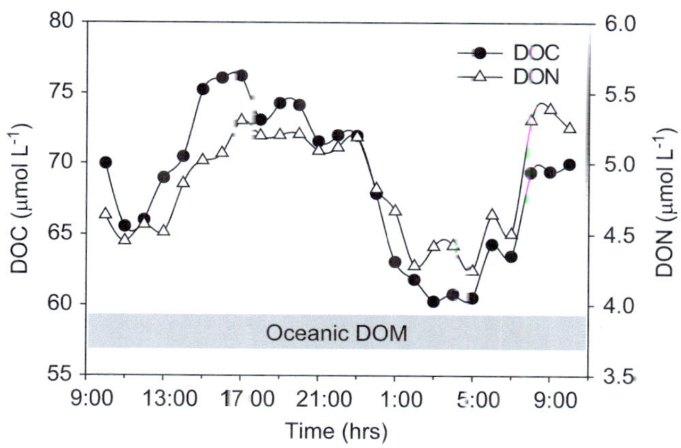


**Fig. 2.2** Diel variations in the DOC and DON concentrations on the reef flat of Shiraho Reef (Obtained and modified from Tanaka et al. 2011b, with permission from ©Springer). DOC: *filled circles*, DON: *open triangles*. The range of oceanic DOM concentrations is shown with *gray* shadows

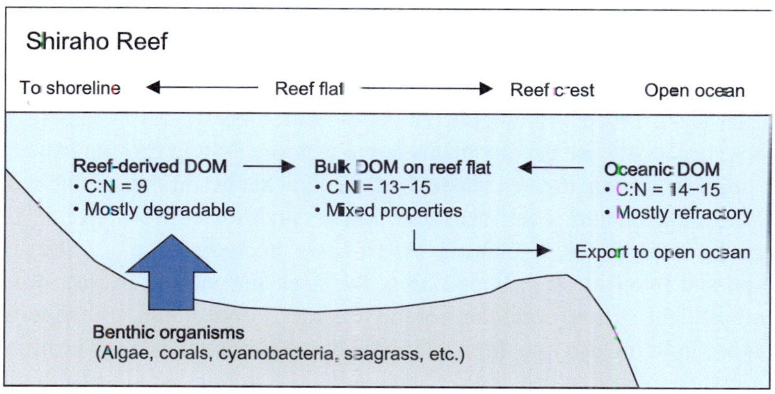


**Fig. 2.3** Diagram of the DOM fluxes at Shiraho Reef. The C:N ratios of DOM were derived from Tanaka et al. (2011a, b)

The observation that the DOM concentrations at Shiraho Reef returned to the offshore oceanic level during high-tide periods indicates that the seawater on the reef flat is almost completely replaced with oceanic waters every day (Tanaka et al. 2011b). In this type of coral reef with a short water residence time, much of DOM that is produced on the reef flat could be exported offshore before this material is decomposed by bacteria (Fig. 2.3). This process actually depends on the water residence time and the degradability of organic matter (Mari et al. 2007). More DOM is exported offshore when the water residence time is shorter and the DOM is less degradable. DOM degradability is discussed later in Subsect. 4.3.

## 2.3 Production Processes of DOM

### 2.3.1 Corals

DOM is released from various organisms and communities in coral reefs. Scleractinian corals have been frequently studied as DOM producers or sometimes as consumers (Crossland 1987; Ferrier-Pagès et al. 1998; Brown and Bythel 2005; Grover et al. 2006, 2008; Tanaka et al. 2009; Nakajima et al. 2010; Naumann et al. 2010; Levas et al. 2015). These organisms continuously release or absorb DOM from the ambient seawater and sometimes release a copious amount of visible mucoid organic matter called coral mucus under stressful conditions such as air exposure (Wild et al. 2004a, b). Several reasons have been proposed to explain the release of DOM and mucus from corals, including excretory pathways for excess organic matter (Davies 1984), defense against sedimentation (Riegl and Branch 1995), and defense against pathogens (Cooney et al. 2002). These functions of coral mucus were reviewed by Brown and Bythel (2005). The production or consumption rate of DOM by corals is usually expressed as a rate that is normalized to the unit area of the coral skeletal surface. The net production rate of DOC showed a very wide range from −22 to 23 mmol $m^{-2}$ $h^{-1}$ (Table 2.2; Fig. 2.4), and the median value was 0.20 mmol $m^{-2}$ $h^{-1}$. This large range of DOC production rates could be caused by various factors, e.g., coral species, the environmental conditions around the coral, and experimental techniques.

When a DOM production rate is measured for corals, these organisms are usually incubated in a closed system for a period by using glass or plastic bottles, and changes in the DOM concentrations are measured. Many conditions differ among these types of cultural experiments, including the type and intensity of provided light, the seawater temperature, and the stirring conditions of the seawater. For example, higher light intensity could increase the photosynthetic rates of endosymbiotic algae (genus *Symbiodinium*) in the coral, which increases the release rates of DOC from the coral colony (Crossland 1987; Naumann et al. 2010). Higher nutrient concentrations in seawater might reduce DOM release rates because the production of endosymbiont cells is promoted with nutrients and less organic matter is available for release (Dubinsky and Berman-Frank 2001; Naumann et al. 2010; Tanaka et al. 2010). Moreover, the concentration of DOM that is available for corals in the initial culture media might affect the net DOM production rates from the corals because these organisms are known to both release and absorb DOM (Grover et al. 2006, 2008; Levas et al. 2013, 2015). The information of these experimental conditions is not available for all studies, so statistically analyzing what factor is most influential to DOM production rates is difficult.

Although absolute release rates of DOM have been reported in many studies as shown in Table 2.2, the relative release rates of DOM to other coral metabolic rates, such as photosynthesis, have scarcely been measured. A few studies showed that 5.4–14% of the net photosynthesis of coral endosymbiotic algae was released as DOC from the corals *Galaxea fascicularis* (Ferrier-Pagès et al. 1998), *Acropora pulchra* (Tanaka et al. 2009), and *Porites lobata* (Haas et al. 2011). The contribution of POC to the TOC that is released from corals was reported to be <40% (Nakajima et al. 2009, 2010), so the ratio of the released TOC to the net photosynthetic rate would be roughly 5–20%. Estimating this type of relative flux would be useful and essential to establish a general concept of coral metabolism and to narrow the extremely wide range of DOM production rates that have been reported.

### 2.3.2 Benthic Algae

Although the number of research articles is still limited, benthic algae have recently been receiving attention as DOM producers in coral reefs (Wild et al. 2010; Haas et al. 2010a, b, 2011, 2013, 2016; Mueller et al. 2014). The net production rates of DOC from macroalgae and turf algae (benthic microalgae) range from −0.50 to 5.5 mmol $m^{-2}$ $h^{-1}$, and the median value is 0.45 mmol $m^{-2}$ $h^{-1}$ (Table 2.2, Fig. 2.4). Haas et al. (2010b) compared the DOC release rates among nine benthic algal species and showed that turf algae (various consortia of the green alga *Cladophora*, red alga *Gelidium*, and associated cyanobacteria assemblages) had the highest DOC release rates (Table 2.2). The release rate was basically activated with elevated temperature and light intensity but inhibited under the highest conditions (28 °C or 500 μmol $m^{-2}$ $s^{-1}$).

The DOC production rates from turf algae and macroalgae (*Amansia rhodantha* and *Turbinaria ornata*) were higher than those from the coral *Porites lobata* (Haas et al. 2011). The ratio of released DOC to the net photosynthetic rate was also higher for benthic algae than the coral: the macroalgae

**Table 2.2** Major DOM producers and their production rates per unit surface area of the producers in coral reefs

| Group | Species | DOC release rate (mmol $m^{-2}$ $h^{-1}$) | References |
|---|---|---|---|
| Corals | *Acropora millepora* | 0.31 | Levas et al. (2015) |
| | *Acropora muricata* (*A. formosa*) | 1.25 | Nakajima et al. (2010) |
| | *Acropora pulchra* | 0.16 | Tanaka et al. (2008a) |
| | *Acropora pulchra* | 0.37 | Tanaka et al. (2009) |
| | *Acropora robusta* (*A. nobilis*) | 2.21 | Nakajima et al. (2009) |
| | *Acropora* sp. | 2.56 | Naumann et al. (2010) |
| | *Euphyllia* sp. | −9.91 to 0.83 (−3.30) | Wild et al. (2012) |
| | *Fungia* sp. | −1.18 | Naumann et al. (2010) |
| | *Goniastrea* sp. | 1.83 | Naumann et al. (2010) |
| | *Madracis mirabilis* | −0.87 | Mueller et al. (2014) |
| | *Manicina* sp. | −13.03 | Haas et al. (2010a) |
| | *Millepora* sp. | 0.77 | Naumann et al. (2010) |
| | *Montipora digitata* | 0.09 | Tanaka et al. (2010) |
| | *Montipora digitata* | 6.74–22.98 (14.68) | Wild et al. (2012) |
| | *Orbicella annularis* | 0.15 | Mueller et al. (2014) |
| | *Pocillopora damicornis* | 0.06 | Haas et al. (2013) |
| | *Pocillopora* sp. | −21.93 | Naumann et al. (2010) |
| | *Porites cylindrica* | 0.14 | Tanaka et al. (2008a) |
| | *Porites lobata* | 0.20 | Haas et al. (2011) |
| | *Porites lobata* | 4.06 | Levas et al. (2013) |
| | *Porites* sp. | 3.17 | Haas et al. (2010a) |
| | *Stylophora* sp. | −1.17 | Naumann et al. (2010) |
| | *Turbinaria reniformis* | 0.25 | Levas et al. (2015) |
| Macroalgae | *Avrainvillea sp.* | −0.50 | Haas et al. (2010a) |
| (Chlorophyta) | *Caulerpa* sp. | 0.56–1.11 (0.83) | Haas et al. (2010b) |
| | *Enteromorpha* | 0.14 | Haas et al. (2010b) |
| | *Halimeda opuntia* | 0.20 | Haas et al. (2011) |
| | *Halimeda* sp. | −0.07 | Haas et al. (2010a) |
| | *Penicillus* sp. | 0.25 | Haas et al. (2010a) |
| | *Rhipocephalus* sp. | 1.01 | Haas et al. (2010a) |
| | *Ulva* sp. | 0.28 | Haas et al. (2010b) |
| (Phaeophyta) | *Dictyota ceylanica* | 0.30 | Haas et al. (2013) |
| | *Hydroclathrus* sp. | 0.40 | Haas et al. (2010b) |
| | *Lobophora* sp. | 0.85 | Haas et al. (2010a) |
| | *Lobophora* sp. | 0.40 | Haas et al. (2010b) |
| | *Sargassum* | 0.47 | Haas et al. (2010b) |
| | *Turbinaria ornata* | 0.50 | Haas et al. (2011) |
| (Rhodophyta) | *Amansia rhodantha* | 0.80 | Haas et al. (2011) |
| | *Hydrolithon reinboldii* | 0.45 | Haas et al. (2011) |
| | *Hydrolithon reinboldii* | 0.10 | Haas et al. (2013) |
| | *Liagora* sp. | 0.40 | Haas et al. (2010b) |
| | *Lithophyllum congestum* | 5.35 | Mueller et al. (2014) |
| | *Peyssonnelia* sp. | 0.24–2.96 (1.0) | Haas et al. (2010b) |
| | Red algae | 1.03 | Haas et al. (2010a) |
| Turf algae | Turf algae | 0.52–5.53 (2.56) | Haas et al. (2010b) |
| | Turf algae | 2.80 | Wild et al. (2010) |
| | Turf algae | 1.4 | Haas et al. (2011) |
| | Turf algae | 0.18 | Haas et al. (2013) |
| Cyanobacteria | *Oscillatoria bonnemaisonii* | 1.36 | Brocke et al. (2015) |
| Seagrass | *Syringodium* sp. | 2.49 | Haas et al. (2010a) |
| | *Thalassia* sp. | 0.18 | Haas et al. (2010a) |

The values in parentheses are the means of several data in the study

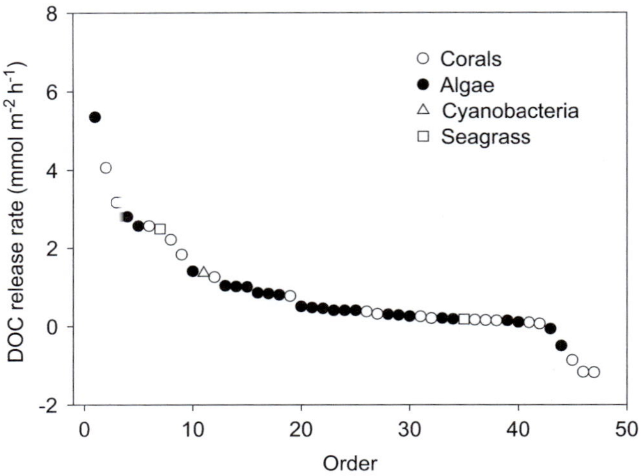


**Fig. 2.4** DOC release rates from benthic organisms in coral reefs. The data in Table 2.2 are ranked in descending order. Four data points (14.68, −3.31, −13.03, and −21.93 mmol $m^{-2}$ $h^{-1}$; see Table 2.2) that lay outside the y-axis range are not shown in the figure

(*A. rhodantha* and *T. ornate*) and turf algae released approximately 20–40% as DOC, while the coral *P. lobata* released 11% (Haas et al. 2011). The ratios of the released DOC to the net photosynthesis of macroalgae in coral reefs were comparable to the ratios for macroalgae in other aquatic ecosystems: for example, the brown alga *Ecklonia cava*, which was collected at Oura Bay (temperate climate zone in Japan), released 18–62% of its net photosynthesis as DOC (Wada et al. 2007). The brown alga *Laminaria hyperborea*, which was collected along the coast of Norway, also released 26% of its net algal production as DOC (Abdullah and Fredriksen 2004). These studies showed a significant contribution of macroalgae to the production and cycling of DOM in aquatic ecosystems, when the macroalgae are one of the dominant benthic communities.

The cover of benthic algae in coral reefs is highly variable and is affected by the environmental conditions. Basically, elevated nutrient concentrations and lower herbivory tend to increase the cover of benthic algae, allowing them to outcompete corals and crustose coralline algae (Smith et al. 1981, 2001, 2005, 2016; Lapointe 1997; Burkepile and Hay 2009; Littler et al. 2010). Macroalgae did not seem to be a dominant benthic community at Shiraho Reef (Nakamura and Nakamori 2009), but turf algae (microphytobenthos) comprised 14–27% of the whole benthic primary production (Suzumura et al. 2002). The cover of macro- and microalgae seemed to be equivalent to what was observed at Kingman Atoll in the Northern Line Islands (20%, Sandin et al. 2008) but seemed to be much lower than what was observed at Tabuaeran and Kiritimati Atolls in the Northern Line Islands (50–70%, Sandin et al. 2008). DOM production rates were not measured in these studies, but the benthic algal communities might considerably contribute to DOM production in the reef (Haas et al. 2011, 2013, 2016).

### 2.3.3 Phytoplankton

Phytoplankton is a major producer of DOM in general open and coastal oceans (Carlson 2002). Coral reefs, particularly fringing coral reefs, often have a relatively shallow water depth, so the contribution of phytoplankton to the primary production or DOM production is considered to be much lower than that of benthic primary producers. The deeper the coral reef is, the greater the contribution of phytoplankton. Charpy-Roubaud et al. (1988) measured the planktonic and benthic primary production rates in Tikehau Atoll (Tuamotu Archipelago, French Polynesia) by using $^{14}C$ and $O_2$ budget methods, respectively, and concluded that the benthic primary production was higher in the area with a depth of 0–10 m, with both production rates being equal at a depth of 10–15 m.

DOM production by phytoplankton has not been measured often in coral reefs. The production rate of DOC by phytoplankton in the lagoon of New Caledonia was approximately 0.005–1.8 mmol $m^{-3}$ $h^{-1}$ under the area's light conditions, including eutrophicated coastal bay stations, and the percentage extracellular release (PER, the percentage of released DOC to the total primary production) varied between 5% and 74% with an average of 35% (Rochelle-Newall et al. 2008). Excluding the coastal bay stations, the DOC production rates of phytoplankton were 0.005–0.4 mmol $m^{-3}$ $h^{-1}$ with a median value of 0.06 mmol $m^{-3}$ $h^{-1}$ (Rochelle-Newall et al. 2008). Strictly comparing this DOC production rate with those of corals and benthic algae from different studies is difficult (Table 2.2) because the rates by these benthos are based on the per unit surface area of the organism and do not consider the reef-scale area (i.e., three-dimensional structure of a sea floor). Nonetheless, if we suppose that a reef flat has a depth of 1 m, the DOC production rate of phytoplankton would be 0.06 mmol $m^{-2}$ $h^{-1}$ according to the median rate in Rochelle-Newall et al. (2008). This DOC production rate constitutes 30% of the median DOC production rate of corals (0.20 mmol $m^{-2}$ $h^{-1}$; Table 2.2) and 13% of benthic algae (0.45 mmol $m^{-2}$ $h^{-1}$; Table 2.2). Phytoplankton at a depth of 25 m in the water column in Takapoto Atoll, French Polynesia, was estimated to produce DOC at 55 mmol $m^{-2}$ $day^{-1}$ (Sakka et al. 2002), which is higher than the DOC production rates of benthic organisms from the unit bottom area.

**Table 2.3** Net primary production (NPP) rates and net DOC production ($NP_{DOC}$) rates at Shiraho Reef

| NPP (mmol $m^{-2}$ $day^{-1}$) | $NP_{DOC}$ (mmol $m^{-2}$ $day^{-1}$) | Season | References |
|---|---|---|---|
| Community scales | | | |
| Microphytobenthos | | | |
| 6.9–24 | – | Mar, Sep | Suzumura et al. (2002) |
| 0.8 | – | Jul, Aug | Nakamura and Nakamori (2009) |
| Seagrass | | | |
| 40 | – | Jul, Aug | Nakamura and Nakamori (2009) |
| Corals | | | |
| 200 | – | Jul, Aug | Nakamura and Nakamori (2009) |
| Reef scales | | | |
| 110 | – | Mar | Kayanne et al. (1995) |
| 130 | – | Sep | Suzuki et al. (1995) |
| 36 ± 12 | 30–36 | Sep | Hata et al. (2002) |
| 96–131 | – | Dec, Mar, Sep | Kayanne et al. (2005) |
| – | 15 ± 57 | Dec, Mar, Jun, Sep | Miyajima et al. (2007a) |
| 11 | – | Jul, Aug | Nakamura and Nakamori (2009) |
| 335–359 | 12–24 | Aug, Sep | Tanaka et al. (2011b) |
| Average of reef-scale studies | | | |
| 125 | 22 | | |

Some values were visually interpreted from figures

### 2.3.4 Production of DOM at Shiraho Reef

The net DOC production rates ($NP_{DOC}$) normalized to the unit horizontal area were reported to be 12–36 mmol $m^{-2}$ $day^{-1}$ at Shiraho Reef (Hata et al. 2002; Miyajima et al. 2007a; Tanaka et al. 2011b; Table 2.3). The DOC production rates at reef scales have never been estimated for other coral reefs around the world but have been measured for subtropical and temperate seagrass-dominated communities, such as 4–25 mmol $m^{-2}$ $day^{-1}$ (Ziegler and Benner 1999), 81 mmol $m^{-2}$ $day^{-1}$ (Barrón et al. 2004), and 2–35 mmol $m^{-2}$ $day^{-1}$ (Barrón and Duarte 2009). These production rates fall within a similar range of DOC production rates that were observed on the reef flat of Shiraho Reef. Considering that $NP_{DOC}$ exceeded the gross primary production rate in the seagrass community (Barrón et al. 2004), a part of the DOM could be released from detritus accumulated in the seagrass community (Smith et al. 1992, Miyajima et al. 1998). When the $NP_{DOC}$ at Shiraho Reef is normalized to a unit water volume, the production rate (12–36 mmol $m^{-3}$ $day^{-1}$) can be compared to general planktonic ecosystems. For example, $NP_{DOC}$ was 0.3 mmol $m^{-3}$ $day^{-1}$ at the surface of the subtropical North Pacific Ocean (Karl et al. 1998) and 0.1–0.7 mmol $m^{-3}$ $day^{-1}$ at a coastal upwelling region of Spain (Marañón et al. 2004). This comparison shows a much higher production rate of DOC on the reef flat, although the total production in the open ocean would become higher when the entire photic zone is considered.

A comparison between the $NP_{DOC}$ and net primary production (NPP) rates shows that the NPP at Shiraho Reef was highly variable compared to $NP_{DOC}$ (Table 2.3). The lowest range of NPP was 11–36 mmol $m^{-2}$ $day^{-1}$ (Hata et al. 2002; Nakamura and Nakamori 2009), which was equivalent to the $NP_{DOC}$ value at the same site. Nakamura and Nakamori (2009) estimated the NPP from a survey on a line transect across the reef flat and described that a large portion of the transect line (64%) was covered by bare carbonate sands, which might have led to the low NPP. Hata et al. (2002) conducted a survey in the summer of 1998, when extensive coral bleaching occurred, so primary production by the coral communities might have not contributed considerably to the entire NPP. These large variations in NPP, which depended on the season, year, and survey area, caused the ratio of $NP_{DOC}$ to NPP to vary greatly from <10% to almost 100% (Table 2.3). By simply averaging the previously obtained data, the $NP_{DOC}$ and NPP at Shiraho Reef were 22 mmol $m^{-2}$ $day^{-1}$ and 125 mmol $m^{-2}$ $day^{-1}$, respectively, and the ratio of $NP_{DOC}$ to NPP was calculated to be 18%. This estimate shows that a considerable amount of DOM is produced on the reef flat and is most likely lost from the reef flat to offshore through drifting seawater.

The DOC and DON fluctuation patterns at Shiraho Reef were consistent (Fig. 2.2), which indicates that the DOM that is produced on the reef flat has a similar chemical composition irrespective of the diel cycle. The relationship between the observed DOC and DON concentrations around the coral reef must be plotted to estimate the C:N ratio of the reef-derived DOM. According to the slope of this relationship, the C:N ratio of the DOM that is produced at Shiraho Reef was estimated to be 9.3 from the surveys in 2008 and 2009 (Fig. 2.5). This C:N ratio for the DOM was slightly higher than that of POM (4.6–7.3; Tanaka et al. 2011a, b) from the same sites at Shiraho Reef. The C:N ratio of bulk DOM on the reef flat (13–15) was slightly lower than that of oceanic DOM (14–15) (Tanaka et al. 2011a, b; Fig. 2.3), which indicates that most of the DOM on the reef flat was derived from the surrounding ocean. Because the bulk DOM on the reef flat was a mixture of oceanic DOM and reef-derived DOM, the properties of the bulk DOM might differ between locations on the reef flat. Similar C:N ratios were also observed at the lagoon of New Caledonia: the C:N ratio of the DOM in the lagoon (12–14) was slightly lower than that of oceanic DOM (15; Mari et al. 2007). These observations indicate that relatively N-rich DOM is produced in coral reefs compared to the ambient open ocean.

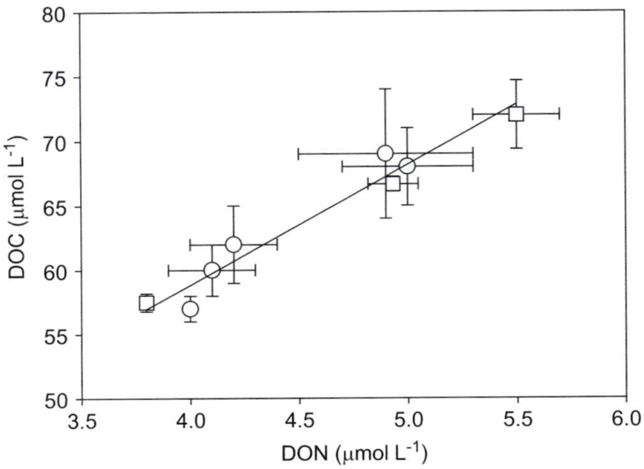


**Fig. 2.5** DOC and DON concentrations at Shiraho Reef, which were measured during the surveys in the summer of 2008 (*squares*) and 2009 (*circles*). The original data were found in Tanaka et al. (2011a, b). A model II regression was fitted to the plot data: $y = 9.3x + 22$ ($r^2 = 0.96$, $p < 0.01$)

### 2.3.5 What Is the Major DOM Producer at Shiraho Reef?

Specifying the major DOM sources in ecosystems is one of the most important and challenging themes in coral reef biogeochemistry. In this subsection, the major DOM sources are estimated for Shiraho Reef as an example study site. Recall that the DOM would mainly be derived from benthic environments, considering the fluctuations in DOM concentrations with the tidal cycle (Fig. 2.2). Because the production rate of DOM was higher at daytime than nighttime (Tanaka et al. 2011b), the production of DOM would be related to photosynthetic activities. The possible DOM producers are mainly corals, seagrass, and benthic macro- or microalgae. Actually, the contribution of these three communities to the total primary production on the reef flat was almost 100% (Nakamura and Nakamori 2009).

Here, we stress that the DOM concentrations were significantly higher at inshore seagrass-dominated sites (SG1 and SG2 in Fig. 2.1) than at offshore coral-dominated sites (CR1 and CR2 in Fig. 2.1) (Tanaka et al. 2011a, b), which suggests that the seagrass-dominated site contributed more to the DOM production on the reef flat. However, the C:N ratio of DOM that was produced on the reef flat was approximately 9.3 (Fig. 2.5), which implies that seagrass and macroalgae are not the major DOM producers because their cells generally have much higher C:N ratios (15–23; Atkinson and Smith 1983; Duarte 1992; Fourqurean et al. 1992). Thus, the remaining possible sources of DOM might be benthic microalgae (turf algae), including cyanobacterial mats, which tend to have similar C:N ratios in their cells to those of the observed DOM, e.g., 7.0 for *Cladophora* sp. (turf algae) and 7.9–9.4 for cyanobacteria (Atkinson and Smith 1983). Benthic cyanobacterial mats were reported to release DOC more than turf algae and contributed to 79% of the total DOC production at the fringing reef of Curaçao (Brocke et al. 2015). At Shiraho Reef, benthic cyanobacteria are extensively observed around the seagrass-dominated zone because of relatively high nutrient concentrations (Blanco et al. 2008) and thus might be a major DOM source on the reef flat alongside other turf algae.

## 2.4 Bacterial Consumption of DOM

The DOM and POM that are produced in coral reefs are mainly removed from the reef by two processes: (1) physical exportation from the reef to the surrounding ocean by water currents (Mari et al. 2007; Wyatt et al. 2010, 2013) and (2) biological decomposition and mineralization in the coral reef (Wild et al. 2010; Haas et al. 2011; Morrow 2011; Tanaka et al. 2011a; Nelson et al. 2013). The former largely depends on the water residence time and hydrodynamics in the reef (Mari et al. 2007; Wyatt et al. 2012). The latter process is mainly driven by heterotrophic bacteria, which rapidly recycle and transfer organic energy to other reef organisms via a microbial loop (Wild et al. 2004a; Naumann et al. 2012; Nakajima et al. 2015). In this section, we focus on the bacterial consumption of DOM in coral reef waters.

### 2.4.1 Distribution of Bacterial Abundance

Bacteria are the most abundant organisms on coral reefs (Rohwer et al. 2002). The abundance of bacteria over coral reefs drastically varies but generally ranges from 2 to 15 × $10^5$ cells $mL^{-1}$, except for some extreme values (Table 2.4). The lowest value (0.7 × $10^5$ cells $mL^{-1}$) was reported in Kingman Atoll, Northern Line Islands, which exhibited no local anthropogenic or terrestrial influence (Dinsdale et al. 2008). In contrast, the highest values, which exceed more than 15 × $10^5$ cells $mL^{-1}$, have often been reported in the lagoons of Tuamotu Atolls, French Polynesia (Torréton and Dufour 1996a, b; Torréton et al. 1997; González et al. 1998; Ferrier-Pagès and Furla 2001; Sakka et al. 2002). The median and average values of bacterial abundance that are observed in coral reef waters around the world are 7.0 × $10^5$ cells $mL^{-1}$ and 9.2 × $10^5$ cells $mL^{-1}$, respectively (Table 2.4).

Several authors have reported that the bacterial abundance is higher in lagoons or reef flats compared to that in the surrounding oceanic waters (Table 2.4). For example, the bacterial abundance in the atoll lagoons of Tuamotu Archipelago was 1.5–5.3 times higher than that in the surrounding ocean (Torréton and Dufour 1996a, b; Torréton et al. 1997). A similar pattern was also found in Great Astrolabe Reef, Fiji, and Majero Atoll, Marshall Islands (Yoshinaga et al. 1991;

**Table 2.4** DOC concentrations, bacterial abundances, and growth rates in various coral reef waters

| Site | References | Specific sampling site | DOC (μM) | Bacterial abundance ($10^5$ cells $ml^{-1}$) | Bacterial growth rate ($day^{-1}$) |
|---|---|---|---|---|---|
| Pacific Ocean | | | | | |
| Bora bay (Miyako Island, Japan) | Ferrier-Pagés and Gattuso (1998) | Reef flat | – | 6.2–13 (7.0) | 1.3–2.8 (2.1) |
| | Casareto et al. (2006) | Reef flat | 70–77 (72) | 1.1–5.8 (3.3) | – |
| Kushimoto (Japan) | Taniguchi et al. (2014) | Reef flat | – | 5.2–8.3 (6.6) | 2.7–6.4 (5.2) |
| Bidong Island (Malaysia) | Nakajima et al. (2013) | Reef flat | 69–73 (71) | 13–14 (14) | 0.72–0.81 (0.76) |
| | | Offshore | 63 | 14 | 0.79 |
| Majuro Atoll (Marshall Islands) | Yoshinaga et al. (1991) | Lagoon | 53–188 (145) | 8.0–14 (11) | 0.04–0.87 (0.35) |
| | | Offshore | 72–143 (109) | 6.2–13 (8.9) | – |
| Ponape Island (Micronesia) | Yoshinaga et al. (1991) | Lagoon | 163–355 (235) | 7.8–16 (11) | 0.17–4.3 (2.5) |
| Lizard Island (GBR, Australia) | Moriarty (1979) | Reef flat/lagoon | – | 1.6–6.1 (3.9) | – |
| | Moriarty et al. (1985) | Reef flat | – | 2.9–6.5 (4.2) | 0.10–6.8 (1.6) |
| | | Reef front | – | 3.7–8.0 (6.1) | 0.05–2.5 (0.68) |
| | | Ocean | – | 3.1–8.0 (6.2) | 0.10–1.5 (0.56) |
| One Tree Island (GBR, Australia) | Linley and Koop (1986) | Reef flat | – | 2.5–6.8 (4.8) | 1.6–4.2 (2.4) |
| Tikehau Atoll (Tuamotu, French Polynesia) | Torréton and Dufour (1996a) | Lagoon | – | 23 | 0.10 |
| | | Reef flat spillway | – | 3.0 | 0.17 |
| | | Offshore | – | 4.5 | 0.051 |
| | Torréton and Dufour (1996b) | Lagoon | – | 18 | 0.18 |
| | Torréton et al. (1997) | Lagoon | 105 | 24 | 0.12 |
| | | Offshore | 87 | 5.0 | 0.090 |
| | González et al. (1998) | Lagoon | – | 12–26 (19) | 0.11 |
| Takapoto Atoll (Tuamotu, French Polynesia) | Torréton and Dufour (1996b) | Lagoon | – | 6.4 | 0.23 |
| | | Offshore | – | 5.0 | 0.10 |
| | Torréton et al. (1997) | Lagoon | 121 | 6.5 | 0.25 |
| | Sakka et al. (2000, 2002) | Lagoon | – | 16 | 0.90 |
| Rangiroa atoll (Tuamotu, French Polynesia) | Ferrier-Pagés and Furla (2001) | Reef flat spillway/channel | – | 4–18 (11) | 0.43–1.4 (0.96) |
| Fakarava atoll (Tuamotu, French Polynesia) | Ferrier-Pagés and Furla (2001) | Reef flat spillway/channel | – | 9–14 (12) | 0.43–0.96 (0.72) |
| Haraiki Atoll (Tuamotu, French Polynesia) | Pagés et al. (1997) and Torréton et al. (2000) | Lagoon | 65–72 (69) | 13–21 (17) | 0.24–0.29 (0.27) |
| Hikueru Atoll (Tuamotu, French Polynesia) | Pagés et al. (1997) and Torréton et al. (2000) | Lagoon | 74–89 (82) | 9.0–9.3 (9.2) | 0.16–0.18 (0.17) |
| Hiti Atoll (Tuamotu, French Polynesia) | Pagés et al. (1997) and Torréton et al. (2000) | Lagoon | 79–80 (80) | 17–24 (21) | 0.10–0.14 (0.12) |
| Kauehi Atoll (Tuamotu, French Polynesia) | Pagés et al. (1997) and Torréton et al. (2000) | Lagoon | 72–78 (75) | 12 | 0.07–0.11 (0.09) |
| Marokau Atoll (Tuamotu, French Polynesia) | Pagés et al. (1997) and Torréton et al. (2000) | Lagoon | 62–73 (68) | 15–16 (15) | 0.070 |
| Nihiru Atoll (Tuamotu, French Polynesia) | Pagés et al. (1997) and Torréton et al. (2000) | Lagoon | 62 | 11–12 (12) | 0.08–0.17 (0.13) |
| Rekareka Atoll (Tuamotu, French Polynesia) | Pagés et al. (1997) and Torréton et al. (2000) | Lagoon | 88–102 (95) | 14–23 (19) | 0.99–1.3 (1.1) |
| Taiaro Atoll (Tuamotu, French Polynesia) | Pagés et al. (1997) and Torréton et al. (2000) | Lagoon | 144–161 (153) | 18–19 (19) | 0.12–0.18 (0.15) |

(continued)

**Table 2.4** (continued)

| Site | References | Specific sampling site | DOC (μM) | Bacterial abundance ($10^5$ cells $ml^{-1}$) | Bacterial growth rate ($day^{-1}$) |
|---|---|---|---|---|---|
| Tekokota Atoll (Tuamotu, French Polynesia) | Pagés et al. (1997) and Torréton et al. (2000) | Lagoon | 61–62 (62) | 2.2–3.0 (2.6) | 0.16–0.22 (0.19) |
| Tepoto Sud Atoll (Tuamotu, French Polynesia) | Pagés et al. (1997) and Torréton et al. (2000) | Lagoon | 64–76 (70) | 6.4–15 (11) | 0.22–0.33 (0.28) |
| Amanu Atoll (Tuamotu, French Polynesia) | Torreton et al. (2002) | Lagoon | – | 9.3 | 0.15 |
| Tuanake Atoll (Tuamotu, French Polynesia) | Torreton et al. (2002) | Lagoon | – | 10 | 0.070 |
| Tuamotu Atolls (Tuamotu, French Polynesia) | Pagés et al. (1997) and Torréton et al. (2000) | Offshore | 64–76 (85) | 6.4–15 (5.5) | 0.028–0.038 (0.033) |
| Great Astrolabe Reef (Fiji) | Torréton et al. (1997) | Lagoon | 114 | 7.5 | 0.28 |
| | | Offshore | 102 | 5.0 | 0.050 |
| | Torréton (1999) | Lagoon | 114 | 7.7 | 0.28 |
| | | Offshore | 110 | 5.1 | 0.033 |
| Maître Island (New Caledonia) | Torréton et al. (2010) | Lagoon | – | 5.5–7.4 (6.4) | 0.12–0.55 (0.28) |
| Kingman Atoll (Northern Line Islands) | Dinsdale et al. (2008) | Offshore | 38–46 (42) | 0.31–0.90 (0.57) | – |
| Palmyra Atoll (Northern Line Islands) | Dinsdale et al. (2008) | Offshore | 47–61 (50) | 1.3–3.5 (2.4) | – |
| Tabuaeran Atoll (Northern Line Islands) | Dinsdale et al. (2008) | Offshore | 42–69 (50) | 2.6–4.6 (3.7) | – |
| Kiritimati Atoll (Northern Line Islands) | Dinsdale et al. (2008) | Offshore | 29–35 (32) | 1.4–8.5 (5.2) | – |
| Atlantic Ocean | | | | | |
| Florida Keys (USA) | Hoch et al. (2008) | Reef flat | 91–130 (104) | 3.4–9.0 (5.5) | 0.57–1.5 (1.1) |
| Dry Tortugas (Florida, USA) | Paul et al. (1986) | Reef flat | – | 2.6–6.3 (4.6) | 0.10–0.99 (0.55) |
| Curaçao (Caribbean Sea) | Gast et al. (1998) | Reef flat | – | 3.3–8.2 (5.7) | 0.13–0.58 (0.26) |
| | Scheffers et al. (2005) | Reef flat | – | 4.4–6.2 (5.3) | 0.11–0.46 (0.20) |
| | van Duyl and Gast (2001) | Reef flat | 142 | – | 0.30 |
| | van Duyl and Gast (2001) | Offshore | 143 | – | 0.13 |
| Belize (Caribbean Sea) | Haas et al. (2016) | | 62–72 (69) | 5.6–6.7 (6.1) | – |
| Tobago (Caribbean Sea) | Haas et al. (2016) | | 47–57 (51) | 6.1–15 (8.5) | – |
| Panama (Caribbean Sea) | Haas et al. (2016) | | 45–52 (48) | 7.0–19 (15) | – |
| Puerto Rico (Caribbean Sea) | Haas et al. (2016) | | 47–54 (50) | 6.4–19 (11) | – |
| Indian Ocean | | | | | |
| Mayotte (Comoros Islands) | Vacelet et al. (1999) | Lagoon | – | 4.3 | – |
| | Houlbrèque et al. (2006) | Lagoon | – | 4.5 | – |
| Sri Lanka | Haas et al. (2016) | | 43–57 (51) | 4.3–19 (11) | |

Some values were visually interpreted from figures. The values in parentheses are the means of several data in the study

Torréton 1999). These bacterial distribution patterns are similar to those of DOM, as reviewed above, and indicate that bacterial growth is enhanced by bioavailable DOM that is produced in the reef.

### 2.4.2 Bacterial Growth and DOM Consumption

#### 2.4.2.1 Bacteria Growth in Natural Environments

The bacterial growth rate is usually defined as the specific growth rate during a period of exponential increase in the bacterial abundance:

$$\text{Specific growth rate} = \frac{\ln N - \ln N_i}{t}$$

where $N$ and $N_i$ indicate the bacterial abundances at the end and beginning of a period ($t$), respectively. The bacterial growth rates in coral reefs have been reported to be 0.03–6.8 $day^{-1}$ (Table 2.4). The median and average growth rates are 0.26 $day^{-1}$ and 0.58 $day^{-1}$, respectively. The median rate indicates that the bacterial abundance takes 2.7 days to double in seawater.

The bacterial growth in a reef water column was not completely supported by DOM that was released from phytoplankton alone (Torréton et al. 2002; Rochelle-Newall et al. 2008). This observation indicates that benthic organisms other than phytoplankton play a major role in providing DOM to pelagic bacteria. In the southwestern lagoon of New Caledonia, only 10–20% of the bacterial C demand in seawater was met by planktonic DOC production (Rochelle-Newall et al. 2008). When seawater becomes more eutrophic, primary production in the seawater is enhanced and more DOM is produced, with the bacterial C demand largely being met by planktonic DOC production, as observed in the coastal bay stations of New Caledonia (Rochelle-Newall et al. 2008).

The bacterial growth rate at Lizard Island, Great Barrier Reef, was higher at daytime than nighttime (Moriarty et al. 1985), which suggests that bacterial production was enhanced as organic matter released from photosynthesizing organisms. This observation corresponds to the DOM production rate on the Shiraho reef flat, which was higher at daytime than nighttime (Tanaka et al. 2011b). Additionally, the bacterial growth rates at Lizard Island were 15 times higher during the summer than those during the winter (Moriarty et al. 1985), which suggests that higher temperatures increased the release rates of organic matter from benthic organisms (Haas et al. 2010b; Naumann et al. 2010).

#### 2.4.2.2 Bacterial Growth Enhanced by Specific Organisms

Bacterial aggregation on coral mucus was first found several decades ago (Sorokin 1973; Ducklow and Mitchell 1979), and a number of studies have reported higher bacterial densities and oxygen consumption rates in coral mucus (Table 2.5). For example, Nakajima et al. (2015) observed that the bacterial abundance in seawater increased much more quickly with the experimental addition of coral mucus than in the control seawater (Fig. 2.6). The other benthic primary producers, e.g., macroalgae, turf algae, and crustose coralline algae, also exuded significant amounts of DOM into the surrounding seawater, which enhanced bacterial growth (Table 2.5).

Measuring bacterial growth rates by using organic matter that is released from coral reef organisms has shown that the growth rates range from 0.33 to 3.5 $day^{-1}$ (Table 2.5). This range of bacterial growth rates seems to be much higher than that observed in natural coral reef waters (Table 2.4), which suggests that bacterial growth in natural reef waters is more or less limited by the availability of organic matter. The bacterial production rates that were measured near the surface of corals were higher than that in the ambient seawater (van Duyl and Gast 2001; Scheffers et al. 2005), which shows that the bacterial growth rates in coral reefs are enhanced if more organic matter is available in the seawater.

#### 2.4.2.3 Relationship Between the DOC Concentration and Bacterial Growth Rates

In theory, a higher DOM concentration supports a higher rate of bacterial DOM consumption and subsequent bacterial cell production until the consumption rate reaches a maximum according to the Michaelis-Menten equation (Kirchman 2012). We tested the relationship between DOC concentrations and bacterial growth rates by using the data in Tables 2.4 and 2.5 to find a general pattern. When only the data from natural reef waters were used (Table 2.4), a significant correlation was not found between these two parameters (Fig. 2.7). This unexpected result may have been caused by differences in the composition of organic matter, such as the C:N ratios, and/or the analytical precision of the determination of DOC concentration (Sharp et al. 2002a) and bacterial abundance (Shibata et al. 2006). Several studies have reported higher bacterial growth rates in seawater with higher DOC concentrations at local scales (Nelson et al. 2011; Rochelle-Newall et al. 2008; Bouvy et al. 2012). However, when the data on incubations with additional organic matter, such as coral mucus (Table 2.5), were combined with the data from natural reef waters, a statistically positive correlation was found between the DOC concentrations and bacterial growth rates (Spearman's rank test, $p < 0.01$; Fig. 2.7). This result shows that bacterial growth rates are clearly enhanced with high DOC concentrations of approximately >100 μM.

The uptake of media from seawater by microbes is limited by kinetic controls or mass transfer controls (Sanford and

**Table 2.5** Bacterial growth rates with the addition of DOM that was released from coral reef organisms

| References | Species or genus | Medium size (μm) | Initial DOC (μM) | Initial bacterial abundance ($10^5$ cells $mL^{-1}$) | Bacterial growth rate ($day^{-1}$) |
|---|---|---|---|---|---|
| Coral | | | | | |
| Herndl and Velimirov (1986) | *Cladocora caespitosa* | <5 | 219 | 5.3 | 1.6 |
| | Seawater control | <5 | 158 | 2.5 | 1.3 |
| van Duyl and Gast (2001) | *Madracis mirabilis/favelota* | <0.7 | 211 | 1.1 | 2.1 |
| | Seawater control | <0.7 | 173 | 1.1 | 2.1 |
| Tanaka et al. (2008a) | *Acropora pulchra* | <0.7 | 63 | 2.0 | 0.33 |
| | *Porites cylindrica* | <0.7 | 63 | 5.0 | 0.62 |
| Tanaka et al. (2008b) | *Acropora pulchra* | <0.7 | 83 | 0.036 | 3.5 |
| | *Porites cylindrica* | <0.7 | 91 | 0.058 | 2.8 |
| Nakajima et al. (2009) | *Acropora robusta* (*A. nobilis*) | <3 | 252 | 7.1 | 2.6 |
| | Seawater control | <3 | 87 | 3.5 | 0.62 |
| | *A. muricata* (*A. formosa*) | <3 | 124 | 5.6 | 3.3 |
| | Seawater control | <3 | 59 | 3.4 | 0.65 |
| Haas et al. (2011) | *Porites* | <0.2 | 80 | 2.0 | 1.0 |
| Nelson et al. (2013) | *Porites* | <0.2 | 78 | – | 0.35 |
| Nakajima et al. (2015) | *A. robusta/A. muricata* | <0.7 | 341 | 21.1 | 2.7 |
| | Seawater control | <0.7 | 82 | 8.2 | 0.72 |
| | Benthic algae | | | | |
| Haas et al. (2011) | *Amansia* | <0.2 | 238 | 2.0 | 0.88 |
| | *Turbinaria* | <0.2 | 192 | 2.5 | 0.91 |
| | *Halimeda* | <0.2 | 123 | 4.0 | 0.84 |
| | Seawater control | <0.2 | 73 | 3.8 | 0.43 |
| Nelson et al. (2013) | *Amansia* | <0.2 | 177 | – | 1.1 |
| | *Turbinaria* | <0.2 | 181 | – | 0.61 |
| | *Halimeda* | <0.2 | 113 | – | 0.65 |
| | Seawater control | <0.2 | 70 | – | 0.41 |

Some values were visually interpreted from figures

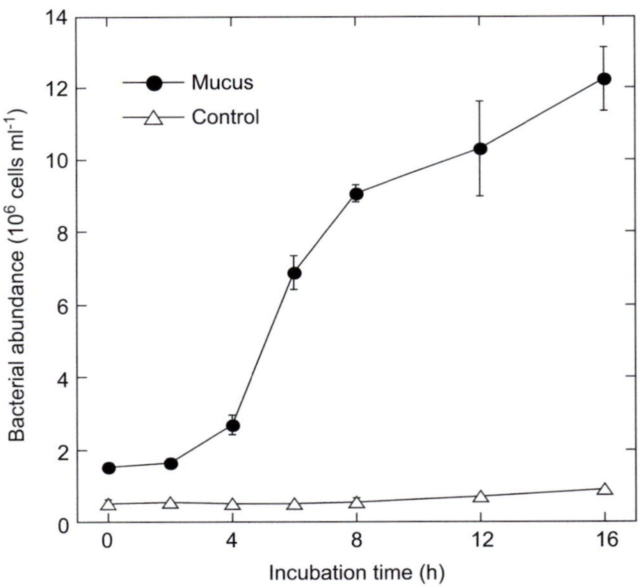


**Fig. 2.6** Increase in bacterial abundance in the seawater with coral mucus (Modified from Nakajima et al. 2015, with permission from ©Blackwell Verlag GmbH)

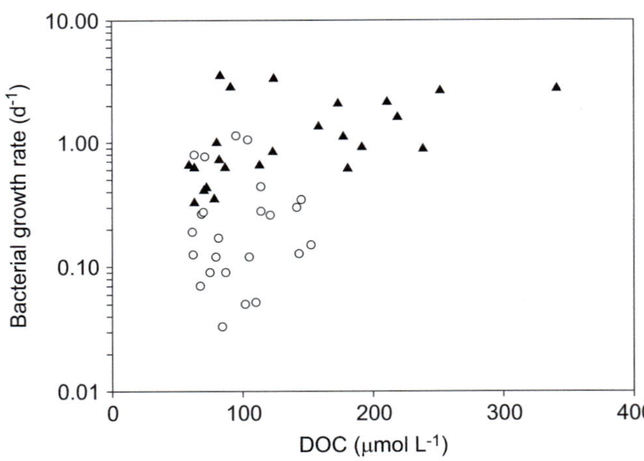


**Fig. 2.7** Relationship between the DOC concentration and bacterial growth rate in coral reefs (The data from Table 2.4 (natural environments) and Table 2.5 (artificial enrichment in organic matter) are shown in *open circles* and *filled triangles*, respectively)

Crawford 2000). When the reaction rates at or below the surface of microbes are faster than the mass transfer rates, the uptake rate is controlled by mass transfer (Thomas and Atkinson 1997). Conversely, when the reaction rates are slower than the mass transfer rates, the uptake rate is kinetically controlled and thus increases with the concentration of the medium up to a maximum rate (Badgley et al. 2006). Nutrient uptake by primary producers in oligotrophic environments such as coral reefs occurs under mass transfer rather than through kinetics (Thomas and Atkinson 1997; Hearn et al. 2001; Atkinson and Falter 2003; Badgley et al. 2006; Atkinson 2011), where nutrients in seawater are not rapidly taken up because of the limitations of diffusion from the seawater to the consumer. Atkinson and Smith (1987) showed that only 11% of $^{32}P$ that was artificially added to seawater on a coral reef flat was removed after the water traveled 450 m across the reef flat. The relationship between DOC concentrations and bacterial growth rates (Fig. 2.7) suggests that DOM consumption by bacteria in natural coral reef waters may be regulated by both the DOM concentration in the seawater and the water flow or turbulence, which controls the mass transfer rates (Peters et al. 1998; Sanford and Crawford 2000).

The median bacterial abundances and growth rates that were observed in natural reef waters ($7.0 \times 10^5$ cells $mL^{-1}$ and 0.26 $day^{-1}$, respectively; Table 2.4) allowed us to calculate a general production rate of bacterial cells of $2.1 \times 10^5$ cells $mL^{-1}$ $day^{-1}$. If we suppose that each bacterial cell contains 20 fg C (Fukuda et al. 1998), the production rate is estimated to be 0.35 μmol C $L^{-1}$ $day^{-1}$. This estimated production rate of bacterial cells in coral reef waters actually covers most of the previously measured rates: 0.04–0.48 μmol C $L^{-1}$ $day^{-1}$ (Torréton et al. 1997), 0.2–2.3 μmol C $L^{-1}$ $day^{-1}$ (Ferrier-Pagès and Gattuso 1998), 0.02 μmol C $L^{-1}$ $day^{-1}$ (Torréton 1999), and 0.29 μmol C $L^{-1}$ $day^{-1}$ (Rochelle-Newall et al. 2008). The bacterial growth efficiency is defined as the ratio of increased bacterial C to consumed organic C (Kirchman 2012). If we suppose that this efficiency is between 15% and 35% as the values for general marine bacteria (Kirchman 2012), the total C consumption by pelagic bacteria in coral reef waters is calculated to be 1.0–2.3 μmol C $L^{-1}$ $day^{-1}$. The reef flat at Shiraho Reef has an approximate average depth of 2 m, so the bacterial consumption rate of organic C in the seawater is 2.0–4.6 mmol $m^{-2}$ $day^{-1}$, which is <21% of the net production rate of DOC on the reef flat (22 mmol $m^{-2}$ $day^{-1}$; Table 2.3).

### 2.4.3 Quantitative Evaluation on DOM Degradability

Quantitatively evaluating the degradability of DOM is important to understand the biogeochemical cycles of DOM in coral reefs. Seawater that is collected in a coral reef (reef flat or lagoon) mainly has two fractions of DOM in terms of degradability: (1) newly produced DOM in the coral reef and (2) refractory DOM that ubiquitously exists in the global oceans (Fig. 2.3). Evaluating the degradability of DOM (1) requires collecting seawater samples inside and outside the reef and comparing the degradability of the DOM in those seawater samples.

The DOC in seawater that was collected on the reef flat at Shiraho Reef was mineralized by 10–18 μM for 1 year under dark conditions, while the offshore oceanic DOC was only mineralized by 4–5 μM (Tanaka et al. 2011a). This result indicates that the coral reef water had more bioavailable DOM than the oceanic seawater. When the mineralization of reef-derived DOC was described by an exponential decay model according to two degradability pools, the degradable fraction comprised 63–94% (average 77%) and the non-degradable fraction comprised 6–37% (average 23%). The degradable fraction had mineralization rates of 0.021–0.077 $day^{-1}$, which means that 63% of the fraction was mineralized within 13–48 days. These mineralization rates (decay constants) of reef-derived DOC are comparable to those of organic C that was released from corals (0.003–0.016 $day^{-1}$; Tanaka et al. 2011c), DOC that was released from temperate macroalgae (0.0058–0.041 $day^{-1}$; Wada et al. 2008), or DOC that was collected at the mid-Atlantic bight (0.22 $day^{-1}$ and 0.018 $day^{-1}$ for the very labile and labile fraction, respectively; Hopkinson et al. 2002). The determination of decay constants is greatly affected by the experimental designs, such as the sampling intervals (e.g., hours, days, or months) or the number of fractions in the decay model (e.g., labile, semi-labile, or refractory). Nonetheless, these comparisons suggest that DOM that is produced in coral reefs has a similar degradability to DOM that is produced in other aquatic ecosystems.

Longer water residence times in a coral reef mean that more bioavailable DOM is decomposed and mineralized to DIC in the reef ecosystem (Mari et al. 2007). In lagoons with extremely long water residence times, such as the atolls in Tuamotu Archipelago, Phoenix, Marshal, and Line Islands (16–2190 days, Delesalle and Sournia 1992), all the labile and semi-labile DOM would be mineralized in the lagoon. The DOM in coral reefs is also transformed into POM via transparent exopolymeric particles (TEP), which causes the organic matter to become trapped within larger particles and to accumulate in the reef (Mari et al. 2007). This transformation of DOM could facilitate the settling of organic matter on the reef sediment and the mineralization of organic matter by bacteria and consequently reduce the export of DOM from the reef to the ambient ocean (Wild et al. 2004a; Mari et al. 2007; Naumann et al. 2012).

Some of the DOM (6–37%) that was produced on the Shiraho reef flat was not mineralized by bacteria in dark incubations over 1 year (Tanaka et al. 2011a). If we suppose that DOC production comprised 18% of the net primary production on Shiraho Reef (Table 2.3), 1–7% (average of 4%) of the net primary production would remain as refractory organic C for at least 1 year. Some fractions of the DOM that are released from corals have been reported to be resistant to bacterial decomposition (Vacelet and Thomassin 1991; Tanaka et al. 2008b, 2011c). These observations showed that some reef-derived DOM is recalcitrant to bacterial decomposition over long periods, and this fraction could be more or less exported offshore irrespective of the water residence time in the reef. Because the degradability tests in Tanaka et al. (2011a) were performed under dark conditions, the existence of sunlight might make these reef-derived DOM more susceptible to bacterial mineralization because of photoreactions (Moran and Zepp 1997; Wada et al. 2015), although only a minor role of photochemical transformation in the cycling of DOM was proposed in a shallow subtropical seagrass-dominated lagoon (Ziegler and Benner 2000).

This section summarized the decomposition of DOM by planktonic bacteria (bacterioplankton). However, the density of bacteria in reef sediment is much higher than that in the water column, and these benthic bacteria also contribute to the decomposition of DOM (Wild et al. 2004a, b, 2005). The relative contributions of benthic and planktonic bacteria to DOM consumption are also important to understanding DOM cycles in coral reefs, which is a subject to be resolved in the future.

## 2.5 Summary

This paper is a first review article that summarized our present understanding of DOM cycles in coral reefs. Data from laboratory experiments, such as DOM that is released from individual organisms, have been more frequently collected from various coral reefs around the world, but elaborate studies on DOM cycles at reef scales are limited to several reefs. In particular, the concentration of DON has scarcely been measured for coral reef waters. Further research is required to understand and generalize the behavior and function of DOM in coral reef ecosystems. The summary of this chapter is as follows:

1. In most coral reefs, the DOM concentrations and bacterial abundance are higher on the reef flat or in the lagoon than in the surrounding ocean. This observation indicates that coral reef ecosystems produce labile DOM for bacteria and likely promote energy flows in microbial food webs.
2. On average, the net production rate of DOC comprised 18% of the net primary production rate at Shiraho Reef. The C:N ratio of the DOM that was produced on the reef flat was estimated to be 9.3, which was much lower than that of DOM in the ambient open ocean. The major DOM producers at this coral reef were inferred to be benthic microalgae, including cyanobacterial mats.
3. Many coral reef organisms release DOM into the ambient seawater. The median DOC release rate was 0.20 mmol $m^{-2}$ $h^{-1}$ for corals and 0.45 mmol $m^{-2}$ $h^{-1}$ for benthic algae. The contribution of phytoplankton to the DOM production on shallow reef flats is much lower than that of corals or benthic algae.
4. The median bacterial abundances and growth rates in the coral reef waters were $7.0 \times 10^5$ cells $mL^{-1}$ and 0.26 $day^{-1}$, respectively. The estimated bacterial consumption rate of organic C in seawater was 1.0–2.3 μmol C $L^{-1}$ $day^{-1}$. The bacterial consumption of DOM might often be controlled by mass transfer.
5. Long-term quantitative evaluations of DOM degradability showed that a considerable portion of the DOM that is produced in coral reefs is not mineralized by bacteria over several months. Thus, coral reef ecosystems more or less provide DOM to the surrounding ocean, irrespective of the water residence time in the reef.

**Acknowledgment** We are grateful to two anonymous reviewers who gave us plenty of valuable comments to improve this article. This study was supported by Brunei Research Council (S&T-14), the Japan Society for the Promotion of Science (JSPS) Asian CORE Program, and JSPS Postdoctoral Fellowships for Research Abroad.

## References

Abdullah MI, Fredriksen S (2004) Production, production, respiration and exudation of dissolved organic matter by the kelp Laminaria hyperborean along the west coast of Norway. J Mar Biol Assoc U K 84:887–894. https://doi.org/10.1017/S002531540401015Xh

Atkinson MJ (2011) Biogeochemistry of nutrients. In: Dubinsky Z, Stambler N (eds) Coral reefs: an ecosystem in transition. Springer, Dordrecht, pp 199–206. https://doi.org/10.1007/978-94-007-0114-4_13

Atkinson MJ, Falter JL (2003) Coral reefs. In: Black K, Shimmield G (eds) Biogeochemistry of marine systems. CRC Press, Boca Raton, pp 40–64

Atkinson MJ, Smith SV (1983) C:N:P ratios of benthic marine plants. Limnol Oceanogr 28:568–574. https://doi.org/10.4319/lo.1983.28.3.0568

Atkinson MJ, Smith DF (1987) Slow uptake of $^{32}P$ over a barrier reef flat. Limnol Oceanogr 32:436–441. https://doi.org/10.4319/lo.1987.32.2.0436

Azam F, Finches T, Gray JC et al (1983) The ecological role of water-column microbes in the sea. Mar Ecol Prog Ser 10:257–263. https://doi.org/10.3354/meps010257

Badgley BD, Lipschultz F, Sebens KP (2006) Nitrate uptake by the reef coral *Diploria strigosa*: effects of concentration, water flow, and irradiance. Mar Biol 149:327–338. https://doi.org/10.1007/s00227-005-0180-5

Barrón C, Duarte CM (2009) Dissolved organic matter release in a *Posidonia oceanica* meadow. Mar Ecol Prog Ser 374:75–84. https://doi.org/10.3354/meps07715

Barrón C, Marbà N, Terrados J et al (2004) Community metabolism and carbon budget along a gradient of seagrass (*Cymodocea nodosa*) colonization. Limnol Oceanogr 49:1642–1651. https://doi.org/10.4319/lo.2004.49.5.1642

Bauer JE, Williams PM, Druffel ERM (1992) $^{14}C$ activity of dissolved organic carbon fractions in the north-central Pacific and Sargasso Sea. Nature 357:67–670. https://doi.org/10.1038/357667a0

Blanchot J, Charpy L, Le Borgne R (1989) Size composition of particulate organic matter in the lagoon of Tikehau atoll (Tuamotu archipelago). Mar Biol 102:329–339. https://doi.org/10.1007/BF00428485

Blanco AC, Nadaoka K, Yamamoto T (2008) Planktonic and benthic microalgal community composition as indicators of terrestrial influence on a fringing reef in Ishigaki Island, Southwest Japan. Mar Environ Res 66:520–535. https://doi.org/10.1016/j.marenvres.2008.08.005

Bouvy M, Combe M, Bettarel Y et al (2012) Uncoupled viral and bacterial distributions in coral reef waters of Tuamotu Archipelago (French Polynesia). Mar Pollut Bull 65:506–515. https://doi.org/10.1016/j.marpolbul.2012.01.001

Brocke HJ, Wenzhoefer F, de Beer D et al (2015) High dissolved organic carbon release by benthic cyanobacterial mats in a Caribbean reef ecosystem. Sci Rep 5:8852. https://doi.org/10.1038/srep08852

Brown BE, Bythel JC (2005) Perspectives on mucus secretion in reef corals. Mar Ecol Prog Ser 296:291–309. https://doi.org/10.3354/meps296291

Burkepile DE, Hay ME (2009) Nutrient versus herbivore control of macroalgal community development and coral growth on a Caribbean reef. Mar Ecol Prog Ser 389:71–84. https://doi.org/10.3354/meps08142

Carlson CA (2002) Production and removal processes. In: Hansel DA, Carlson CA (eds) Biogeochemistry of marine dissolved organic matter. Academic, San Diego, pp 91–151. https://doi.org/10.1016/B978-012323841-2/50006-3

Casareto BE, Charpy L, Blanchot J, Suzuki Y, Kurosawa K, Ishikawa Y (2006) Phototrophic prokaryotes in Bora Bay, Miyako Island, Okinawa, Japan. In: Proceedings of 10th international coral reef symposium, pp 844–853

Cawley KM, Ding Y, Fourqurean J et al (2012) Characterising the sources and fate of dissolved organic matter in Shark Bay, Australia: a preliminary study using optical properties and stable carbon isotopes. Mar Freshw Res 63:1098–1107. https://doi.org/10.1071/MF12028

Charpy-Roubaud CJ, Charpy L, Lemasson L (1988) Benthic and planktonic primary production of an open atoll lagoon (Tikehau, French Polynesia). Proc 6th Int Coral Reef Symp 2:551–556

Cho BC, Azam F (1988) Major role of bacteria in biogeochemical fluxes in the ocean's interior. Nature 332:441–443. https://doi.org/10.1038/332441a0

Cooney RP, Pantos O, Le Tissier MDA et al (2002) Characterisation of the bacterial consortium associated with black band disease in coral using molecular microbiological techniques. Environ Microbiol 4:401–413. https://doi.org/10.1046/j.1462-2920.2002.00308.x

Crossland CJ (1987) In situ release of mucus and DOC-lipid from the corals *Acropora variabilis* and *Stylophora pistillata* in different light regimes. Coral Reefs 6:35–42. https://doi.org/10.1007/BF00302210

Crossland CJ, Hatcher BG, Smith SV (1991) Role of coral reefs in global ocean production. Coral Reefs 10:55–64. https://doi.org/10.1007/BF00571824

Davies PS (1984) The role of zooxanthellae in the nutritional energy requirements of *Pocillopora eydouxi*. Coral Reefs 2:181–186. https://doi.org/10.1007/BF00263571

Delesalle B, Sournia A (1992) Residence time of water and phytoplankton biomass in coral reef lagoons. Cont Shelf Res 12:939–949. https://doi.org/10.1016/0278-4343(92)90053-M

de Goeij JM, van den Berg H, van Oostveen MM et al (2008) Major bulk dissolved organic carbon (DOC) removal by encrusting coral reef cavity sponges. Mar Ecol Prog Ser 357:139–151. https://doi.org/10.3354/meps07403

de Goeij JM, van Oevelen D, Vermeij MJA et al (2013) Surviving in a marine desert: the sponge loop retains resources within coral reefs. Science 342:108–110. https://doi.org/10.1126/science.1241981

Dinsdale EA, Rohwer F (2011) Fish or germs? Microbial dynamics associated with changing trophic structures on coral reefs. In: Dubinsky Z, Stambler N (eds) Coral reefs: an ecosystem in transition. Springer Science + Business Media B.V., pp 231–240. doi:https://doi.org/10 1007/978-94-007-0114-4_16

Dinsdale EA, Pantos O, Smriga S et al (2008) Microbial ecology of four coral atolls in the Northern Line Islands. PLoS One 3:e1584. https://doi.org/10.1371/journal.pone.0001584

Duarte CM (1992) Nutrient concentration of aquatic plants: patterns across species. Limnol Oceanogr 37:882–889. https://doi.org/10.4319/lo.1992.37.4.0882

Duarte CM, Cebrián J (1996) The fate of marine autotrophic production. Limnol Oceanogr 41:1758–1766. https://doi.org/10.4319/lo.1996.41.8.1758

Dubinsky Z, Berman-Frank I (2001) Uncoupling primary production from population growth in photosynthesizing organisms in aquatic ecosystems. Aquat Sci 63:4–17. https://doi.org/10.1007/PL00001343

Ducklow HW, Mitchell R (1979) Composition of mucus released by coral reef coelenterates. Limnol Oceanogr 24:706–714. https://doi.org/10.4319/lo.1979 24.4.0706

Ferrier-Pagès C, Furla P (2001) Pico- and nanoplankton biomass and production in the two largest atoll lagoons of French Polynesia. Mar Ecol Prog Ser 211:63–76. https://doi.org/10.3354/meps211063

Ferrier-Pagès C, Gattuso J-P (1998) Biomass, production and grazing rates of pico- and nanoplankton in coral reef waters (Miyako Island, Japan). Microb Ecol 35:46–57. https://doi.org/10.1007/s002489900059

Ferrier-Pagès C, Gattuso J-P, Cauwet G et al (1998) Release of dissolved organic carbon and nitrogen by the zooxanthellate coral *Galaxea fascicularis*. Mar Ecol Prog Ser 172:265–274. https://doi.org/10.3354/meps172265

Ferrier-Pagès C, Leclercq N, Jaubert J et al (2000) Enhancement of pico- and nanoplankton growth by coral exudates. Aquat Microb Ecol 21:203–209. https://doi.org/10.3354/ame021203

Fourqurean JW, Zieman JC, Powell GVN (1992) Phosphorus limitation of primary production in Florida Bay: evidence from C:N:P ratios of the dominant seagrass *Thalassia testudinum*. Limnol Oceanogr 37:162–171. https://doi.org/10.4319/lo.1992.37.1.0162

Fuhrman JA, Sleeter TD, Carlson CA, Proctor LM (1989) Dominance of bacterial biomass in the Sargasso Sea and its ecological implications. Mar Ecol Prog Ser 57:207–218. https://doi.org/10.3354/meps057207

Fukuda R, Ogawa H, Nagata T, Koike I (1998) Direct determination of carbon and nitrogen contents of natural bacterial assemblages in marine environments. Appl Environ Microb 64:3352–3358

Furnas M, Mitchell A, Skuza M, Brodie J (2005) In the other 90%: phytoplankton responses to enhanced nutrient availability in the Great Barrier Reef Lagoon. Mar Pollut Bull 51:253–265. https://doi.org/10.1016/j.marpolbul.2004.11.010

Gast GJ, Wiegman S, Wieringa E et al (1998) Bacteria in coral reef water types: removal of cells, stimulation of growth and mineralization. Mar Ecol Prog Ser 167:37–45. https://doi.org/10.3354/meps167037

Gattuso J-P, Pichon M, Delesalle B et al (1996) Carbon fluxes in coral reefs. I. Lagrangian measurement of community metabolism and resulting air-sea $CO_2$ disequilibrium. Mar Ecol Prog Ser 145:109–121. https://doi.org/10.3354/meps145109

Gattuso J-P, Frankignoulle M, Wollast R (1998) Carbon and carbonate metabolism in coastal aquatic ecosystems. Annu Rev Ecol Syst 29:405–434. https://doi.org/10.1146/annurev.ecolsys.29.1.405

González JM, Torréton J-P, Dufour P et al (1998) Temporal and spatial dynamics of the pelagic microbial food web in an atoll lagoon. Aquat Microb Ecol 16:53–64. https://doi.org/10.3354/ame016053

Grover R, Maguer J-F, Allemand D et al (2006) Urea uptake by the scleractinian coral *Stylophora pistillata*. J Exp Mar Biol Ecol 332:216–225. https://doi.org/10.1016/j.jembe.2005.11.020

Grover R, Maguer J-F, Allemand D et al (2008) Uptake of dissolved free amino acids by the scleractinian coral *Stylophora pistillata*. J Exp Biol 211:860–865. https://doi.org/10.1242/jeb.012807

Haas AF, Jantzen C, Naumann S et al (2010a) Organic matter release by the dominant primary producers in a Caribbean reef lagoon: implication for *in situ* $O_2$ availability. Mar Ecol Prog Ser 409:27–39. https://doi.org/10.3354/meps08631

Haas AF, Naumann MS, Struck U et al (2010b) Organic matter release by coral reef associated benthic algae in the Northern Red Sea. J Exp Mar Biol Ecol 389:53–60. https://doi.org/10.1016/j.jembe.2010.03.018

Haas AF, Nelson CE, Kelly LW et al (2011) Effects of coral reef benthic primary producers on dissolved organic carbon and microbial activity. PLoS One 6:e27973. https://doi.org/10.1371/journal.pone.0027973

Haas AF, Nelson CE, Rohwer F et al (2013) Influence of coral and algal exudates on microbially mediated reef metabolism. Peer J 1:e108. https://doi.org/10.7717/peerj.108

Haas AF, Fairoz MFM, Kelly LW et al (2016) Global microbialization of coral reefs. Nat Microbiol 1:16042. https://doi.org/10.1038/nmicrobiol.2016.42

Hata H, Kudo S, Yamano H et al (2002) Organic carbon flux in Shiraho coral reef (Ishigaki Island, Japan). Mar Ecol Prog Ser 232:129–140. https://doi.org/10.3354/meps232129

Hearn CJ, Atkinson MJ, Falter JL (2001) A physical derivation of nutrient-uptake rates in coral reefs: effects of roughness and waves. Coral Reefs 20:347–356. https://doi.org/10.1007/s00338-001-0185-6

Hedges JI (1992) Global biogeochemical cycles; progress and problems. Mar Chem 39:67–93. https://doi.org/10.1016/0304-4203(92)90096-S

Herndl GJ, Velimirov B (1986) Microheterotrophic utilization of mucus released by the Mediterranean coral *Cladocora cespitosa*. Mar Biol 90:363–369. https://doi.org/10.1007/BF00428560

Hoch MP, Dillon KS, Coffin RB et al (2008) Sensitivity of bacterioplankton nitrogen metabolism to eutrophication in sub-tropical coastal waters of Key West, Florida. Mar Pollut Bull 56:913–926. https://doi.org/10.1016/j.marpolbul.2008.01.030

Hopkinson CS Jr, Vallino JJ, Amy N (2002) Decomposition of dissolved organic matter from the continental margin. Deep-Sea Res II 49:4461–4478. https://doi.org/10.1016/S0967-0645(02)00125-X

Houlbrèque F, Delesalle B, Blanchot J et al (2006) Picoplankton removal by the coral reef community of La Prévoyante, Mayotte Island. Aquat Microb Ecol 44:59–70. https://doi.org/10.3354/ame044059

Iryu Y, Nakamori T, Matsuda S et al (1995) Distribution of marine organisms and its geological significance in the modern reef complex of the Ryukyu Islands. Sediment Geol 99:243–258. https://doi.org/10.1016/0037-0738(95)00047-C

Karl DM, Hebel DV, Björkman K et al (1998) The role of dissolved organic matter release in the productivity of the oligotrophic North Pacific Ocean. Limnol Oceanogr 43:1270–1286. https://doi.org/10.4319/lo.1998.43.6.1270

Kawahata H, Yukino I, Suzuki A (2000) Terrestrial influences on the Shiraho fringing reef, Ishigaki Island, Japan: high carbon input relative to phosphate. Coral Reefs 19:172–178. https://doi.org/10.1007/s003380000093

Kayanne H, Suzuki A, Saito H (1995) Diurnal changes in the partial pressure of carbon dioxide in coral reef water. Science 269:214–216. https://doi.org/10.1126/science.269.5221.214

Kayanne H, Hata H, Kudo S et al (2005) Seasonal and bleaching-induced changes in coral reef metabolism and $CO_2$ flux. Glob Biogeochem Cycles 19:GB3015. https://doi.org/10.1029/2004GB002400

Kennedy DM, Woodroffe CD (2002) Fringing reef growth and morphology: a review. Earth-Sci Rev 57:255–277. https://doi.org/10.1016/S0012-8252(01)00077-0

Kirchman DL (2012) Processes in microbial ecology. Oxford University Press, Oxford

Lapointe BE (1997) Nutrient thresholds for bottom-up control of macroalgal blooms on coral reefs in Jamaica and southeast Florida. Limnol Oceanogr 42:1119–1131. https://doi.org/10.4319/lo.1997.42.5_part_2.1119

Lapointe BE, Langton R, Bedford BJ et al (2010) Land-based nutrient enrichment of the Buccoo reef complex and fringing coral reefs of Tobago, West Indies. Mar Pollut Bull 60:334–343. https://doi.org/10.1016/j.marpolbul.2009.10.020

Lee SH, Kang YC, Fuhrman JA (1995) Imperfect retention of natural bacterioplakton cells by glass fiber filters. Mar Ecol Prog Ser 119:285–290

Levas S, Grottoli AG, Hughes A et al (2013) Physiological and biogeochemical traits of bleaching and recovery in the mounding species of coral *Porites lobata*: implications for resilience in mounding corals. PLoS One 8:e63267. https://doi.org/10.1371/journal.pone.0063267

Levas S, Grottoli AG, Warner ME et al (2015) Organic carbon fluxes mediated by corals at elevated $pCO_2$ and temperature. Mar Ecol Prog Ser 519:153–164. https://doi.org/10.3354/meps11072

Linley E, Koop K (1986) Significance of pelagic bacteria as a trophic resource in a coral reef lagoon, One Tree Island, Great Barrier Reef. Mar Biol 92:457–464. https://doi.org/10.1007/BF00392505

Littler MM, Littler DS, Brooks BL (2010) The effects of nitrogen and phosphorus enrichment on algal community development: artificial mini-reefs on the Belize Barrier Reef sedimentary lagoon. Harmful Algae 9:255–263. https://doi.org/10.1016/j.hal.2009.11.002

Marañón E, Cermeño P, Fernández E et al (2004) Significance and mechanisms of photosynthetic production of dissolved organic carbon in a coastal eutrophic ecosystem. Limnol Oceanogr 49:1652–1666. https://doi.org/10.4319/lo.2004.49.5.1652

Mari X, Rochelle-Newall E, Torréton J-P et al (2007) Water residence time: a regulatory factor of the DOM to POM transfer efficiency. Limnol Oceanogr 52:808–819. https://doi.org/10.4319/lo.2007.52.2.0808

Meyers-Schulte KH, Hedges JI (1986) Molecular evidence for a terrestrial component of organic matter dissolved in ocean water. Nature 321:61–63. https://doi.org/10.1038/321061a0

Miyajima T, Koike I, Yamano H et al (1998) Accumulation and transport of seagrass-derived organic matter in reef flat sediment of Green Island, Great Barrier Reef. Mar Ecol Prog Ser 175:251–259. https://doi.org/10.3354/meps175251

Miyajima T, Hata H, Umezawa Y et al (2007a) Distribution and partitioning of nitrogen and phosphorus in a fringing reef lagoon of Ishigaki Island, northwestern Pacific. Mar Ecol Prog Ser 341:45–57. https://doi.org/10.3354/meps341045

Miyajima T, Tanaka Y, Koike I et al (2007b) Evaluation of spatial correlation between nutrient exchange rates and benthic biota in a reef-flat ecosystem by GIS-assisted flow-tracking. J Oceanogr 63:643–659. https://doi.org/10.1007/s10872-007-0057-y

Moran MA, Zepp RG (1997) Role of photoreactions in the formation of biologically labile compounds from dissolved organic matter. Limnol Oceanogr 42:1307–1316. https://doi.org/10.4319/lo.1997.42.6.1307

Moriarty DJW (1979) Biomass of suspended bacteria over coral reefs. Mar Biol 53:193–200. https://doi.org/10.1007/BF00389189

Moriarty DJW, Pollard PC, Hunt WG (1985) Temporal and spatial variation in bacterial production in the water column over a coral reef. Mar Biol 85:285–292. https://doi.org/10.1007/BF00393249

Morrow KM (2011) Allelochemicals produced by Caribbean macroalgae and cyanobacteria have species-specific effects on reef coral microorganisms. Coral Reefs 30:309–320. https://doi.org/10.1007/s00338-011-0747-1

Mostofa KMG, Liu C-Q, Mottaleb MA et al (2013) Dissolved organic matter in natural waters. In: Mostofa KMG et al (eds) Photobiogeochemistry of organic matter. Springer, Berlin/Heidelberg, pp 1–137. https://doi.org/10.1007/978-3-642-32223-5_1

Mueller B, van der Zande RM, van Leent PJM (2014) Effect of light availability on dissolved organic carbon release by Caribbean reef algae and corals. Bull Mar Sci 90:875–893. https://doi.org/10.5343/bms.2013.1062

Muller-Parker G, Cook CB, D'Elia CF (1994) Elemental composition of the coral *Pocillopora damicornis* exposed to elevated seawater ammonium. Pac Sci 48:234–246

Nakajima R, Yoshida T, Azman BAR et al (2009) In situ release of coral mucus by *Acropora* and its influence on the heterotrophic bacteria. Aquat Ecol 43:815–823. https://doi.org/10.1007/s10452-008-9210-y

Nakajima R, Yoshida T, Fujita K et al (2010) Release of particulate and dissolved organic carbon by the scleractinian coral *Acropora formosa*. Bull Mar Sci 86:861–870. https://doi.org/10.5343/bms.2009.1069

Nakajima R, Yoshida T, Shibata A, Othman BHR, Toda T (2011) Quality and quantity of particulate organic carbon in a coral reef at Tioman Island, Malaysia. Sains Malaysiana 40:1375–1382

Nakajima R, Tsuchiya K, Nakatomi N et al (2013) Enrichment of microbial abundance in the sea-surface microlayer over a coral reef: implications for biogeochemical cycles in reef ecosystems. Mar Ecol Prog Ser 490:11–22. https://doi.org/10.3354/meps10481

Nakajima R, Tanaka Y, Yoshida T et al (2015) High inorganic phosphate concentration in coral mucus and its utilization by heterotrophic bacteria in a Malaysian coral reef. Mar Ecol 36:835–841. https://doi.org/10.1111/maec.12158

Nakamura T, Nakamori T (2009) Estimation of photosynthesis and calcification rates at a fringing reef by accounting for diurnal variations and the zonation of coral reef communities on reef flat and slope: a case study for the Shiraho reef, Ishigaki Island, southwest Japan. Coral Reefs 28:229–250. https://doi.org/10.1007/s00338-008-0454-8

Naumann MS, Haas A, Struck U et al (2010) Organic matter release by dominant hermatypic corals of the Northern Red Sea. Coral Reefs 29:649–659. https://doi.org/10.1007/s00338-010-0612-7

Naumann MS, Richter C, Mott C et al (2012) Budget of coral-derived organic carbon in a fringing coral reef of the Gulf of Aqaba, Red Sea. J Mar Syst 105–108:20–29. https://doi.org/10.1016/j.jmarsys.2012.05.007

Nelson CE, Alldredge AL, McCliment EA et al (2011) Depleted dissolved organic carbon and distinct bacterial communities in the water column of a rapid-flushing coral reef ecosystem. ISME J 5:1374–1387. https://doi.org/10.1038/ismej.2011.12

Nelson CE, Goldberg SJ, Kelly LW et al (2013) Coral and macroalgal exudates vary in neutral sugar composition and differentially enrich reef bacterioplankton lineages. ISME J 7:962–979. https://doi.org/10.1038/ismej.2012.161

Ogawa H, Tanoue E (2003) Dissolved organic matter in oceanic waters. J Oceanogr 59:129–147. https://doi.org/10.1023/A:1025528919771

Ogawa H, Fukuda R, Koike I (1999) Vertical distributions of dissolved organic carbon and nitrogen in the Southern Ocean. Deep-Sea Res 46:1809–1826. https://doi.org/10.1016/S0967-0637(99)00027-8

Opsahl S, Benner R (1997) Distribution and cycling of terrigenous dissolved organic matter in the ocean. Nature 386:480–482. https://doi.org/10.1038/386480a0

Pagès J, Torreton JP, Sempere R (1997) Dissolved organic carbon in coral-reef lagoons by high temperature catalytic oxidation and UV spectrometry. C R Acad Sci Ser II A Sci Terre Planetes 324:915–922. https://doi.org/10.1016/S1251-8050(97)82505-0ro

Paul JH, DeFlaun MF, Jeffrey WH (1986) Elevated levels of microbial activity in the coral surface microlayer. Mar Ecol Prog Ser 33:29–40. https://doi.org/10.3354/meps033029

Pavés HJ, González HE (2008) Carbon fluxes within the pelagic food web in the coastal area off Antofagasta (23°S), Chile: the significance of the microbial versus classical food webs. Ecol Model 212:218–232. https://doi.org/10.1016/j.ecolmodel.2007.10.004

Peters F, Marrasé C, Gasol JM et al (1998) Effects of turbulence on bacterial growth mediated through food web interactions. Mar Ecol Prog Ser 172:293–303. https://doi.org/10.3354/meps172293

Pomeroy LR (1974) The ocean's food web, a changing paradigm. Bioscience 24:499–504. https://doi.org/10.2307/1296885

Riegl B, Branch GM (1995) Effects of sediment on the energy budgets of four scleractinian (Bourne 1900) and five alcyonacean (Lamouroux 1816) corals. J Exp Mar Biol Ecol 186 259–275. https://doi.org/10.1016/0022-0981(94)00164-9

Rix L, de Goeij JM, Mueller CE et al (2016) Coral mucus fuels the sponge loop in warm- and cold-water coral reef ecosystems. Sci Rep 6:18715. https://doi.org/10.1038/srep18715

Rochelle-Newall EJ, Torreton J-P, Mari X et al (2008) Phytoplankton–bacterioplankton coupling in a subtropical South Pacific coral reef lagoon. Aquat Microb Ecol 50:221–229. https://doi.org/10.3354/ame01158

Rohwer F, Seguritan V, Azam F et al (2002) Diversity and distribution of coral-associated bacteria. Mar Ecol Prog Ser 243:1–10. https://doi.org/10.3354/meps243001

Sakka A, Legendre L, Gosselin M, Delesalle B (2000) Structure of the oligotrophic planktonic food web under low grazing of heterotrophic bacteria: Takapoto Atoll, French Polynesia. Mar Ecol Prog Ser 197:1–17. https://doi.org/10.3354/meps197001

Sakka A, Legendre L, Gosselin M et al (2002) Carbon budget of the planktonic food web in an atoll lagoon (Takapoto, French Polynesia). J Plankton Res 24:301–320. https://doi.org/10.1093/plankt/24.4.301

Sandin SA, Smith JE, DeMartini EE et al (2008) Baselines and degradation of coral reefs in the northern Line Islands. PLoS One 3:e1548. https://doi.org/10.1371/journal.pone.0001548

Sanford LP, Crawford SM (2000) Mass transfer versus kinetic control of uptake across solid-water boundaries. Limnol Oceanogr 45:1180–1186. https://doi.org/10.4319/lo.2000.45.5.1180

Scheffers SR, Bak RPM, Duyl FC (2005) Why is bacterioplankton growth in coral reef framework cavities enhanced? Mar Ecol Prog Ser 299:89–99. https://doi.org/10.3354/meps299089

Seitzinger SP, Sanders RW (1999) Atmospheric inputs of dissolved organic nitrogen stimulate estuarine bacteria and phytoplankton. Limnol Oceanogr 44:721–730. https://doi.org/10.4319/lo.1999.44.3.0721

Sharp JH, Carlson CA, Peltzer ET et al (2002a) Final dissolved organic carbon broad community intercalibration and preliminary use of DOC reference materials. Mar Chem 77:239–253. https://doi.org/10.1016/S0304-4203(02)00002-6

Sharp JH, Rinker KR, Savidge KB et al (2002b) A preliminary methods comparison for measurement of dissolved organic nitrogen in seawater. Mar Chem 78:171–184. https://doi.org/10.1016/S0304-4203(02)00020-8

Sharp JH, Beauregard AY, Burdige D et al (2004) A direct instrument comparison for measurement of total dissolved organic nitrogen in seawater. Mar Chem 84:181–193. https://doi.org/10.1016/j.marchem.2003.07.003

Sherr EB, Sherr BF (1987) High rates of consumption of bacteria by pelagic ciliates. Nature 325:710–711. https://doi.org/10.1038/325710a0

Shibata A, Goto Y, Saito H et al (2006) Comparison of SYBR green I and SYBR gold stains for enumerating bacteria and viruses by epifluorescence microscopy. Aquat Microb Ecol 43:223–231. https://doi.org/10.3354/ame043223

Smith DC, Simon M, Alldredge AL et al (1992) Intense hydrolytic enzyme activity on marine aggregates and implications for rapid particle dissolution. Nature 359:139–142. https://doi.org/10.1038/359139a0

Smith JE, Smith CM, Hunter CL (2001) An experimental analysis of the effects of herbivory and nutrient enrichment on benthic community dynamics on a Hawaiian reef. Coral Reefs 19:332–342. https://doi.org/10 1007/s003380000124

Smith JE, Runcie JW, Smith CM (2005) Characterization of a large-scale ephemeral bloom of the green alga Cladophora sericea on the coral reefs of West Maui, Hawaii. Mar Ecol Prog Ser 302:77–91. https://doi.org/10.3354/meps302077

Smith JE, Brainard R, Carter A, Grillo S, Edwards C, Harris J, Lewis L, Obura D, Rohwer F, Sale E, Vroom PS, Sandin S (2016) Re-evaluating the health of coral reef communities: baselines and evidence for human impacts across the central Pacific. Proc R Soc B Biol Sci 283:20151985. https://doi.org/10.1098/rspb.2015.1985

Smith SV, Kimmerer WJ, Laws EA, Brock RE, Walsh TW (1981) Kaneohe Bay sewage diversion experiment: perspectives on ecosystem responses to nutritional perturbation. Pac Sci 35:279–395

Sorokin YI (1973) Trophical role of bacteria in the ecosystem of the coral reef. Nature 242:415–417. https://doi.org/10.1038/242415a0

Suzuki A, Nakamori T, Kayanne H (1995) The mechanism of production enhancement in coral reef carbonate systems: model and empirical results. Sediment Geol 99:259–280. https://doi.org/10.1016/0037-0738(95)00048-D

Suzuki Y, Casareto BE, Kurosawa K (2000) Import and export fluxes of HMW-DOC and LMW-DOC on a coral reef at Miyako Island, Okinawa. Proc 9th Int Coral Reef Symp 1:555–560

Suzumura M, Miyajima T, Hata H et al (2002) Cycling of phosphorus maintains the production of microphytobenthic communities in carbonate sediments of a coral reef. Limnol Oceanogr 47:771–781. https://doi.org/10.4319/lo.2002.47.3.0771

Tanaka Y, Miyajima T, Koike I et al (2007) Imbalanced coral growth between organic tissue and carbonate skeleton caused by nutrient enrichment. Limnol Oceanogr 52:1139–1146. https://doi.org/10.4319/lo.2007.52.3.1139

Tanaka Y, Miyajima T, Koike I et al (2008a) Production of dissolved and particulate organic matter by the reef-building corals, *Porites cylindrica* and *Acropora pulchra*. Bull Mar Sci 82:237–245

Tanaka Y, Miyajima T, Ozawa H (2008b) Bacterial degradability of dissolved organic carbon in coral mucus. In: Proceedings of 11th international coral reef symposium, pp 945–949

Tanaka Y, Miyajima T, Umezawa Y et al (2009) Net release of dissolved organic matter by the scleractinian coral *Acropora pulchra*. J Exp Mar Biol Ecol 377:101–106. https://doi.org/10.1016/j.jembe.2009.06.023

Tanaka Y, Ogawa H, Miyajima T (2010) Effects of nutrient enrichment on the release of dissolved organic carbon and nitrogen by the scleractinian coral *Montipora digitata*. Coral Reefs 29:675–682. https://doi.org/10.1007/s00338-010-0639-9

Tanaka Y, Ogawa H, Miyajima T (2011a) Production and bacterial decomposition of dissolved organic matter in a fringing coral reef. J Oceanogr 67:427–437. https://doi.org/10.1007/s10872-011-0046-z

Tanaka Y, Miyajima T, Watanabe A et al (2011b) Distribution of dissolved organic carbon and nitrogen in a coral reef. Coral Reefs 30:533–541. https://doi.org/10.1007/s00338-011-0735-5

Tanaka Y, Ogawa H, Miyajima T (2011c) Bacterial decomposition of coral mucus as evaluated by long-term and quantitative observation. Coral Reefs 30:443–449. https://doi.org/10.1007/s00338-011-0729-3

Taniguchi A, Yoshida T, Eguchi M (2014) Bacterial production is enhanced by coral mucus in reef systems. J Eex Mar Biol Ecol 461:331–336. https://doi.org/10.1016/j.jembe.2014.09.004

Tedetti M, Cuet P, Guigue C et al (2011) Characterization of dissolved organic matter in a coral reef ecosystem subjected to anthropogenic pressures (La Réunion Island, Indian Ocean) using multi-dimensional fluorescence spectroscopy. Sci Total Environ 409:2198–2210. https://doi.org/10.1016/j.scitotenv.2011.01.058

Thibodeau B, Miyajima T, Tayasu I et al (2013) Heterogeneous dissolved organic nitrogen supply over a coral reef: first evidence from nitrogen stable isotope ratios. Coral Reefs 32:1103–1110. https://doi.org/10.1007/s00338-013-1070-9

Thomas FIM, Atkinson MJ (1997) Ammonium uptake by coral reefs: effects of water velocity and surface roughness on mass transfer. Limnol Oceanogr 42:81–88. https://doi.org/10.4319/lo.1997.42.1.0081

Torréton J-P (1999) Biomass, production and heterotrophic activity of bacterioplankton in the Great Astrolabe Reef lagoon (Fiji). Coral Reefs 18:43–53. https://doi.org/10.1007/s003380050152

Torréton J-P, Dufour P (1996a) Temporal and spatial stability of bacterioplankton biomass and productivity in an atoll lagoon. Aquat Microb Ecol 11:251–261. https://doi.org/10.3354/ame011251

Torréton J-P, Dufour P (1996b) Bacterioplankton production determined by DNA synthesis, protein synthesis, and frequency of dividing cells in Tuamotu atoll lagoon and surrounding ocean. Microb Ecol 32:185–202. https://doi.org/10.1007/BF00185887

Torréton J-P, Pagès J, Dufour P et al (1997) Bacterioplankton carbon growth yield and DOC turnover in some coral reef lagoons. Proc 8th Int Coral Reef Sym 1:942–952

Torréton J-P, Talbot V, Garcia N (2000) Nutrient stimulation of bacterioplankton growth in Tuamotu atoll lagoons. Aquat Microb Ecol 21:125–137. https://doi.org/10.3354/ame021125

Torréton J-P, Pagès J, Talbot V (2002) Relationships between bacterioplankton and phytoplankton biomass, production and turnover rate in Tuamotu atoll lagoons. Aquat Microb Ecol 28:267–277. https://doi.org/10.3354/ame028267

Torréton J-P, Rochelle-Newall E, Pringault O et al (2010) Variability of primary and bacterial production in a coral reef lagoon (New Caledonia). Mar Pollut Bull 61:335–348. https://doi.org/10.1016/j.marpolbul.2010.06.019

Umezawa Y, Miyajima T, Kayanne H et al (2002a) Significance of groundwater nitrogen discharge into coral reefs at Ishigaki Island, southwest of Japan. Coral Reefs 21:346–335. https://doi.org/10.1007/s00338-002-0254-5

Umezawa Y, Miyajima T, Yamamuro M et al (2002b) Fine-scale mapping of land-derived nitrogen in coral reefs by $\delta^{15}N$ in macroalgae. Limnol Oceanogr 47:1405–1406. https://doi.org/10.4319/lo.2002.47.5.1405

Vacelet E, Thomassin BA (1991) Microbial utilization of coral mucus in long term in situ incubation over a coral reef. Hydrobiologia 211:19–32. https://doi.org/10.1007/BF00008613

Vacelet E, Arnoux A, Thomassin BA et al (1999) Influence of freshwater and terrigenous material on nutrients, bacteria and phytoplankton in a high island lagoon: Mayotte, Comoro Archipelago, Indian Ocean. Hydrobioligia 380:165–178. https://doi.org/10.1023/A:1003432829602

van Duyl FC, Gast GJ (2001) Linkage of small-scale spatial variations in DOC, inorganic nutrients and bacterioplalnkton growth with different coral reef water types. Aquat Microb Ecol 24:17–26

Volkman JK, Tanoue E (2002) Chemical and biological studies of particulate organic matter in the ocean. J Oceanogr 58:265–279. https://doi.org/10.1023/A:1015809708632

Wada S, Aoki MN, Tsuchiya Y et al (2007) Quantitative and qualitative analyses of dissolved organic matter released from *Ecklonia cava* Kjellman, in Oura Bay, Shimoda, Izu Peninsula, Japan.

J Exp Mar Biol Ecol 349:344–358. https://doi.org/10.1016/j.jembe.2007.05.024

Wada S, Aoki MN, Mikami A et al (2008) Bioavailability of macroalgal dissolved organic matter in seawater. Mar Ecol Prog Ser 370:33–44. https://doi.org/10.3354/meps07645

Wada S, Omori Y, Kayamyo Y et al (2015) Photoreactivity of dissolved organic matter from macroalgae. Reg Stud Mar Sci 2:12–18. https://doi.org/10.1016/j.rsma.2015.08.018

Watanabe A, Yamamoto T, Nadaoka K et al (2013) Spatiotemporal variations in $CO_2$ flux in a fringing coral reef simulated using a novel carbonate system dynamics model. Coral Reefs 32:239–254. https://doi.org/10.1007/s00338-012-0964-2

Wild C, Huettel M, Klueter A et al (2004a) Coral mucus functions as an energy carrier and particle trap in the reef ecosystem. Nature 428:66–70. https://doi.org/10.1038/nature02344

Wild C, Rasheed M, Werner U et al (2004b) Degradation and mineralization of coral mucus in reef environments. Mar Ecol Prog Ser 267:159–171. https://doi.org/10.3354/meps267159

Wild C, Woyt H, Huettel M (2005) Influence of coral mucus on nutrient fluxes in carbonate sands. Mar Ecol Prog Ser 287:87–98. https://doi.org/10.3354/meps287087

Wild C, Niggl W, Naumann MS, Haas AF (2010) Organic matter release by Red Sea coral reef organisms – potential effects on microbial activity and in situ $O_2$ availability. Mar Ecol Prog Ser 411:61–71. https://doi.org/10.3354/meps08653

Wild C, Laforsch C, Mayr C et al (2012) Effect of water currents on organic matter release by two scleractinian corals. Aquat Ecol 46:335–341. https://doi.org/10.1007/s10452-012-9404-1

Williams PM, Druffel ERM (1987) Radiocarbon in dissolved organic matter in the central North Pacific Ocean. Nature 330:246–248. https://doi.org/10.1038/330246a0

Wyatt ASJ, Lowe RJ, Humphries S et al (2010) Particulate nutrient fluxes over a fringing coral reef: relevant scales of phytoplankton production and mechanisms of supply. Mar Ecol Prog Ser 405:113–130. https://doi.org/10.3354/meps08508

Wyatt ASJ, Falter JL, Lowe RJ et al (2012) Oceanographic forcing of nutrient uptake and release over a fringing coral reef. Limnol Oceanogr 57:401–419. https://doi.org/10.4319/lo.2012.57.2.0401

Wyatt ASJ, Lowe RJ, Humphries S et al (2013) Particulate nutrient fluxes over a fringing coral reef: source-sink dynamics inferred from carbon to nitrogen ratios and stable isotopes. Limnol Oceanogr 58:409–427. https://doi.org/10.4319/lo.2013.58.1.0409

Yahel G, Sharp JH, Marie D et al (2003) In situ feeding and element removal in the symbiont-bearing sponge *Theonella swinhoei*: bulk DOC is the major source for carbon. Limnol Oceanogr 48:141–149. https://doi.org/10.4319/lo.2003.48.1.0141

Yoshinaga I, Fukami K, Ishida Y (1991) Comparison of DNA and protein synthesis rates of bacterial assemblages between coral reef waters and pelagic waters in tropical ocean. Mar Ecol Prog Ser 76:167–174. https://doi.org/10.3354/meps076167

Ziegler S, Benner R (1999) Dissolved organic carbon cycling in a subtropical seagrass-dominated lagoon. Mar Ecol Prog Ser 180:149–160. https://doi.org/10.3354/meps180149

Ziegler S, Benner R (2000) Effects of solar radiation on dissolved organic matter cycling in a subtropical seagrass meadow. Limnol Oceanogr 45:257–266. https://doi.org/10.4319/lo.2000.45.2.0257

# 3 Studies on Stress Responses of Corals in Japan

Tomihiko Higuchi

**Abstract**

Coral reefs in Japan are threatened by multiple environmental stresses at both the global and the local scales. Declining water quality in coastal reefs has been reported in the Okinawa region due to red soil runoff, agricultural fertilizers, and antifouling chemicals. Many studies on the stress responses of corals have been conducted by Japanese researchers. For example, metabolic changes due to stresses such as high temperature and chemical discharge have been reported in quantitative terms. Antioxidant enzyme activities and mycosporine-like amino acids have been studied as possible defense mechanisms against environmental stress. Moreover, coral bleaching has been frequently reported and actively studied since the 1980s in Japan. The synergistic effects of multiple stressors have also been studied, with several studies reporting accelerated bleaching under conditions of high seawater temperature and low water quality (high nitrate concentration and high bacterial abundance). The importance of water flow has been extensively studied, and it has been suggested to allow more rapid recovery from bleaching and a higher survival rate. To mitigate environmental stresses on coral reefs, it is important to evaluate risks due not only to global warming but also to local stresses.

**Keywords**

Global warming • Eutrophication • Chemical discharge • Coral bleaching • Oxidative stress • Antioxidant • Cold stress

## 3.1 Environmental Stresses on Coral Reefs in Japan

Coral reefs are threatened by several serious environmental problems, including global warming, coral bleaching, coastal development, increased sedimentation, tourism, overfishing, hurricanes, ocean acidification, and coral diseases (Brown 1997; Hughes et al. 2003; Fabricius 2005; Orr et al. 2005; Hoegh-Guldberg et al. 2007). Many studies have reported that elevated seawater temperature is the most notable environmental change affecting the symbiotic system between the host coral and its symbiotic algae, zooxanthellae (e.g., Brown 1997). In Okinawa, Japan, extensive bleaching due to high seawater temperature was observed in 1998 (Yamazato 1999). Seawater temperature was nearly 30 °C from the surface to a depth of 50 m in July 1998, compared with around 27 °C from the surface to a depth of 30 m in July 1997 (Yamazato 1999). The period of elevated temperature, during which the sea surface temperature (SST) exceeded 30 °C, corresponded with the period of bleaching, which confirmed that the major driver of coral bleaching in the Ryukyu Islands was high SST (Kayanne et al. 2002). On the shallow reef flat, seawater was 2–3 °C warmer than the measured SST in the outer ocean during the daytime and reached 35 °C in the middle of August 1998 (Kayanne et al. 1999). On the other hand, especially in northerly parts of Japan, low temperature

T. Higuchi (✉)
Atmosphere and Ocean Research Institute, The University of Tokyo, Tokyo, Japan
e-mail: thiguchi@aori.u-tokyo.ac.jp


A. Iguchi, C. Hongo (eds.), *Coral Reef Studies of Japan*, Coral Reefs of the World 13,
https://doi.org/10.1007/978-981-10-6473-9_3

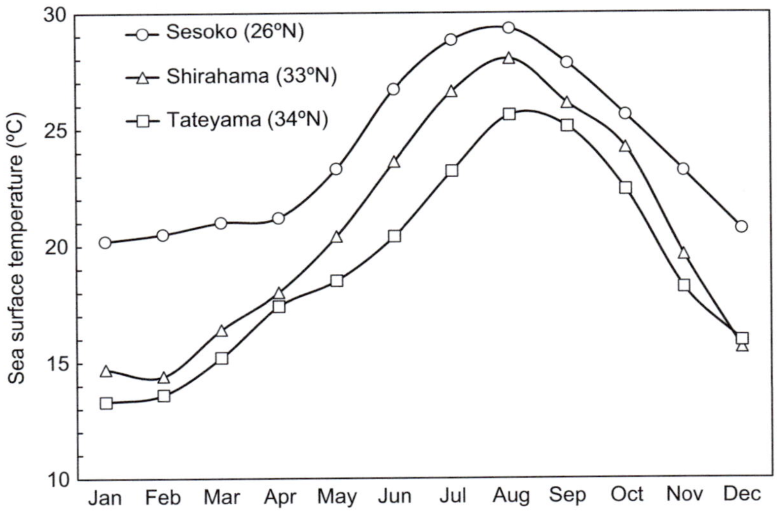


**Fig. 3.1** Monthly averages of sea surface temperature around Japan in 2013. ○, at Sesoko (26°N); △, at Shirahama (33°N); □, at Tateyama (34°N). Temperature data was extracted from Higuchi et al. (2015a)

is a serious stress determining coral survival (Higuchi et al. 2015a). SST close to the northern limit of the coral distribution in the temperate zone of Japan can fall as low as 13 °C in winter (e.g., Tateyama: 34°N, Higuchi et al. 2015a). Notably, corals survive winter SST below 13 °C around Iki Island (33.5°N) (Yamano et al. 2001). Thus, the coral distribution in Japan includes a wide range of temperatures across different geographic zones (Fig. 3.1).

Declining water quality in the coastal environment due to coastal development is considered to be one of the primary causes of the global decline in coral reefs (Hoegh-Guldberg 1999; Pandolfi et al. 2003). In Japan, red soil runoff from developed land has been a major environmental issue since the 1980s (Omija 2004). High turbidity after rainfall can often be observed in coral reef areas (Fig. 3.2). Terrestrial runoff of sediments, nutrients, and pollutants can adversely impact corals due to burial and smothering and can also reduce recruitment, decrease calcification, and decrease photosynthetically active radiation (PAR) (Fabricius 2005). The effects of short-term sedimentation with red soil on the corals *Coelastrea aspera*, *Porites lobata*, and *Pavona frondifera* were investigated in laboratory experiments (Ismail and Tsuchiya 2005). The results showed that massive corals, such as *C. aspera* and *P. lobata*, were more resistant to sediment stress than the platelike coral *P. frondifera*. In addition to the direct effects of red soil particles covering the corals, photochemical studies of red soil-polluted seawater have shown that photoformation of hydrogen peroxide ($H_2O_2$), a reactive oxygen species (ROS), was higher in red soil-polluted seawater than in unpolluted seawater (Nakajima et al. 2004). This increase in $H_2O_2$ photoformation was correlated with the total dissolved iron (Fe) concentration, which was higher in red soil-polluted seawater than in clean seawater (Nakajima et al. 2004; Okada et al. 2005). Furthermore, Nakajima et al. (2004) reported that $H_2O_2$ photochemical formation was further enhanced by strong UV radiation, suggesting that the decrease in the stratospheric ozone layer may increase concentrations of $H_2O_2$ in seawater in the future. In the Red Sea, $H_2O_2$ concentrations are 0.10–0.25 μM during the daytime in the lagoon at low tide (Shaked and Armoza-Zvuloni 2013). The concentration of seawater $H_2O_2$ in coral reef areas of Okinawa Island can reach 0.16 μM on sunny days (Arakaki et al. 2005), and higher concentrations (ca. 0.5 μM) of $H_2O_2$ were also observed when seawater was kept in a quartz test tube in the sunlight (Arakaki et al. 2007). This suggests that seawater that is not well mixed, such as in tidal pools at low tide, could have much higher concentrations of $H_2O_2$, reaching levels toxic to corals (Higuchi et al. 2009a).

Chemical discharge to coral reefs through rivers is also important to consider when assessing coral reef stressors. Pollution from domestic agricultural, industrial, and shipping activities has not been investigated as a problem affecting coral reefs around the Ryukyu Archipelago (Kitada et al. 2008; Sheikh et al. 2007; Tanabe et al. 2008). Terrestrial discharge onto coral reefs can be linked to changes in coral population and community dynamics in Japan (West and van Woesik 2001; Dikou and van Woesik 2006). Based on 15 years of data, it appeared that high turbidity influenced by recent terrestrial runoff reduced the resilience of *Acropora* species to severe thermal stress events because the corals could not adapt to the decline in water quality (Hongo and Yamano 2013). In coastal areas, pollution with nutrients, especially nitrate, via inflow from rivers and groundwater has increased over the last few decades (Umezawa et al. 2002; Higuchi et al. 2014). Nutrient pollution due to human activity has increased due to several factors but most notably the increased use of chemical fertilizers in agriculture (Umezawa et al. 2002). Reefs that receive groundwater discharge, such as the Sesoko reef in Okinawa, exhibit increased nitrate concentrations of about 10 μM (Meekaew et al. 2014). This increase is most notable at low tide due to the concen-

**Fig. 3.2** Effect of red soil pollution on coastal reef. (**a**) Sunny day and (**b**) after rainfall

trated effect of groundwater at that time (Higuchi et al. 2014). Inoue et al. (2014) indicated that coral skeletons of *Porites* sp. recorded clear seasonal variations in manganese (Mn), which is a proxy for river discharge. Tanahara et al. (2013) studied temporal changes in heavy metal concentrations in sediment cores from tidal land downstream of the main rivers in Naha, Okinawa, a densely populated area. Analysis of the soil revealed contamination by heavy metals from development upstream that had flowed into the tidal area. Diuron, [*N*′-(3,4-dichlorophenyl)-*N*,*N*-dimethyl-urea] (DCMU), is a photosystem II herbicide and is considered a priority hazardous substance by the European Commission (Malato et al. 2002). In 2004, around 11 tons of diuron was used for agricultural purposes in Okinawa Prefecture, which had the highest usage of this toxic substance in Japan outside of the Tokyo metropolitan region (Sheikh et al. 2009a). Diuron concentrations were periodically measured between 2007 and 2009 using 93 transects of the Shiraho coral reef waters of the Ryukyu Archipelago and the Todoroki River, which receives runoff from sugarcane production and discharges to the reefs (Sheikh et al. 2009a). Average DCMU concentrations in the river and reefs were 60.2 ± 157 and 2.3 ± 7 ng L$^{-1}$, respectively, with concentrations on the Shiraho reef ranging from below detection to 90 ng L$^{-1}$. The maximum concentration of DCMU was detected in winter after a period of intense rainfall. The DCMU contamination of Ryukyu Archipelago coral reefs is higher than that of other tropical reefs, as measured in Hong Kong (1.9–6.3 ng L$^{-1}$; Shaw et al. 2008) and in the Great Barrier Reef, Australia (0.2–1.6 ng L$^{-1}$; Shaw and Mueller 2005). In contrast, the maximum DCMU concentration in Okinawa sediment (0.22 μg kg$^{-1}$) was low relative to that in the Great Barrier Reef World Heritage Area, Australia ($<$0.1–10.1 μg kg$^{-1}$) (Haynes et al. 2000; Kitada et al. 2008; McMahon et al. 2005). Although DCMU is frequently found at high concentrations close to farmland and not cities, high concentrations were detected in both rural and moderately urbanized areas (Kitada et al. 2008). DCMU concentrations in urban areas were sometimes higher than in rural areas, which might be caused by greater usage in the gardens of urban homes (Kitada et al. 2008). Irgarol 1051 is used in antifouling paint, and high concentrations were observed in ports with shipping activity (Sheikh et al. 2009b). Kitada et al. (2008) reported that Irgarol 1051 can be detected in river areas far downstream of the source pier or fishery harbor (0.6–3.2 km). These results suggested that Irgarol 1051 could be transported from river mouths to the coastal ocean (Kitada et al. 2008). The average concentration of Irgarol detected in commercial and fisheries ports was 24.70 ± 9.88 ng $l^{-1}$, and in Naha Bay, its average concentration was 10.00 ± 12.98 ng $l^{-1}$, approaching the environmental risk limit for marine organisms (Sheikh et al. 2009b). Irgarol 1051 has been used since around 2003 instead of tributyltin (TBT), which is on the list of compounds legally restricted by the International Maritime Organization (IMO) (Gatidou et al. 2007). Sheikh et al. (2007) reported the presence of organotin compounds [TBT and triphenyltin (TPhT)] in water and sediments of the Manko and Okukubi estuaries with high boating and agricultural activities. Although high TBT concentrations were not detected in reef areas, the concentrations of TBT reported in upper streams may pose a risk to marine life (>1 ng $L^{-1}$; Tong et al. 1996). Bisphenol A (BPA) is classified as an endocrine disruptor and is released into the environment as a result of its use in polycarbonate plastic and epoxy resin (Krishnan et al. 1993). Concentrations of BPA in river water were low in reef areas in Okinawa (Kawahata et al. 2004; Kitada et al. 2006). Although BPA was detected in river sediment, its concentration remained too low for effects such as estrogenic stimulation of corals and coral reef organisms (Kitada et al. 2008). Therefore, the BPA concentrations presently observed in Okinawa are not likely to damage coral reefs. Benzophenone-3 (BP-3; oxybenzone) is used in sunscreen lotions as it protects against the damaging effects of ultraviolet radiation (Downs et al. 2016). Downs et al. (2016) suggested that oxybenzone may be a hazard to coral reef conservation by threatening the resiliency of coral reefs to climate change. In Okinawa, BP-3 levels on coral reefs 300–600 m away from public swimming beaches ranged from 0.4 to 3.8 ng $L^{-1}$. BP-3 was detected up to 600 m from the public beach, suggesting that this distance represents its maximum dispersion from the source (Tashiro and Kameda 2013). Although the concentration of BP-3 in Okinawa is not higher than those in other areas such as Hawaii (Downs et al. 2016), the threat of BP-3 from swimmers to corals and coral reefs remains important, as a concentration of 6.5 μg $L^{-1}$ has been shown to affect coral planulae. Coral reefs in Japan are affected by multiple environmental stresses, many of which can be linked to anthropogenic activities. For other effects of anthropogenic activities such as the impact of ocean acidification on coral reefs, please see Chap. 5. Predation of corals by organisms such as crown-of-thorns starfish is discussed in Chap. 9.

## 3.2 Stress Responses of Corals in Japan

Stress responses of hermatypic corals in Japan have been the subject of numerous studies (Table 3.1). Many indicators have been used to evaluate coral stress levels. Coral metabolism, an indicator of the physiological condition of coral colonies, has been studied using an apparatus called the continuous-flow, complete-mixing (CFCM) experimental system (Fujimura et al. 2008), which can quantitatively measure changes in coral metabolism under increasing stress levels. This system can simultaneously control temperature, light intensity, and chemical composition of the seawater. The CFCM system can acquire quantitative information during long-term experiments while continuously supplying fresh seawater (Fujimura et al. 2008). Using this experimental system, Fujimura et al. (2008) reported that high temperature increased respiration rate and decreased both photosynthesis and calcification rates in corals. Agostini et al. (2016) proposed that mitochondrial electron transport system activity (ETSA) could be used as an indicator of coral health, especially for the health of the host animal. ETSA showed positive correlations with metabolic rates and physiological variables such as zooxanthella density, biomass, photosynthesis, respiration, and calcification. Thus, measurement of all of the different metabolic rates during incubation could be avoided by measuring ETSA as a global indicator of coral health.

Contamination by the new antifouling biocide Irgarol is prevalent in the coastal areas around the coral reefs of Okinawa Island. However, current concentrations in the environment are <1 to 35 ng $L^{-1}$ and thus do not seriously impact the survival of reef-building corals in Japan (Sheikh et al. 2009b). Regarding other chemical contaminants, Watanabe et al. (2006) reported tissue detachment and decreases in symbiont density in *Acropora tenuis* juvenile polyps at a concentration of 1 μg $L^{-1}$ TBT or 10 μg $L^{-1}$ DCMU. A long-term experiment (about 50 days) showed adverse effects on coral growth at lower concentrations (0.4 μg $L^{-1}$ TBT and 1 μg $L^{-1}$ DCMU; Watanabe et al. 2007). Sheikh et al. (2012) indicated that DCMU inhibited photosynthesis and calcification of corals at a concentration of 10 μg $L^{-1}$. Yuyama et al. (2012b) examined the effects of thermal and chemical (DCMU and TBT) stresses on *A. tenuis* juvenile polyps. Oxidative stress-responsive protein was commonly expressed by stressed corals, regardless of the stressor.

ROS can lead to significant oxidative damage to the coral-algae symbiotic system (Lesser 2006; Weis 2008). Smith et al. (2005) hypothesized that $H_2O_2$ is the most important ROS associated with coral bleaching because it can act as a signaling molecule between zooxanthellae and their symbiotic hosts. Considering the high concentration of $H_2O_2$ in red soil-polluted areas, the physiological effects of $H_2O_2$ on reef-building corals were tested. Higuchi et al. (2009b) reported the effects of various concentrations of $H_2O_2$ on carbon metabolism by the coral *Coelastrea aspera* (formerly *Goniastrea aspera*). Higher concentrations of $H_2O_2$ decreased photosynthesis and calcification but did not affect respiration

**Table 3.1** List of studies in Japan on responses of corals against environmental changes

| Parameter | Response | References |
|---|---|---|
| High temperature | 1. Photosynthesis, respiration, and calcification of *Coelastrea aspera* were maximum in 29 °C. Photosynthesis and calcification were decreased from 31–33 °C, and respiration was increased with 31–33 °C | Fujimura et al. (2008) |
| | 2. Under 32 °C a large number of *Symbiodinium* were expelled, and loss of *Symbiodinium* from coral tissues led to coral bleaching. Six species of scleractinian corals, *Acropora selago*, *Acropora muricata*, *Heliofungia actiniformis*, *Ctenactis echinata*, *Oxypora lacera*, and *Pocillopora eydouxi*, were used | Fujise et al. (2013) |
| UV radiation | 3. UV-reduced condition (UV cut 78%) decreased 40% photoformation of OH radical in tissue extract of *Galaxea fascicularis* | Higuchi et al. (2010) |
| Hydrogen peroxide ($H_2O_2$) | 4. 0.3 μM does not affect on coral metabolisms. >3 μM $H_2O_2$ decreased photosynthesis (−18% with 3 μM) and calcification (−23% with 3 μM) of *Coelastrea aspera* | Higuchi et al. (2009b) |
| | 5. 0.3 and 3 μM $H_2O_2$ increased activities of antioxidant enzyme catalase | Higuchi et al. (2008) |
| $H_2O_2$ plus high temperature | 6. 3 μM $H_2O_2$ decreased photosynthesis (−21%) and calcification (−7%) of *Galaxea fascicularis*. The synergistic effect of 3 μM H2O2 combined with high seawater temperature (31 °C) increased respiration rate (+134%). | Higuchi et al. (2009a) |
| Red soil | 7. The massive corals, *Coelastrea aspera* and *Porites lobata*, are more resistant to sediment stress than the platelike coral *Pavona frondifera* | Ismail and Tsuchiya (2005) |
| Nitrate and phosphate plus high temperature | 8. The corals *Acropora tenuis* in high nutrient (4.8 μM $NO_3$, 0.28 μM $PO_4$) lost more zooxanthellae per unit surface area than those in low nutrient (0.2 μM $NO_3$, 0.06 μM $PO_4$), but the lost percentages were not significantly different between low and high nutrient under thermal stress of 31 °C | Tanaka et al. (2014) |
| Nitrate plus high temperature or high light | 9. 10 μM nitrate alone does not induce stress with ambient temperature (27 °C) and low light (200 μmol $m^{-2}$ $s^{-1}$). However, high nitrate increased bleaching severity (about 70% of zooxanthellae lost) under increased light (600 μmol $m^{-2}$ $s^{-1}$) or high temperature (32 °C) within 6 days | Higuchi et al. (2015b) |
| High light | 10. Corals containing clade C showed a greater decline in Fv/Fm (−74%), compared to decline in corals associated with clade D (−50%) with 500 μmol $m^{-2}$ $s^{-1}$ after 3 h | Yuyama et al. (2016) |
| Low temperature | 11. Measuring photosynthesis and respiration of *Acropora pruinosa* under 3 h exposure | Nakamura et al. (2004) |
| | 12. 13 °C stress led to death on *Acropora solitaryensis* and *Acropora hyacinthus* before 5 days. *A. pruinosa* survived with bleaching for 10 days under 13 °C. After water temperature increased, *A. pruinosa* recover from cold bleaching | Higuchi et al. (2015a) |
| Bacteria plus high temperature | 13. Increased abundance of bacteria such as *Vibrio coralliilyticus* and *Sulfitobacter* sp. did not affect coral metabolisms and zooxanthellae density under ambient temperature (27 °C). However, *Sulfitobacter* sp. accelerated coral bleaching and severely decreased photosynthesis and calcification under high temperature (32 °C) | Higuchi et al. (2013) |
| DCMU | 14. Tissue detachment and decreased in the symbiont density of *Acropora tenuis* at the concentrations of 10 μg $L^{-1}$ within 10 days exposure | Watanabe et al. (2006) |
| | 15. Adverse effects on coral growth at 1 μg $L^{-1}$with 50 days exposure | Watanabe et al. (2007) |
| | 16. Decreased photosynthesis and calcification of corals with 10 μg $L^{-1}$ for 4 days | Sheikh et al. (2012) |
| Water velocity | 17. Reduced water flow amplified photodamage of zooxanthellae under strong irradiance. Higher water-flow rates strongly reduced the degree of photoinhibition and the severity of bleaching in *Acropora digitifera* | Nakamura et al. (2005) |

(Higuchi et al. 2009b). The synergistic effect of high $H_2O_2$ with high temperature significantly increased respiration rates, which surpassed the effect of either $H_2O_2$ or high temperature alone (Higuchi et al. 2009a). High temperature alone decreased photosynthesis and calcification and increased respiration. These results suggest that the current level of $H_2O_2$ in Okinawa seawater is not likely to cause significant acute effects on the metabolic activities of corals. However, if the ozone layer in the stratosphere becomes thinner in the future, the amount of UV radiation reaching the surface of the ocean may increase, which in turn will increase the photochemical formation of $H_2O_2$. It is therefore important to continue monitoring the concentrations of seawater $H_2O_2$ in coral reef areas. Incidentally, high concentrations of

$H_2O_2$ (5 mM) induced spawning in Okinawan staghorn corals (Hayashibara et al. 2004). This spawning technique is used for many studies of larval and juvenile corals (Suzuki et al. 2011; Yamashita et al. 2013).

Environmental stresses such as high temperature can also induce increased production of ROS, including $H_2O_2$, by the coral host and its symbiont. The production of $H_2O_2$ was induced by inhibition of the chlorophyll a–chlorophyll $c_2$–peridinin protein complex (acpPC) in *Symbiodinium* (Takahashi et al. 2008). One defense mechanism against increased ROS is an increase in the activity of free radical scavenger enzymes such as superoxide dismutase (SOD) and catalase (CAT). SOD and CAT act in concert to inactivate superoxide radicals and $H_2O_2$. Under high-temperature conditions, both SOD and CAT activities increased in coral tissue and zooxanthellae (Yakovleva et al. 2004; Higuchi et al. 2008). Agostini et al. (2016) reported that high temperature increased Mn SOD, which may correlate with the production of ROS in the mitochondria. Yakovleva et al. (2004) suggested that coral bleaching triggered by high temperature was caused by the production of ROS, including $H_2O_2$, within the zooxanthellae. In $H_2O_2$ exposure experiments, CAT activities in both coral tissue and zooxanthellae increased (Higuchi et al. 2008). However, CAT activities in corals were not significantly affected by even tenfold increased concentrations of $H_2O_2$. It was thus suggested that corals have a limit to their ability to increase CAT activity (Higuchi et al. 2008, 2012). Thus, both high temperature and high $H_2O_2$ affect the metabolic and antioxidant activities of corals.

Strong UV radiation is a direct stress on the coral symbiotic system (Gleason and Wellington 1993). For many coral reefs, the overlying water column allows UV radiation to penetrate to depths of 20 m or deeper because oligotrophic waters are dominated by the optical properties of the water itself and not by dissolved or particulate constituents in the water column (Banaszak and Lesser 2009). Higuchi et al. (2010) reported photochemical production of hydroxyl radical (•OH) in extracts of both coral hosts and endosymbiotic zooxanthellae. Their results indicated that, whether or not coral bleaching occurred, coral tissues and symbiotic zooxanthellae had the potential to produce •OH photochemically under exposure to sunlight. Moreover, reducing the amount of UV radiation reduced •OH formation rates by 40%, indicating that UV radiation strongly affects •OH formation in coral tissue and zooxanthellae in addition to its formation through photochemical processes. Gleason and Wellington (1993) showed that reef-building corals neutralize the damaging effects of UV light through the synthesis or accumulation of UV-absorbing compounds, known as mycosporine-like amino acids (MAAs). Shick et al. (1999) indicated that MAAs originate from the algae in algal-invertebrate symbioses during a bleaching event. As more zooxanthellae are lost from coral tissue, more UV radiation can penetrate into the corals, enhancing the formation of •OH in the coral tissues and causing more damage. On the other hand, the whole genome sequence of *Acropora digitifera* indicated the presence of homologues encoding all four shinorines, a group of MAAs. Thus, MAA synthesis in corals is not symbiont dependent (Shinzato et al. 2011). Yakovleva and Hidaka (2009) reported that MAA concentration and composition varied among color morphotypes of the coral *Galaxea fascicularis*. Yakovleva et al. (2004) proposed that mycosporine-glycine, an MAA, also plays a role in providing rapid protection against oxidative stress, functioning as a biological antioxidant in the host coral and zooxanthellae under thermal stress.

Responses of host corals to stressors can vary depending on the types of zooxanthellae associated with them. Using monoclonal cultures of *Symbiodinium* clades, Yuyama et al. (2012a) observed changes in the expression of stress marker genes in the coral depending on the symbiont type. The expression levels of four genes, encoding green and red fluorescent proteins, an oxidative stress-responsive protein, and an ascorbic acid transporter, decreased at high temperatures when juvenile corals were associated with clade A zooxanthellae but increased when the zooxanthellae were in clade D. Yuyama et al. (2016) reported that juvenile corals hosting *Symbiodinium* clade C showed a lower survival rate than those with clade D symbionts during thermal stress, although photosynthetic efficiency ($F_v/F_m$) was lower at 30 °C than at 25 °C in both clades. In addition, after 3 h of strong light exposure (500 μmol $m^{-2}$ $s^{-1}$), corals containing clade C1 showed a greater decline in $F_v/F_m$ (−74%) than corals associated with clade D (−50%), indicating that an association with clade D leads to a greater tolerance to strong light than with clade C (Yuyama et al. 2016).

Under natural conditions, a single source of stress, such as elevated water temperature, rarely occurs in isolation, and instead multiple stressors often occur simultaneously. While most studies have focused on responses to individual stressors, the synergistic interactions between increased temperature and other factors have also been examined, including UV radiation (Lesser et al. 1990), salinity (Porter et al. 1999), copper (Nyström et al. 2001), and nitrate (Nordemar et al. 2003). Researchers in Japan have investigated the combined effects of multiple stresses on corals for better elucidation of the impacts of environmental changes on Japanese coral reefs (Higuchi et al. 2009a, 2013; Tanaka et al. 2014). Higuchi et al. (2013) reported the effects of high seawater temperature and bacterial inoculation (a mixture of five potential coral pathogens naturally found in coastal regions of Okinawa, Japan) on coral bleaching, metabolism, and antioxidant defenses in *Montipora digitata*. At ambient temperature (27 °C), the added bacteria did not cause coral bleaching and did not affect metabolism or antioxidant enzyme activities. However, bacterial inoculations in addition to high-temperature stress (32 °C) resulted in severe coral bleaching, decreased photosynthetic efficiency ($F_v/F_m$), and decreased photosynthesis and calcification rates. Corals under thermal stress alone showed signs of bleaching, but the changes in zooxanthella density and metabolic rates were less severe than for those under combined thermal stress and bacterial challenge. Among the bacteria examined, the sulfite-oxidizing bacteria *Sulfitobacter* sp. had the greatest capacity to enhance and accelerate the bleaching process under thermal stress. *Sulfitobacter* sp. was isolated from the water surrounding bleached *M. digitata*. These results suggest that bacterial challenge and thermal stress can have synergistic negative effects on living corals (Higuchi et al. 2013). As shown above, high concentrations of nitrate caused by agricultural fertilizer runoff (>10 μM) have been measured in coastal areas (Umezawa et al. 2002; Higuchi et al. 2014). Higuchi et al. (2015b) reported the combined effects of elevated nitrate concentrations with physical stresses (temperature and light intensity) on coral bleaching and antioxidant enzyme activity. Two nitrate concentration levels (<1, 10 μM) were tested along with two temperature conditions (27, 32 °C) and two levels of photon intensity (200, 600 μmol $m^{-2}$ $s^{-1}$). The density of zooxanthellae did not significantly change when subjected to only high nitrate or only high light. On the other hand, the combined stresses of high light or high temperature with high nitrate significantly decreased zooxanthella density. Moreover, there was a negative correlation between zooxanthella density and activities of the antioxidant enzymes SOD and CAT, leading to the conclusion that the observed coral bleaching occurred due to high ROS levels (Higuchi et al. 2015b). High nitrate concentration also affects the resilience of the zooxanthella-coral relationship at high temperatures (Chumun et al. 2013). Taken together, these results suggested that both high temperature and high light intensity affected coral bleaching when nitrate concentrations were high, but their combined effect cannot be estimated by simply summing the effects of the individual stressors.

## 3.3 Studies on Coral Bleaching in Japan

Massive coral bleaching incidents have recently increased around the world due to changes in the environment with both natural and anthropogenic causes. The main causes of coral bleaching are believed to be high seawater temperatures, high irradiance, and strong UV radiation (Fitt et al. 2001). High seawater temperature disrupts the symbiotic association between the animal host and the symbiotic algae, leading to the bleaching of the coral. During the 1997–1998 El Niño, sea surface temperatures rose well above normal, and many tropical coral reefs were bleached or experienced high mortality (Glynn et al. 2001). In the past, several bleaching events have been reported on Japanese coral reefs (e.g., 1980, 1983, 1987, 1991, 1993, 1994, 1998, 2001, 2003, 2005, 2007, 2010, 2013; Yamazato 1999; Yara et al. 2009; Muko et al. 2013; Harii et al. 2014). Uniquely, Japanese divers and researchers used the Sango (coral) Map Project (Namizaki et al. 2013), a citizen science project, to collect bleaching information in 2013. The most severe bleaching event in Japan occurred in 1998 (Fig. 3.3), when many corals were bleached and died in Okinawa (Yamazato 1999). Bleaching was first observed in early July, peaked in August and September, and ceased in early October (Hasegawa et al. 1999, Yamazato 1999). Seawater temperature in 1998 was 1–2 °C higher than in normal years (Kayanne et al. 2002). Harii et al. (2014) showed that a shift in dominant coral taxa, from branching *Montipora* and *Acropora* to *Porites* and *Heliopora*, resulted from multiple disturbances, such as the bleaching events in 1998 and 2007, physical damage from strong typhoons, and inputs of sediment after heavy rain.

As observed in mesocosm experiments, coral bleaching severely affects carbon metabolism including photosynthesis and calcification (Fujimura et al. 2001). While Takahashi et al. (2004) showed that photosynthesis in zooxanthellae is very sensitive to water temperature and is impaired when temperature is high, the exact mechanism of the loss of zooxanthellae during bleaching remains unclear. In one thermal stress experiment, a large number of *Symbiodinium* were expelled, and this loss from coral tissues led to coral bleaching (Fujise et al. 2013). However, in another investigation of corals during a bleaching event, degraded zooxanthellae were found in high numbers in the coral tissue (Mise and Hidaka 2003). Fujise et al. (2014) proposed that the expulsion mechanisms differed depending on temperature. Under ambient temperature conditions, the expulsion of *Symbiodinium* was undertaken as part of a regulatory mechanism to maintain constant zooxanthella density. However, at 30 °C (moderate thermal stress), *Symbiodinium* were damaged, and corals selectively digested the damaged cells or expelled them without digestion. Under high-temperature (32 °C) stress, the abundance of healthy zooxanthellae in coral tissues decreased, the proportion of abnormal cells increased, and the rate of zooxanthella expulsion was reduced (Suzuki et al. 2015). Within the shrunken cells, a chlorophyll-like pigment, $13^2$, $17^3$-cyclopheophorbide a enol (cPPB-aE), which is not ordinarily found in healthy zooxanthellae, was detected. The presence of cPPB-aE in zooxanthellae suggests that chlorophyll was degraded to cPPB-aE, a compound that is not fluorescent. The lack of fluorescence reduced the rate of ROS formation in zooxanthellae. Therefore, it has been suggested that the formation of cPPB-aE in zooxanthellae is a defense mechanism by the host coral to avoid oxidative stress (Suzuki et al. 2015).

**Fig. 3.3** Mass bleaching event on 1998 in Okinawa, Japan (Photo credit: Yoshikatsu Nakano (Univ. of the Ryukyus))

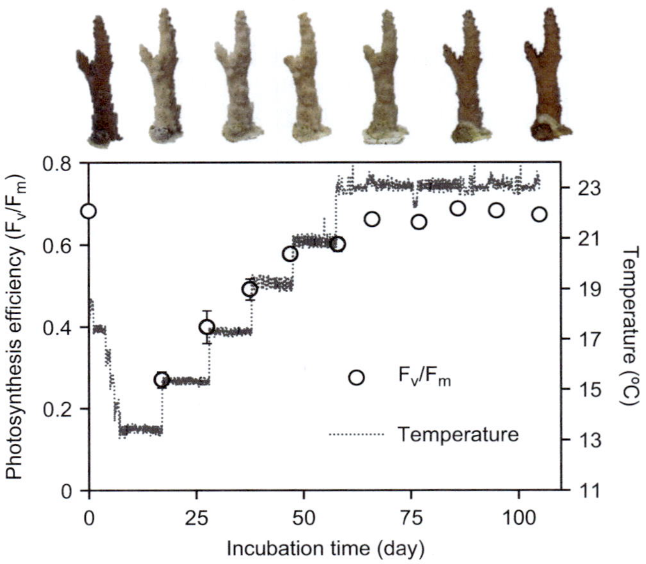


**Fig. 3.4** Variation in the photosynthetic quantum yield $F_v/F_m$ and photograph of *Acropora pruinosa* under cold stress and recovery process (Modified from Higuchi et al. 2015a)

On the other end of the temperature range experienced by corals in Japan, Nakamura et al. (2004) investigated the effect of cold stress on the coral *Acropora pruinosa* found in the warm temperate zone of Japan and found that photosynthesis and respiration were reduced under low-temperature conditions. Higuchi et al. (2015a) also reported that acroporid corals incubated at low temperature (13 °C) showed serious bleaching symptoms (Fig. 3.4). Photosynthetic efficiency, respiration, calcification, and activities of the antioxidant enzymes SOD and CAT all decreased at low temperature. Decreases in photosynthesis and calcification were also observed with high-temperature bleaching stress, but the decreases in respiration and antioxidant enzyme activities were the opposite of the responses to heat stress. *A. pruinosa*, which is distributed up to the northernmost limit for hermatypic corals in the world, did not die after 10 days under cold stress and was able to recover when temperature was increased (Fig. 3.4). Cold stress bleaching might be a result of the disposal of zooxanthellae that lost their photosynthetic function due to low temperature. Cold stress bleaching could also serve to limit the energy consumption of zooxanthellae with low photosynthetic efficiency. We suggested that cold bleaching is an important mechanism for *A. pruinosa* to survive low winter temperatures (Higuchi et al. 2015a).

The effect of water flow on the response of corals to high-temperature stress and the bleaching mechanism has been extensively studied in Japan (Nakamura 2010). Limited water flow lowers the temperature threshold for inhibition (Nakamura and Yamasaki 2013), leading to enhanced bleaching and increased damage to zooxanthella photosystems. On the other hand, high water flow allows faster recovery from bleaching in terms of chlorophyll concentration and zooxanthella density and a higher survival rate for colonies exposed to high temperature, both experimentally and in the field (Nakamura and van Woesik 2001; Nakamura et al. 2005).

In a global study, Downs et al. (2002) suggested that bleaching is the final defense of corals against oxidative stress. Several studies in Japan have investigated this bleaching mechanism under high-temperature conditions (Yakovleva et al. 2004; Higuchi et al. 2008; etc.). If antioxidant activity is insufficient, the final defense mechanism (i.e., bleaching) might be activated. However, bleaching in cold conditions might not be due to oxidative stress (Higuchi et al. 2015a) and was not triggered by the addition of a high concentration of $H_2O_2$ (Higuchi et al. 2008). It may be that because the oxidative stress was not caused by zooxanthellae, corals did not expel them (Fig. 3.5). Furthermore, coral bleaching was inhibited by simultaneous $H_2O_2$ and high-temperature stresses (Higuchi et al. 2012). It is likely that the bleaching mechanism was not triggered because the high

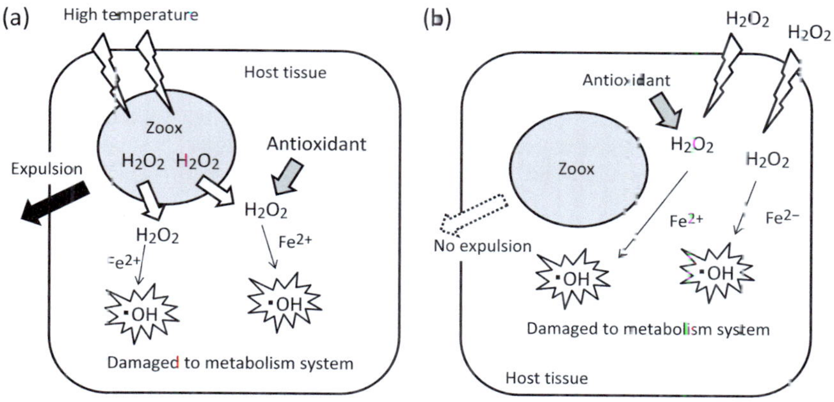


**Fig. 3.5** Reactive oxygen species in zooxanthellae (zoox) and host tissue (**a**) under high-temperature stress and (**b**) under $H_2O_2$ stress. Both conditions cause oxidative stresses in coral cells and damage to metabolic system. However, under $H_2O_2$ stress only, coral does not expel zoox

$H_2O_2$ concentration was in the coral cytosol during $H_2O_2$ exposure, rather than by the zooxanthellae.

## 3.4 Future Directions

As discussed above, many studies on the effects of environmental stresses on corals have been conducted in Japan. The response to each stressor has been tested quantitatively, and this information will help to determine conservation standards for coral reefs in Japan. However, stress response studies were mainly conducted under manipulated conditions in the laboratory, and more field-based quantitative research is still required. Coral bleaching was studied under both hot- and cold-temperature stresses. In Japan, Yara et al. (2009) suggested that both the frequency and the areal extent of coral bleaching around the Ryukyu Islands are expected to increase in the second half of the twenty-first century. Moreover, as multiple stressors enhance the bleaching of corals in a synergistic manner (Higuchi et al. 2013, 2015b), we need to account not only for global warming but also for local stresses such as nutrient discharge in reef assessments. These stresses should be avoided to mitigate environmental impacts on coral reefs.

## References


Agostini S, Fujimura H, Hayashi H, Fujita K (2016) Mitochondrial electron transport activity and metabolism of experimentally bleached hermatypic corals. J Exp Mar Biol Ecol 475:100–107

Arakaki T, Fujimura H, Hamdun AM, Okada K, Kondo H, Oomori T, Tanahara A, Taira H (2005) Simultaneous measurement of hydrogen peroxide and Fe species (Fe(II) and Fe(tot)) in Okinawa island seawater: impact of red soil pollution. J Oceanogr 61:561–568

Arakaki T, Ikota H, Okada K, Kuroki Y, Nakajima H, Tanahara A (2007) Behavior of hydrogen peroxide between atmosphere and coastal seawater around Okinawa Island. Chikyukagaku Geochem 41:35–41. (in Japanese with English abstract)

Banaszak AT, Lesser MP (2009) Effects of solar ultraviolet radiation on coral reef organisms. Photochem Photobiol 8:1276–1294

Brown BE (1997) Coral bleaching: causes and consequences. Coral Reefs 16:129–138

Chumun PK, Casareto BE, Higuchi T, Irikawa A, Bhagooli R, Ishikawa Y, Suzuki Y (2013) High nitrate levels exacerbate thermal photo-physiological stress of zooxanthellae in the reef-building coral *Pocillopora damicorn s*. Eco-Eng 25:75–83

Dikou AR, van Woesik R (2006) Partial colony mortality reflects coral community dynamics: a fringing reef study near a small river in Okinawa, Japan. Mar Pollut Bull 52:269–280

Downs CA, Fauth JE, Halas JC, Dustan P, Bemiss J, Woodley CM (2002) Oxidative stress and seasonal coral bleaching. Free Radic Biol Med 33(4):533–543

Downs CA, Kramarsky-Winter E, Segal R, Fauth J, Knutson S, Bronstein O, Ciner FR, Jeger R, Lichtenfeld Y, Woodley CM, Pennington P, Cadenas K, Kushmaro A, Loya Y (2016) Toxicopathological effects of the sunscreen UV filter, Oxybenzone (benzophenone-3), on coral planulae and cultured primary cells and its environmental contamination in Hawaii and the U.S. Virgin Islands. Arch Environ Contam Toxicol 70:265–288

Fabricius KE (2005) Effects of terrestrial runoff on the ecology of corals and reefs: review and synthesis. Mar Pollut Bull 50:125–146

Fitt WK, Brown BE, Warner ME, Dunne RP (2001) Coral bleaching: interpretation of thermal tolerance limits and thermal thresholds in tropical corals. Coral Reefs 20:51–65

Fujimura H, Oomori T, Maehira T, Miyahira K (2001) Changes of coral carbon metabolism influenced by coral bleaching. Galaxea, JCRS 3:41–50

Fujimura H, Higuchi T, Shiroma K, Arakaki T, Hamdun AM, Nakano Y, Oomori T (2008) Continuous-flow complete-mixing system for assessing the effects of environmental factors on colony-level coral metabolism. J Biochem Bioph Meth 70:865–872

Fujise L, Yamashita H, Suzuki G, Koike K (2013) Expulsion of zooxanthellae (*Symbiodinium*) from several species of scleractinian corals: comparison under non-stress conditions and thermal stress conditions. Galaxea, JCRS 15:29–36

Fujise L, Yamashita H, Suzuki G, Sasaki K, Liao LM, Koike K (2014) Moderate thermal stress causes active and immediate expulsion of photosynthetically damaged zooxanthellae (*Symbiodinium*) from corals. PLoS One 9:e114321

Gatidou G, Thomaidis NS, Zhou JL (2007) Fate of Irgarol 1051, diuron and their main metabolites in two UK marine systems after restrictions in antifouling paints. Environ Int 33 70–77

Gleason DF, Wellington GM (1993) Ultraviolet radiation and coral bleaching. Nature 365:836–838

Glynn PW, Mate JL, Baker AC, Calderon MO (2001) Coral bleaching and mortality in Panama and Equador during the 1997–1998 El Nino Southern Oscillation event: spatial/temporal patterns and comparisons with the 1982–1983 event. Bull Mar Sci 69:79–109

Harii S, Hongo C, Ishihara M, Ide Y, Kayanne H (2014) Impacts of multiple disturbances on coral communities at Ishigaki Island, Okinawa, Japan, during a 15 year survey. Mar Ecol Prog Ser 509:171–180

Hasegawa H, Ichikawa K, Kobayashi M, Kobayashi T, Hoshino M, Mezaki S (1999) The mass-bleaching of coral reefs in the Ishigaki lagoon, 1998. Galaxea, JCRS 1:31–39. (in Japanese with English abstract)

Hayashibara T, Iwao K, Omori M (2004) Induction and control of spawning in Okinawan staghorn corals. Coral Reefs 23:406–409

Haynes D, Muller J, Carter S (2000) Pesticide and herbicide residues in sediments and sea grasses from the great barrier reef world heritage area and Queensland coast. Mar Pollut Bull 41:287–297

Higuchi T, Fujimura H, Arakaki T, Oomori T (2008) Activities of antioxidant enzymes (SOD and CAT) in the coral *Galaxea fascicularis* against increased hydrogen peroxide concentrations in seawater. Int Coral Reef Symp 11:926–930

Higuchi T, Fujimura H, Arakaki T, Oomori T (2009a) The synergistic effects of hydrogen peroxide and elevated seawater temperature on the metabolic activities of the coral, *Galaxea fascicularis*. Mar Biol 156:589–596

Higuchi T, Fujimura H, Ikota H, Arakaki T, Oomori T (2009b) The effects of hydrogen peroxide on metabolism in the coral *Goniastrea aspera*. J Exp Mar Biol Ecol 370:48–55

Higuchi T, Fujimura H, Hitomi Y, Arakaki T, Oomori T, Suzuki Y (2010) Photochemical formation of hydroxyl radicals in tissue extracts of the coral *Galaxea fascicularis*. Photochem Photobiol 86:1421–1426

Higuchi T, Fujimura H, Suzuki Y (2012) Multiple effects of hydrogen peroxide and temperature on antioxidants and bleaching. Int Coral Reef Symp 12:4C_3

Higuchi T, Agostini S, Casareto BE, Yoshinaga K, Suzuki T, Nakano Y, Fujimura H, Suzuki Y (2013) Bacterial enhancement of bleaching and physiological impacts on the coral *Montipora digitata*. J Exp Mar Biol Ecol 440:54–60

Higuchi T, Takagi KK, Matoba K, Kobayashi S, Tsurumi R, Arakaki S, Nakano Y, Fujimura H, Oomori T, Tsuchiya M (2014) The nutrient and carbon dynamics that mutually benefit coral and seagrass in mixed habitats under the influence of groundwater at Bise coral reef, Okinawa, Japan. Int J Mar Sci 4:1–15

Higuchi T, Agostini S, Casareto BE, Suzuki Y, Yuyama I (2015a) The northern limit of corals of the genus *Acropora* in temperate zones is determined by their resilience to cold bleaching. Sci Rep 5:18467

Higuchi T, Yuyama I, Nakamura T (2015b) The combined effects of nitrate with high temperature and high light intensity on coral bleaching and antioxidant enzyme activities. Reg Stud Mar Sci 2:27–31

Hoegh-Guldberg O (1999) Climate change, coral bleaching and the future of the world's coral reefs. Mar Freshwater Res 50(8):839

Hoegh-Guldberg O, Mumby PJ, Hooten AJ, Steneck RS et al (2007) Coral reefs under rapid climate change and ocean acidification. Science 318:1737–1742

Hongo C, Yamano H (2013) Species-specific responses of corals to bleaching events on anthropogenically turbid reefs on Okinawa Island, Japan, over a 15-year period (1995–2009). PLoS One 8:e60952

Hughes TP, Baird AH, Bellwood DR, Card M et al (2003) Climate change, human impacts, and the resilience of coral reefs. Science 301:929–933

Inoue M, Ishikawa D, Miyaji T, Yamazaki A, Suzuki A, Yamano H, Kawahata H, Watanabe T (2014) Evaluation of Mn and Fe in coral skeletons (*Porites* spp.) as proxies for sediment loading and reconstruction of fifty years of land use on Ishigaki Island, Japan. Coral Reefs 33:363–373

Ismail M, Tsuchiya M (2005) Oxygen consumption rate of the corals *Goniastrea aspera*, *Porites lobata*, and *Pavona frondifera* under red soil stress in the Ryukyu Islands, Okinawa, Japan. Galaxea, JCRS 7:11–22

Kawahata H, Ohta H, Inoue M, Suzuki A (2004) Endocrine disrupter nonylphenol and bisphenol A contamination in Okinawa and Ishigaki Islands, Japan-within coral reefs and adjacent river mouths. Chemosphere 55:1519–1527

Kayanne H, Harii S, Yamano H, Tamura M, Ide Y, Akimoto F (1999) Changes in living coral coverage before and after the 1998 bleaching event on coral reef flats of Ishigaki Island, Ryukyu Islands. Galaxea, JCRS 1:73–82. (in Japanese with English abstract)

Kayanne H, Harii S, Ide Y, Akimoto F (2002) Recovery of coral populations after the 1998 bleaching on Shiraho Reef, in the southern Ryukyus, NW Pacific. Mar Ecol Prog Ser 239:93–103

Kitada Y, Kawahata H, Suzuki A, Oomori T (2006) Concentrations of nonylphenol and bisphenol A in the Okinawa Island using liquid chromatography-electrospray tandem mass spectrometry. Int Coral Reef Symp 10:810–818

Kitada Y, Kawahata H, Suzuki A, Oomori T (2008) Distribution of pesticides and bisphenol A in sediments collected from rivers adjacent to coral reefs. Chemosphere 71:2082–2090

Krishnan A, Stathis P, Permceth S, Tokes L, Feldman D (1993) Bisphenol A: an estrogenic substance is released from polycarbonate flasks during autoclaving. Endocrinology 132:2279–2286

Lesser MP (2006) Oxidative stress in marine environment: biochemistry and physiological ecology. Annu Rev Physiol 68:253–278

Lesser MP, Stochaj WR, Tapley DW, Shick JM (1990) Bleaching in coral reef anthozoans: effects of irradiance, ultraviolet radiation, and temperature on the activities of protective enzymes against active oxygen. Coral Reefs 8:225–232

Malato S, Blanco J, Fernandez-Alba R, Aguera A, Rodrigues A (2002) Photocatalytic treatment of water-soluble pesticides by photo Fenton and $TiO_2$ using solar energy. Catal Today 76:209–220

McMahon K, Bengston Nash SM, Eagelshman G, Muller JF, Duke N, Winderrich S (2005) Herbicide contamination and the potential impact to seagrass meadows in Harvey Bay, Queensland, Australia. Mar Pollut Bull 51:325–334

Meekaew A, Casareto BE, Higuchi T, Chumun PK, Suzuki Y (2014) Dissolved organic carbon cycling and the roles of the microbial community in coexistence of corals and seagrasses in Bise, Okinawa, Japan. Eco-Eng 26:81–88

Mise T, Hidaka M (2003) Degradation of zooxanthellae in the coral *Acropora nasuta* during bleaching. Galaxea, JCRS 5:33–39

Muko S, Arakaki S, Nagao M, Sakai K (2013) Growth form dependent response to physical disturbance and thermal stress in *Acropora* corals. Coral Reefs 32:269–280

Nakajima H, Okada K, Fujimura H, Arakaki T, Tanahara A (2004) Photochemical formation of peroxides in coastal seawater around Okinawa Island. Bunseki Kagaku 53:891–897. (in Japanese with English abstract)

Nakamura T (2010) Importance of water-flow on the physiological responses of reef-building corals. Galaxea, JCRS 12:1–14

Nakamura T, van Woesik R (2001) Water-flow rates and passive diffusion partially explain different survival of corals during 1998 bleaching event. Mar Ecol Prog Ser 212:301–304

Nakamura T, Yamasaki H (2013) Differential influence of water-flow on photoinhibition of photosynthesis in endosymbiotic algae among shallow-inhabiting scleractinian corals. Galaxea, JCRS 15S:160–165

Nakamura E, Yokohama Y, Tanaka J (2004) Photosynthetic activity of a temperate coral *Acropora pruinosa* (Scleractinia, Anthozoa) with symbiotic algae in Japan. Phycol Res 52:38–44

Nakamura T, vanWoesik R, Yamasaki H (2005) Water flow reduces photoinhibition of photosynthesis in endosymbiotic algae of reef-building coral *Acropora digitifera* (Scleractinia, Anthozoa). Mar Ecol Prog Ser 301:109–118
Namizaki N, Yamano H, Suzuki R, Oohori K, Onaga H, Kishimoto T, Sagawa T, Machida Y, Yasumura S, Satoh T, Shigiya T, Shibata T, Tsuchikawa M, Miyamoto Y, Harukawa K, Hirate K, Furuse K, Hokoyama K, Yamanaka Y, Wagatsuma T (2013) The potential of citizen monitoring programs for marine areas: activities of the two-year Sango (Coral) Map Project. Galaxea, JCRS 15:391–395
Nordemar I, Nyström M, Dizon R (2003) Effects of elevated seawater temperature and nitrate enrichment on the branching coral *Porites cylindrica* in the absence of particulate food. Mar Biol 142:669–677
Nyström M, Nordemar I, Tedengren M (2001) Simultaneous and sequential stress from increased temperature and copper on the metabolism of hermatypic coral *Porites cylindrica*. Mar Biol 138:1225–1231
Okada K, Nakajima H, Fujimura H, Arakaki T, Tanahara A, Oomori T (2005) Photochemical behavior of Fe (II) in coastal seawater around Okinawa Island by HPLC with a Ferrozine reagent. Bunseki Kagaku 54:861–867. (in Japanese with English abstract)
Omija T (2004) Terrestrial inflow of soils and nutrients. In: The Japanese Coral Reef Society and Ministry of the Environment (ed) Coral reefs of Japan. Ministry of the Environment, Tokyo, pp 64–68
Orr JC, Fabry VJ, Aumont O, Bopp L, Doney SC, Feely RA, Gnanadesikan A, Gruber N, Ishida A, Joos F, Key RM, Lindsay K, Maier-Reimer E, Matear R, Monfray P, Mouchet A, Najjar RG, Plattner GK, Rodgers KB, Sabine CL, Sarmiento JL, Schlitzer R, Slater RD, Totterdell IJ, Weirig MF, Yamanaka Y, Yool A (2005) Anthropogenic ocean acidification over the twenty-first century and its impact on calcifying organisms. Nature 437:681–668
Pandolfi JM, Bradbury RH, Sala E, Hughes TP, Bjorndal KA, Cooke RG, McArdle D, McClenachan L, Newman MJH, Paredes G, Warner RR, Jackson JBC (2003) Global trajectories of the long-term decline of coral reef ecosystems. Science 301:955–958
Porter JW, Lewis SK, Porter KG (1999) The effect of multiple stressors on the Florida keys coral reef ecosystem: a landscape hypothesis and a physiological test. Limnol Oceanogr 44:941–949
Shaked Y, Armoza-Zvuloni R (2013) Dynamics of hydrogen peroxide in a coral reef: sources and sinks. J Geophys Res Biogeosci 118:1793–1801
Shaw C, Mueller JF (2005) Preliminary evaluation of the occurrence of herbicides and PAHs in the wet tropics region of the great barrier reef, Australia, using passive samplers. Mar Pollut Bull 51:876–881
Shaw C, Lam KS, Mueller JF (2008) Photosystem II herbicide pollution in Hong Kong and its potential photosynthetic effects on corals. Mar Pollut Bull 57:473–478
Sheikh MA, Tsuha K, Wang X, Sawano K, Imo ST, Oomori T (2007) Spatial and seasonal behaviour of organotin compounds in protected subtropical estuarine ecosystems in Okinawa. Jpn. Int J Environ Anal Chem 87:847–861
Sheikh MA, Fujimura H, Miyagi T, Uechi Y, Yokota T, Yasumura S, Oomori T (2009a) Detection and ecological threats of PSII herbicide diuron on coral reefs around the Ryukyu Archipelago, Japan. Mar Pollut Bull 58:1922–1926
Sheikh MA, Higuchi T, Fujimura H, Imo TS, Miyagi T, Oomori T (2009b) Contamination and impacts of the new antifouling biocide Irgarol-1051 on subtropical coral reef waters. Int J Environ Sci Technol 6:353–358
Sheikh MA, Oomori T, Fujimura H, Higuchi T, Imo T, Akamatsu A, Miyagi T, Yokota T, Yasumura S (2012) Distribution and potential effects of novel antifouling herbicide diuron on coral reefs herbicides. In: Alvarez-Fernandez R (ed) Environmental impact studies and management approaches. InTech, Rijeka, pp p83–p91
Shick JM, Romaine-Lioud S, Ferrier-Pagès C, Gattuso JP (1999) Ultraviolet-B radiation stimulates shikimate pathway-dependent accumulation of mycosporine-like amino acids in the coral *Stylophora pistillata* despite decreases in its population of symbiotic dinoflagellates. Limnol Oceanogr 44:1667–1682
Shinzato C, Shoguchi E, Kawashima T, Hamada M, Hisata K, Tanaka M, Fujie M, Fujiwara M, Koyanagi R, Ikuta T Fujiyama A, Miller DJ, Satoh N (2011) Using the *Acropora digitifera* genome to understand coral responses to environmental change. Nature 476:320–323
Smith DJ, Suggett DJ, Baker NR (2005) Is photoinhibition of zooxanthellae photosynthesis the primary cause of thermal bleaching in corals? Glob Chan Biol 11:1–11
Suzuki G, Kai S, Yamashita H, Suzuki K, Iehisa Y, Hayashibara T (2011) Narrower grid structure of artificial reef enhances initial survival of in situ settled coral. Mar Pollut Bull 62:2803–2812
Suzuki T, Casareto BE, Shioi Y, Ishikawa Y, Suzuki Y (2015) Finding of $13^2$, $17^3$-cyclopheophorbide a enol as a degradation product of chlorophyll in shrunk zooxanthellae of the coral *Montipora digitata*. J Phycol 51:37–45
Takahashi S, Nakamura T, Sakamizu M, van Woesik R, Yamasaki H (2004) Repair machinery of symbiotic photosynthesis as the primary target of heat stress for reef-building corals. Pla Cell Physiol 45:251–255
Takahashi S, Whitney S, Itoh S, Maruyama T, Badger M (2008) Heat stress causes inhibition of the de novo synthesis of antenna proteins and photobleaching in cultured *Symbiodinium*. Proc of the Natl Acad Sci USA 105:4203–4208
Tanabe S, Takahashi S, Malarvannan G, Ikemoto T, Anan Y, Kunisue T, Isobe T, Agusa T, Nakamura M (2008) Survey on hazardous Chemicals in Aquatic Organisms Inhabiting Nansei Shoto Islands: report on the contamination status of fish and shellfish. Wildl Contam Assessement Nansei Shoto Islands 2005–2007:25–46
Tanahara A, Nakaema F, Suzuki H, Kinjyo Y (2013) Chronological changes in heavy metal concentration in sediment collected from Manko tideland. J Jap Coral Reef Soc 15:79–89. (in Japanese with English abstract)
Tanaka Y, Inoue M, Nakamura T, Suzuki A, Sakai K (2014) Loss of zooxanthellae in a coral under high seawater temperature and nutrient enrichment. J Exp Mar Biol Ecol 457:220–225
Tashiro Y, Kameda Y (2013) Concentration of organic sun-blocking agents in seawater of beaches and coral reefs of Okinawa Island, Japan. Mar Pollut Bull 77:333–340
Tong SL, Panga FY, Phanga SM, Laib HC (1996) Tributyltin distribution in the coastal environment of Peninsular Malaysia. Environ Pollut 91:209–216
Umezawa Y, Miyajima T Kayanne H, Koike I (2002) Significance of groundwater nitrogen discharge into coral reefs at Ishigaki Island, southwest of Japan. Coral Reefs 21:346–356
Watanabe T, Yuyama I, Yasumura S (2006) Toxicological effects of biocides on symbiotic and aposymbiotic juveniles of the hermatypic coral Acropora Tenuis. J Exp Mar Biol Ecol 339:177–188
Watanabe T, Utsunomiya Y, Yuyama (2007) Long-term laboratory culture of symbiotic coral juveniles and their use in eco-toxicological study. J Exp Mar Biol Ecol 352(1):177–186
Weis VM (2008) Cellular mechanisms of cnidarian bleaching: stress causes the collapse of symbiosis. J Exp Biol 211:3059–3066
West K, van Woesik R (2001) Spatial and temporal variance of river discharge on Okinawa (Japan): inferring the temporal impact on adjacent coral reef. Mar Pollut Bull 42:864–872

Yakovleva I, Hidaka M (2009) Survey of mycosporine-like amino acids in different morphotypes of the coral *Galaxea fascicularis* from Okinawa, Japan. Galaxea, JCRS 11:109–118

Yakovleva I, Bhagooli R, Takemura A, Hidaka M (2004) Differential susceptibility to oxidative stress of two scleractinian corals: antioxidant functioning of mycosporine-glycine. Comp Biochem Physiol B 139:721–730

Yamano H, Hori K, Yamauchi M, Yamagawa O, Ohmura A (2001) Highest-latitude coral reef at Iki Island, Japan. Coral Reefs 20:9–12

Yamashita H Suzuki G, Hayashibara T, Koike K (2013) *Acropora* recruits harbor "rare" *Symbiodinium* in the environmental pool. Coral Reefs 32:355–366

Yamazato K (1999) Coral bleaching in Okinawa, 1980 vs. 1998. Galaxea, JCRS 1:83–87

Yara Y, Fujii M, Yamanaka Y, Okada N, Yamano H, Oshima K (2009) Projected effects of global warming on coral reefs in sea close to Japan. J Jpn Coral Reef Soc 11:131–140. (in Japanese with English abstract)

Yuyama I, Harii S, Hidaka M (2012a) Algal symbiont type affects gene expression in juveniles of the coral *Acropora tenuis* exposed to thermal stress. Mar Environ Res 76:41–47

Yuyama I, Ito Y, Watanabe T, Hidaka M, Suzuki Y, Nishida M (2012b) Differential gene expression in juvenile polyps of the coral *Acropora tenuis* exposed to thermal and chemical stresses. J Exp Mar Biol Ecol 430-431:17–24

Yuyama I, Nakamura T, Higuchi T, Hidaka M (2016) Different stress tolerances of juveniles of the coral *Acropora tennis* associated with Claude C1 and D *Symbiodinium*. Zool Stud 55:19

# Coral Disease in Japan

# 4

Naohisa Wada, Aki Ohdera, and Nobuhiro Mano

**Abstract**

While 40 different coral diseases have been described globally since the first report in 1973, the causative agents for most cases have yet to be identified. In Japan, a total of ten coral diseases, including black band disease, brown band disease, white syndrome, pigmentation response, and growth anomalies have been confirmed in the field by epidemiological surveys and monitoring projects. However, disease-induced coral mortality has yet to be reported. A national government-based monitoring project for coral reef conservation has recorded the three major diseases (black band disease, white syndrome, and growth anomalies) in ten areas (Kushimoto, Shikoku, Amakusa, Yakushima, Chichijima, Setouchi, Kerama, Miyako, Ishigaki, and Sekisei Lagoon/Iriomote) of Japan since 2003. Furthermore, a substantial number of studies have contributed to the understanding of coral diseases, identifying ten additional diseases in six coral genera, including a new disease discovered in the temperate coral *Turbinaria peltata* in 2009. In order to elucidate the causative mechanism of coral disease, researchers have utilized a variety of approaches including lipid measurements, molecular technique, and new histopathology methods. However, despite the efforts, further research is required to fully understand the mechanism and impacts of coral disease. This chapter summarizes current knowledge on coral pathogens and discusses the future of coral disease research in Japan.

**Keywords**

Coral disease • Black band disease • Growth anomalies • White syndrome • White spot syndrome

## 4.1 Introduction

In the past several decades, the world's coral reefs have shown a decreasing trend in total coral cover, (Hughes et al. 2003; Bellwood et al. 2004; Bruno and Selig 2007), with 32.8% of the 845 species found in reefs currently considered to be in danger of extinction (Carpenter et al. 2008). The drivers of coral reef decline has been attributed to coral bleaching caused by climate change, pollution linked to tourism and industrial development, sediment depositions from estuarine outflow following heavy rainfall, overfishing, and predation by the crown-of-thorns starfish, as well as coral disease (Dustan 1999; Porter et al. 2001; Harvell et al. 2007; Wilkinson 2008; Maynard et al. 2015).

N. Wada (✉)
Biodiversity Research Center, Academia Sinica, No. 128, Sec. 2, Academia Rd, Nangang, Taipei, 11529, Taiwan

Department of Marine Science and Resources, College of Bioresource Science, Nihon University, Fujisawa, Kanagawa, Japan
e-mail: naohisa0308.nw@gmail.com

A. Ohdera
Department of Biology, Eberly College Science, Pennsylvania State University, University Park, PA, USA

N. Mano
Department of Marine Science and Resources, College of Bioresource Science, Nihon University, Fujisawa, Kanagawa, Japan


A. Iguchi, C. Hongo (eds.), *Coral Reef Studies of Japan*, Coral Reefs of the World 13,
https://doi.org/10.1007/978-981-10-6473-9_4

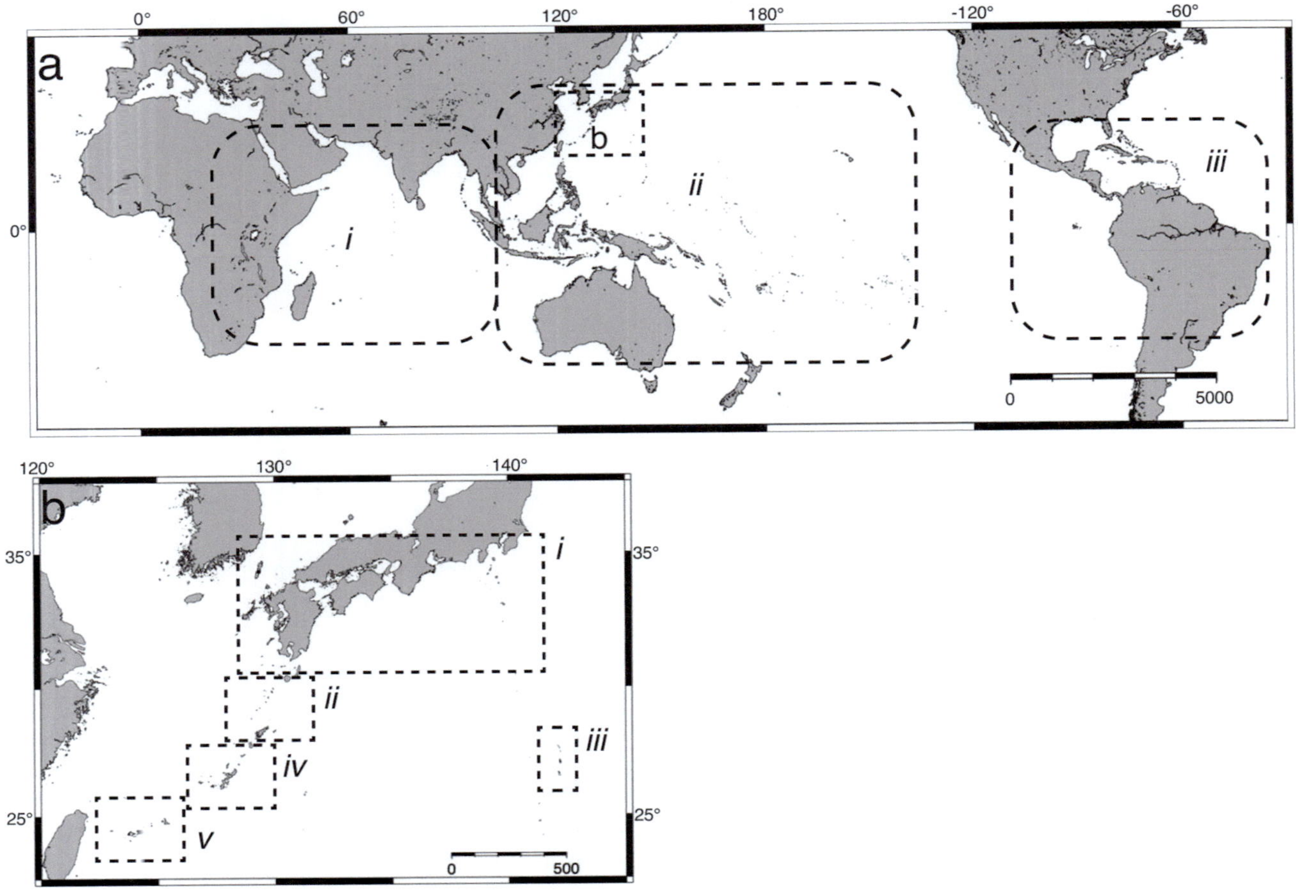


**Fig. 4.1** World coral reef distributions are separated into three regions (**a**-*i*) Indian Ocean, (**a**-*ii*) Caribbean Ocean, (**a**-*iii*) Pacific Ocean. Japanese coral reefs are found in five regions within the seas surrounding (**b**-*i*) Honshu, Shikoku, and Kyushu, (**b**-*ii*) Yakushima to Amami, (**b**-*iii*) Ogasawara Islands, (**b**-*iv*) Okinawa Main, (**b**-*v*) Yaeyama Islands

Since the identification of black band disease (BBD) in Belize in 1973 (Antonius 1973), over 40 new coral diseases and syndromes have been reported from the Caribbean, Indo-Pacific, and Red Sea (Fig. 4.1, Table 4.1; Bruckner 2015), including reports of severe damages to reefs caused by coral disease. Aronson and Precht (2001) suggested white band disease to be responsible for the 95% decline in the Caribbean coral *Acropora palmata* and *A. cervicornis* populations from 1980 to 1990. Due to difficulties of predicting the timing of occurrence and severity of outbreaks, understanding the factors that can lead to and exacerbate the severity of the outbreaks is imperative.

Woodley et al. (2008) has classified diseases of both stony and gorgonian corals into the following six categories:

1. Bleaching through loss or degradation of the dinoflagellate symbiont caused by biotic (bacteria) and abiotic (temperature, UV radiation, salinity, toxicants, etc.) factors
2. Physiological and morphological abnormalities caused by noninfectious environmental stressors such as toxicants, sedimentation, and pollution
3. Physical damage or trauma caused by predators
4. Parasitic infestation caused by protozoans, metazoans, or sponges
5. Growth abnormalities caused by hypertrophy, hyperplasia, neoplasia, or tumors
6. Partial or whole colony infections by bacteria, fungi, virus, or other microorganisms potentially leading to colony mortality

To date, there have been only several cases in which the causative agent has been experimentally linked to an identifiable disease sign, thereby satisfying Koch's postulate.

White pox (WP) found in *A. palmata* has been linked to a rod-shaped gram-negative bacterium, *Serratia marcescens*, which causes distinct white patches throughout an afflicted colony (Patterson et al. 2002). The rod-shaped

**Table 4.1** Common coral diseases identified worldwide

| Reported year | | Disease (acronym) | Coral affected | Location | Location in Fig. 4.1 | Source |
|---|---|---|---|---|---|---|
| 1970s | 1973 | Black band disease (BBD) | Massie corals | Belize | A-iii | Antonius (1973) |
| | 1977 | White plague type I (WPL I) | *Agaricia agaricites* | Florida | A-iii | Dustan (1977) |
| | | Shut down reaction (SDR) | *Acropora cervicornis*, *Montastrea annularis*, and *Siderastrea siderea* | Belize and Florida | A-iii | Antonius (1977) |
| 1980s | 1982 | White band disease type I (WBD I) | *Acr. palmata* | Caribbean Sea | A-iii | Gladfelter (1982) |
| | 1983 | Growth anomalies (GAs) | *Acr. palmata* | Venezuela | A-iii | Bak (1983) |
| 1990s | 1994 | Dark spot disease (DSD) | *Orbicella annularis* | Colombian Caribbean | A-iii | Garzón-Ferreira et al. (2001) |
| | | Caribbean yellow band disease (CYBD) | *Orb. annularis* and *Orb. Faveolata* | Florida | A-iii | Reeves (1994) |
| | 1996 | White pox (WP) | *Acr. palmata* and *Acr. cervicornis* | Florida Keys | A-iii | Holden (1996) |
| | 1997 | Vibrio shiloi-induced bleaching (VSB) | *Oculina patagonica* | Mediterranean coast, Israel | A-i | Kushmaro et al. (1997) |
| | | Yellow blotch/band (YBL) | *Orb. faveolata* | Florida Keys | A-iii | Santavy and Peters (1997) |
| | | White plague type II (WPL II) | *Millepora alcicornis* and *Dichocoenia stokesi* | Florida Keys | A-iii | Richardson et al. (1998) |
| | 1998 | White band disease type II (WBD II) | *Acr. cervicornis* | San Salvador and Bahamas | A-iii | Ritchie and Smith (1998) |
| | | Arabian yellow band disease (AYBD) | *Acr. clathrata*, *Acr. pharaonis*, *Acr. tenuis*, *Acr. valida*, *Acr. flouria*, *Porites lutea*, *P. lichen*, *P. nodfera*, *Turbinaria reniformis*, and *Cyphastrea* | Gulf | A-i | Korrubel and Riegl (1998) |
| | 1999 | Skeletal eroding band (SEB) | Massive and branching corals | Indo-Pacific | A-i and ii | Antonius (1999) |
| 2000s | 2001 | White plague type III (WPL III) | *Orb. annularis* and *Colpophyllia natans* | Florida Keys | A-iii | Richardson et al. (2001) |
| | 2002 | Pink-line syndrome (PLS) | *P. lutea* | Lakshadweep island | A-i | Ravindran and Raghukumar (2002) |
| | 2003 | *Vibrio coralliilyticus*-induced bleaching (VCB) | *Pocillopora damicornis* | Red Sea | A-i | Ben-Haim and Zicherman-Keren (2003) |
| | | *Porites* ulcerative white spot disease (PUWS) | *Porites* spp. | Philippines | A-ii | Raymundo et al. (2003) |
| | 2004 | Brown band disease (BrBD) | Acroporidae | Great Barrier Reef | A-ii | Willis et al. (2004) |
| | | White syndrome (WS) | Acroporidae and Poritidae | Great Barrier Reef | A-ii | Willis et al. (2004) |
| | 2006 | Red band disease (RBD) | *Pachyseris speciosa* and *Porites* sp. | Palau | A-ii | Sussman et al. (2006) |

gram-negative bacterium *Aurantimonas coralicida* is responsible for white plague type II (WPL II), which causes whitening and tissue mortality at the base of *Dichocoenia* and *Orbicella* colonies (Denner et al. 2003). White plague type I (WPL I) in *Favia* spp. is caused by the rod-shaped gram-negative bacterium. *Thalassomonas loyana*, which exhibits

**Table 4.2** Coral diseases identified from Japanese waters

| Reported year | | Disease (acronym) | Coral affected | Location | Location in Fig. 4.1 | Source |
|---|---|---|---|---|---|---|
| 2000s | 2000 | GAs | *Montipora informis* | Sesoko, Okinawa | B-iv | Yamashiro et al. (2000) |
| | 2004 | WS | Table *Acropora* spp. | Sekisei Lagoon, Okinawa | B-v | Biodiversity Center of Japan (2004) |
| | | BBD | n.d. | Sekisei Lagoon, Okinawa | B-v | Biodiversity Center of Japan (2004) |
| | 2009 | Tissue loss | *Porites cylindrica* | Sekisei Lagoon, Okinawa | B-v | Asoh (2009) |
| | | White spot syndromee (WPS) | *Turbinaria peltata* | Miyazaki | B-i | Yamashiro and Fukuda (2009) |
| 2010s | 2012 | Pigmenation response | *P. lutea* and *P. lobata* | Zamami and Ginowan, Okinawa | B-iv | Weil et al. (2012) |
| | | PUWS | *P. lutea* and *P. lobata* | Zamami and Ginowan, Okinawa | B-iv | Weil et al. (2012) |
| | | BrBD | *A. nobilis* | Zamami and Ginowan, Okinawa | B-iv | Weil et al. (2012) |
| | | Compromised health conditions (CH) | *Oxypora lacera* | Zamami and Ginowan, Okinawa | B-iv | Weil et al. (2012) |
| | 2017 | Skeletal eroding band (SEB) | *Montipora* spp. | Aka, Okinawa | B-iv | Wada et al. (in press) |

*n.d.* no data

similar signs to WPL II (Thompson et al. 2006). White syndrome (WS), which displays rapid loss of tissue leading to exposure of the skeleton beneath, is caused by an infection by a gram-negative *Vibrio* species, affecting *Acropora*, *Montipora*, and *Pachyseris* species (Sussman et al. 2008). Other *Vibrios* have also been identified to cause signs similar to coral bleaching. *V. shiloi*-induced bleaching (VSB; Kushmaro et al. 2001) was identified in *Oculina patagonica*, while *V. coralliilyticus*-induced bleaching (VCB; Ben-Haim and Rosenberg 2002) has been found in *Pocillopora damicornis*. Finally, an infection by a cyanobacterium *Phormidium valderianum*, leads to a pink discoloration of the tissue in *Porites*, known as pink-line syndrome (PLS; Ravindran and Raghukumar 2002).

Since the first identification of growth anomalies in 2000 (GAs; Yamashiro et al. 2000), ten diseases have been identified in Japan (Table 4.2). This chapter reviews the coral diseases identified through studies conducted by the Ministry of Environment, Government of Japan, and Okinawa Prefecture. We further focus on current researches concerning confirmed growth anomalies and infectious diseases, events and their causes, and the future of disease research on Japanese coral reefs.

## 4.2 Monitoring Projects for Coral Reef Conservation

Conservation of coral reefs and major habitats in Japanese waters has been mainly led by large-scale national government survey projects, while small-scale projects covering smaller reef areas have been led by local to regional monitoring efforts. For most of these projects, in addition to coral cover and species composition, bleaching severity assessment, crown-of-thorns starfish and mollusk predation, as well as disease prevalence have also been monitored. In order to provide an overview of the disease outbreaks observed in Japanese coral reefs, this section summarizes the abovementioned surveys and their contributions to the coral conservation efforts.

### 4.2.1 National Government-Based Surveys

Coral reefs are increasingly threatened by global climate change as well the environment destruction caused by anthropogenic impacts. The goal of the national government surveys is to determine the biological and ecological state of Japanese coral reefs and associated coral communities in order to better develop conservation policies. Since 2003, the Ministry of Environment has been conducting surveys in 24 regions where coral reefs exist, as part of the Monitoring Site 1000 project (Biodiversity Center of Japan 2003–2015). The results of the surveys are summarized according to two geographical regions (Fig. 4.2). The southern distribution of Japanese coral reefs can be encompassed by seven areas (Setouchi, Daito, Okinawa, Kerama, Miyako, Ishigaki, and Sekisei Lagoon/Iriomote) at south of the Tokara Archipelago. High- latitude coral reef communities can be found at north of Yakushima and Tanegashima and encompassed by nine areas (Tateyama, Kushimoto, Shikoku southwest shore, Iki, Amakusa, Southern seacoast of Kagoshima Pref., Yakushima, Kodakarajima, and Chichijima).

Monitoring projects were conducted with between 5 and 125 sites surveyed per area, utilizing the "spot check" method (Nomura 2004): sites in which corals exhibiting signs of growth anomalies (GAs), black band disease (BBD), and white syndrome (WS) have been recorded, in addition to locations in which signs deviating from the abovementioned diseases have also been reported.

The results of the surveys related to disease occurrence have been summarized and shown in Table 4.3. While there are areas in which diseases have not been reported, GAs have been reported from eight areas (Kushimoto, Shikoku southwest shore, Amakusa, Yakushima, Chichijima, Kerama, Miyako, and Sekisei Lagoon/Iriomote). BBD has been reported from two areas (Kerama and Sekisei Lagoon/Iriomote), and WS from eight areas (Kushimoto, Shikoku southwest shore, Amakusa, Chichijima, Setouchi, Miyako, Ishigaki, and Sekisei Lagoon/Iriomote) (Table 4.3). Disease occurrence was highest at Sekisei Lagoon/Iriomote, with all three disease categories being identified from the site since 2003. In 2004, WS was identified in species of *Acropora*; BBD in encrusting *Montipora*, *Echinopora lamellosa*, and *Pachyseris*; and GAs in tabulate or corymbose *Acropora* and *Montipora* species. In 2014, of the 125 sites surveyed, WS was identified from 115 of those (92%), GAs from 35, and BBD from 14. GAs were reported from 2006 in the Chichijima area of the Ogasawara Islands. In 2007 and 2008, a high number of cases of GAs were reported for Acroporid species, including *A. hyacinthus* and *A. florida*. In the Miyako area, only one case of WS was reported in 2006. However, by 2008, WS was identified in most surveyed sites within the area. Since 2008, WS was reported from one or two sites, highlighting the large shift in disease occurrence in some areas.

Other diseases and disease-like (DL) signs distinguishable from the aforementioned diseases have also been reported from survey areas. Since 2004, patch-like patterns of coral tissue necrosis have been reported from *A. hyacinthus* in the Kushimoto area. The occurrence of the disease increased between 2009 and 2014, with the patch-like signs appearing in approximately half of the survey areas. Observations of the lesion front with a microscope revealed a high abundance of a rod-shaped protozoan (length = 500 μm, width 80 μm). This disease may be related to skeletal eroding band (Antonius and Lipscomb 2000) or brown band disease (Willis et al. 2004) but further investigation is required. Colonies of *Turbinaria peltata* in the same area were also discovered exhibiting small white patches (<1 cm circular spots) in 2012 and identified as white spot syndrome (Yamashiro and Fukuda 2009).

### 4.2.2 Local to Regional-Scale Monitoring

From the summer of 2009 to the spring of 2010, Okinawa Prefecture conducted a coastal survey of the islands, encompassed by nine areas (Fig. 4.3). Each area was surveyed once, employing the manta method or the spot check method. Data on coral cover and abundance of major species, severity of bleaching and disease, and presence/absence of predatory species were collected. A summary of the data is provided in Table 4.4.

Disease surveys were conducted via the manta method (English and Baker 1994), with diseases categorized into BBD, WS, GAs, and others. BBD was only identified in the Miyako area, in 0.1% of corals. WS was identified from six areas, with percent diseased corals ranging from 0.2% in Ihea and Izena Islands to as high as 3.1% in Ie, Aguni, and Tonaki areas. GAs were identified in two regions, Ie/Aguni and Tonaki/Hateruma. 0.6% and 10.9% of corals in the two regions exhibited signs of GAs, respectively.

## 4.3 Current States

The goal of this section is to outline the current state research concerning coral disease in Japan. It summarizes the five epidemiological surveys that have been conducted. In addition, this section introduces the three pathological studies conducted mainly on three types of coral disease current found in Japan.

### 4.3.1 Epidemiological Surveys

Of the five epidemiological surveys of Japanese coral reefs that have been conducted, two comprehensive studies focusing on all diseases currently known have been conducted by Weil et al. (2012) and Wada et al. (in press). Three have reported on a subset of known diseases (Asoh 2008, 2009; Yamashiro and Fukuda 2009; Irikawa et al. 2011).

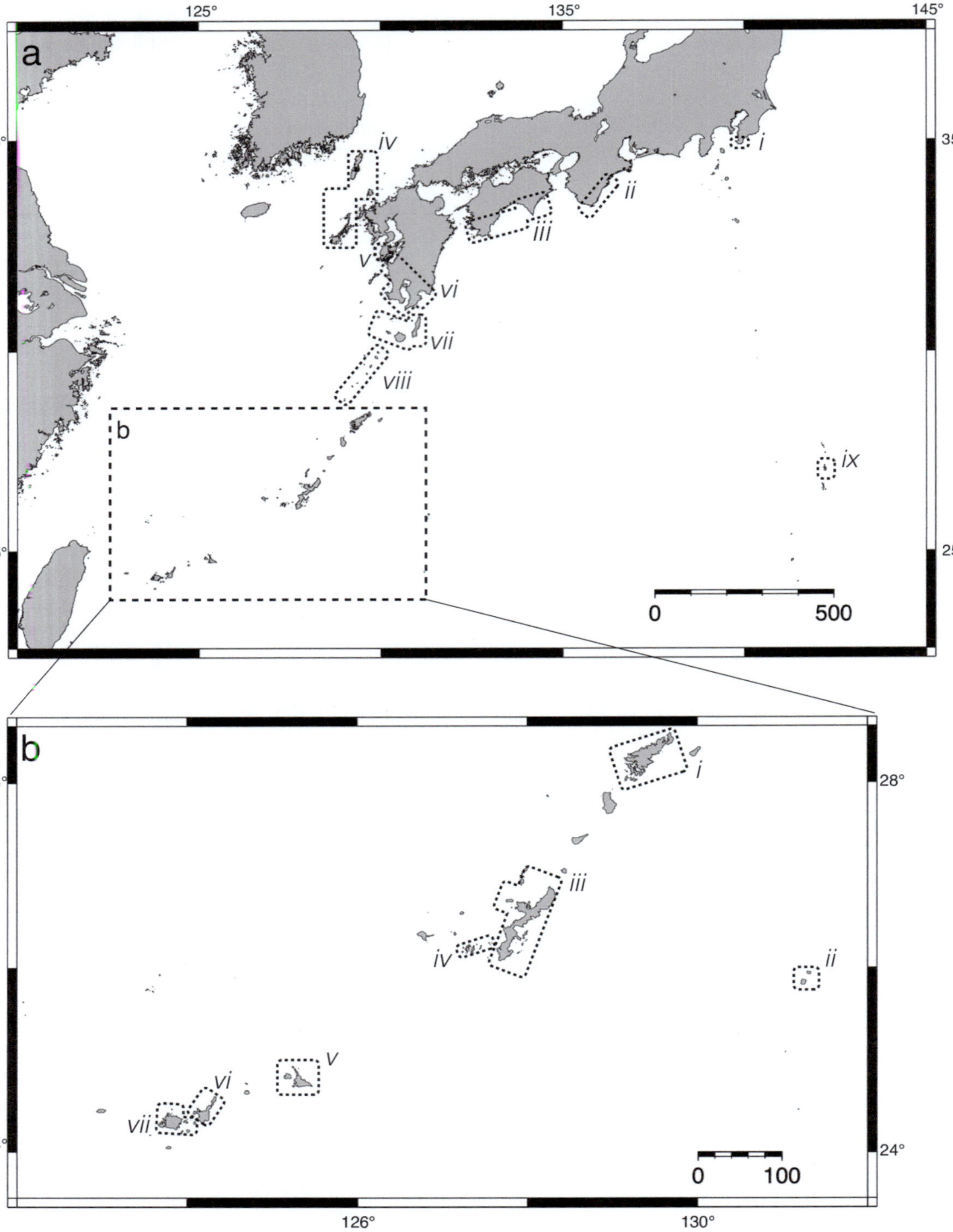


**Fig. 4.2** Area surveyed as part of the monitoring sites, 1000 (national government-based survey). (**a**) High latitude coral community areas: (*i*) Tateyama, (*ii*) Kushimoto, (*iii*) Shikoku southwest shore, (*iv*) Iki, (*v*) Amakusa, (*vi*) Southern seacoast of Kagoshima Pref., (*vii*) Yakushima, (*viii*) Kodakarajima, and (**a**-*ix*) Chichijima. (**b**) Lower latitude survey area: (*i*) Setouchi, (*ii*) Daito, (*iii*) Okinawa, (*iv*) Kerama, (*v*) Miyako, (*vi*) Ishigaki, and (*vii*) Sekisei Lagoon/ Iriomote

**Table 4.3** Confirmed diseases from sites within areas surveyed under the Monitoring Sites 1000 Project (National government survey)

| Area (Location in Fig. 4.2) | States | 2003 | 2004 | 2005 | 2006 | 2007 | 2008 | 2009 | 2010 | 2011 | 2012 | 2013 | 2014 |
|---|---|---|---|---|---|---|---|---|---|---|---|---|---|
| Tateyama (a-i) | Total | n.d. | 5 | 5 | 5 | 6 | 6 | 6 | 6 | 6 | 6 | 5 | 6 |
| | BBD | | | | | | | | | | | | |
| | GAs | | | | | | | | | | | | |
| | WS | | | | | | | | | | | | |
| | Others | | | | | | | | | | | | |
| Kushimoto (a-ii) | Total | 15 | 14 | 14 | 17 | 17 | 16 | 16 | 14 | 18 | 17 | 18 | 15 |
| | BBD | | | | | | | | | | | | |
| | GAs | | | | | | | | | | [a] | 2 | |
| | WS | | | | | | | | | | [a] | 1 | |
| | Others | | 1 | 1 | 1 | 1 | 2 | 8< | 1< | 11 | [a] | 15[b] | 9 |
| Shikoku southwest shore (a-iii) | Total | 11 | 52 | 30 | 29 | 29 | 30 | 30 | 31 | 33 | 31 | 31 | 32 |
| | BBD | | | | | | | | | | | | |
| | GAs | | | | | | | | | | | [d] | |
| | WS | | | | | | 2[c] | | | | | [d] | |
| | Others | | | | | | | | | | | [d] | |
| Iki (a-iv) | Total | n.d. | 4 | 10 | 10 | 14 | 14 | 15 | 15 | 15 | 15 | 15 | 12 |
| | BBD | | | | | | | | | | | | |
| | GAs | | | | | | | | | | | | |
| | WS | | | | | | | | | | | | |
| | Others | | | | 2 | | 1[e] | | 1 | | | | |
| Amakusa (a-v) | Total | 12 | 15 | 15 | 15 | 15 | 15 | 15 | 15 | 15 | 15 | 15 | 15 |
| | BBD | | | | | | | | | | | | |
| | GAs | | | | | | | | | | | | |
| | WS | | | | | | 8 | 8 | [f] | [f] | [f] | [f] | 11 |
| | Others | | | | | | | | | | | | |
| Southern seacoast of Kagoshima prof. (a-vi) | Total | n.d. | 16 | 18 | 18 | 18 | 18 | 17 | 17 | 16 | 15 | 16 | 15 |
| | BBD | | | | | | | | | | | | |
| | GAs | | | | | | | | | | | | |
| | WS | | | | | | | | | | | | |
| | Others | | | | | | | 3 | | | | | |
| Yakushima (a-vii) | Total | n.d. | 17 | 18 | 18 | 19 | 19 | 19 | 19 | 19 | 19 | 19 | 19 |
| | BBD | | | | | | | | | | | | |
| | GAs | | | | | | | | | | 3 | | |
| | WS | | | | | | | | | | | | |
| | Others | [g] | [g] | [g] | [g] | [g] | [g] | | | | | | |
| Kodakarajima (a-viii) | Total | n.d. | n.d. | 13 | n.d. | n.d. | n.d. | n.d. | 23 | n.d. | n.d. | n.d. | n.d. |
| | BBD | | | | | | | | | | | | |
| | GAs | | | | | | | | | | | | |
| | WS | | | | | | | | | | | | |
| | Others | | | | | | | | | | | | |

(continued)

**Table 4.3** (continued)

| | | | | | | | | | | | | | |
|---|---|---|---|---|---|---|---|---|---|---|---|---|---|
| Chichijima (a-ix) | Total | 10 | 12 | 12 | 12 | 12 | 12 | 12 | 12 | 12 | 12 | 12 | 12 |
| | BBD | | | | | | | | | | | | |
| | GAs | | | | 1 | [h] | [h] | [h] | [h] | [h] | [h] | [h] | [h] |
| | WS | | | | | | | | | | | | |
| | Others | | | | | [h] | [h] | [h] | [h] | [h] | [h] | [h] | [h] |
| Setouchi (b-i) | Total | n.d. | 15 | 15 | 15 | 15 | 15 | 15 | 15 | 15 | 15 | 15 | 15 |
| | BBD | | | | | | | | | | | | |
| | GAs | | | | | 1 | 1 | 1 | 1 | | | | |
| | WS | | | | | | | | | | | | |
| | Others | | | | | | | | | | | | |
| Daito (b-ii) | Total | n.d. | n.d. | n.d. | n.d. | 15 | n.d. | n.d. | n.d. | 18 | n.d. | n.d. | n.d. |
| | BBD | | | | | | | | | | | | |
| | GAs | | | | | | | | | | | | |
| | WS | | | | | | | | | | | | |
| | Others | | | | | | | | | | | | |
| Okinawa (b-iii) | Total | 17 | 94 | 60 | 60 | 65 | 67 | 55 | 69 | 71 | 67 | 51 | 69 |
| | BBD | | | | | | | | | | | | |
| | GAs | | | | | | | | | | | | |
| | WS | | | | | | | | | | | | |
| | Others | | | | | | | | | | | | |
| Kerama (b-iv) | Total | 10 | 10 | 10 | 10 | 10 | 10 | 10 | 12 | 12 | 12 | 12 | 12 |
| | BBD | | | | | | | | | | | j | |
| | GAs | | | | 1 | | | | | | | | |
| | WS | | | | | | | | | | | | |
| | Others | | | | [i] | | | | | | 1 | [j] | 7 |
| Miyako (b-v) | Total | n.d. | 12 | 14 | 14 | 14 | 14 | 14 | 14 | 14 | 14 | 14 | 14 |
| | BBD | | | | | | | | | | | | |
| | GAs | | | | 3 | | | | | | | | |
| | WS | | | | 1 | 8 | 9 | 2< | 1 | 1 | 1 | 1 | 2 |
| | Others | | | | | | | | | | | | |
| Ishigaki (b-vi) | Total | 75 | 75 | 75 | 75 | 75 | 77 | 77 | 77 | 77 | 77 | 77 | 77 |
| | BBD | | | | | | | | | | | | |
| | GAs | | | | | | | | | | | | |
| | WS | | | | | [k] | [k] | [k] | | | | | |
| | Others | | | | | | | | | | | [l] | [l] |
| Sekisei Lagoon/Iriomote (b-vii) | Total | 123 | 123 | 123 | 123 | 123 | 125 | 125 | 125 | 125 | 125 | 125 | 125 |
| | BBD | 4 | 5 | 4 | 6 | 1 | 3 | 8 | 10 | 12 | 11 | 14 | 14 |
| | GAs | 9 | 41 | 34 | 57 | 37 | 50 | 48 | 27 | 49 | 41 | 49 | 35 |
| | WS | 31 | 80 | 103 | 113 | 112 | 110 | 109 | 101 | 102 | 106 | 115 | 118 |
| | Others | | | | | | | | | | | | |

(continued)

**Table 4.3** (continued)

*Total* total number of sites surveyed, *BBD* black band disease, *GAs* growth anomalies, *WS* white syndrome, *n.d.* no data
[a]WS, GAs, white spot syndrome (WSS), and disease-like (DL) were observed at many sites
[b]DL was found in 12 sites, WSS at 2 sites, what is believed to be *Porites* pink spot syndrome (PPSS) was observed at 1 site
[c]What was thought to be WS was observed at two sites
[d]Of the ten sites in which diseases were confirmed, WS and GA were observed in most. PPSS and white patch syndrome were observed on several coral colonies at a portion of the surveyed sites
[e]Brown band disease-like sign was observed at one site (BrBD)
[f]WS was observed at most sites between 2010 and 2011, but only at a few sites between 2012 and 2013
[g]Observations of DL were continued from 2004, with an increase seen in 2009
[h]GAs and DL were observed on colonies at 9 sites in 2007, 12 sites in 2008, 11 sites in 2009, 11 sites in 2010, 10 sites in 2011, 11 sites in 2012, 10 sites in 2013, and 9 spots in 2014
[i]BBD and DL were observed at eight sites
[j]While DL was observed for several sites, no additional information available
[k]WS identified from 44% of sites in 2007, but numbers were showing decline in 2008 and 2009
[l]DL was observed in 2013, and increases were seen in 2014

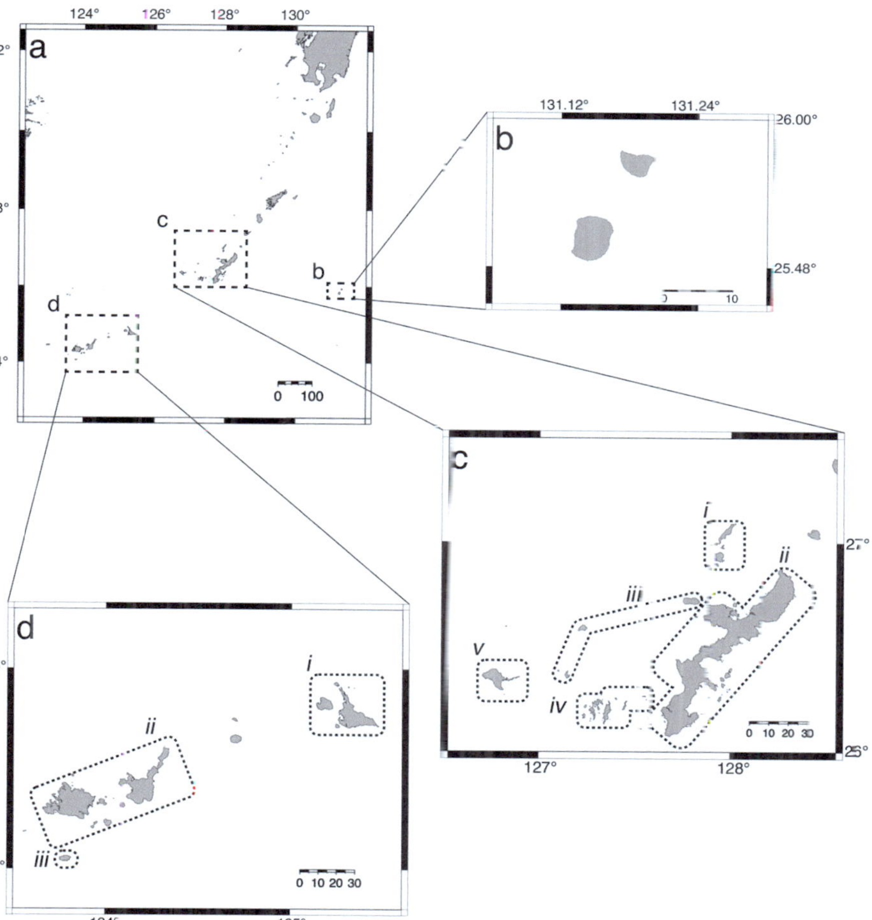


**Fig. 4.3** (**a**) Local to regional-scale monitoring survey areas (conducted by the Okinawa Prefecture). (**b**) Daito Island region, and (**c**) Okinawa Islands (*i.* Iheya/Izena, *ii.* Okinawa Island, *iii.* Ie/Aguni/Tonaki, *iv.* Kerama Islands, *v.* Kume Island), and (**d**) Yaeyama Islands region (*i.* Miyako Island, *ii.* Yaeyama Islands, and *iii.* Hateruma)

The study conducted by Weil et al. (2012) surveyed scleractinian corals, soft corals, sponges, as well as crustose coralline algae for signs of disease and was conducted across four reef sites, two at Ginowan-Oyama, Okinawa Island, and two at Zamami, Kerama Islands, in spring and autumn of 2010. The survey identified seven diseases: black band disease (BBD), white syndrome (WS), pigmentation response (PR), *Porites* ulcerative white spots (PUWS), brown band disease (BrBD), crustose coralline white syndrome (CCWS), and compromised health condition (CH) (Fig. 4.4). Of the seven, PR, PUWS, BrBD, and CCWS have never been reported in Japanese reefs previously. First described in the Philippines in 1996, PUWS is characterized by small ovoid areas of bleached or necrotic tissue, leading to a multifocal pattern of skeleton exposure (Raymundo et al. 2003, 2003). Weil et al. (2012) reported colonies of both *Porites lutea* and

**Table 4.4** Summarized results of the local to regional-scale monitoring (conducted by Okinawa Prefecture) by area

| Area (location in Fig. 4.3) | | Okinawa Island (C-ii) | Kerama Islands (C-iv) | Kume Islands (C-v) | Yaeyama Islands (D-ii) | Iheya/Izena (C-i) | Ie/Agunii/Tonaki (C-iii) | Miyako Island (D-i) | Daito (B) | Hateruma (D-iii) |
|---|---|---|---|---|---|---|---|---|---|---|
| Survey period | | Sep. 2009–Jan. 2010 | Jun. 2010–Nov. 2010 | Sep. 2010–Feb. 2011 | Jul. 2010–Mar. 2011 | Sep. 2011 | Feb. 2010–Jul. 2011 | Jun. 2011–Feb. 2012 | Aug. 2011 | Apr. 2010 |
| Number of survey point | | 154 | 56 | 17 | 107 | 18 | 16 | 30 | 2 | 7 |
| Composition of coral cover | | | | | | | | | | |
| | <5% | 62.3 | 5.3 | 11.8 | 1.9 | | | | | 14.3 |
| | 5–10% | 27.9 | 35.1 | 17.6 | 21.5 | 61.1 | 31.3 | 16.7 | 50.0 | |
| | 10–25% | 9.0 | 42.1 | 41.2 | 27.1 | 38.9 | 50.0 | 63.3 | 50.0 | 57.1 |
| | >25% | 0.6 | 17.6 | 29.4 | 49.5 | | 18.8 | 20.0 | | 28.6 |
| Dominant composition of coral species (%) | | | | | | | | | | |
| | *Porites* spp. | 29.0 | 18.6 | 9.9 | 7.2 | 3.0 | 4.3 | 6.9 | 12.9 | |
| | *Pocillopora* spp. | 18.9 | 11.7 | 19.4 | 2.9 | 4.4 | 18.7 | 0.3 | 43.0 | 2.0 |
| | *Acropora* spp. | 14.2 | 16.2 | 28.5 | 44.0 | 33.5 | 28.9 | 46.8 | | 10.3 |
| | *Montipora* spp. | 1.8 | 2.0 | | 2.9 | 2.2 | | 2.0 | | 7.3 |
| | Faviidae | 1.5 | | 0.9 | 0.1 | | | 0.5 | | |
| | Multiple species | 4.4 | 38.0 | 38.5 | 30.7 | 45.3 | 45.3 | 29.4 | 38.4 | 63.0 |
| | Others | 0.8 | 7.5 | 1.6 | 6.5 | 1.0 | 0.8 | 1.9 | 1.0 | 5.8 |
| | Non-dominant[a] | 29.4 | 6.1 | 1.3 | 5.7 | 10.6 | 39.9 | 12.2 | 5.7 | 11.6 |
| Disease states (%) | | | | | | | | | | |
| | Healthy | 96.1 | 98.4 | 97.3 | 97.4 | 99.8 | 96.3 | 98.3 | 100.0 | 89.1 |
| | BBD | | | | | | | 0.1 | | |
| | WS | | 0.3 | 0.7 | 1.8 | 0.2 | 3.1 | 0.5 | | |
| | GA | | | | | | 0.6 | | | |
| | Others | | 1.1 | | | | | | | 10.9 |
| | n.d. | 3.9 | 0.3 | 2.0 | 0.7 | | | 1.1 | | |

[a]Areas in which a dominant species or coral colonies cannot be identified

*P. lobata* in their survey sites to be affected with PUWS. Colonies of *P. lutea* were detected the disease at a frequency of 23.1 ± 10.4% in Ginowan-Oyama.

Wada et al. (in press) surveyed all coral colonies found within an approximately 70,000 m$^2$ area of reef off of Akajima Island, Kerama Islands from May to July in 2010 and 2011, generating a map of disease distribution. BBD, skeletal eroding band (SEB), GAs, WS, as well as coral bleaching were confirmed between the two survey sites (Figs. 4.5 and 4.6). BBD was predominantly confirmed in encrusting *Montipora* spp., GAs in *Acropora* sp. and encrusting *Montipora* spp., and WS in *Acropora* sp. SEB, which was observed in *Montipora* spp. and *Pocillopora* spp., has not been reported in Japan previously. The most commonly encountered disease for both sites was BBD, with the characteristic black band patterns being formed by an aggregation of microbes, including cyanobacteria. In order to determine the pattern of occurrence for BBD, spatial analysis for the distribution of infected encrusting *Montipora* spp. was conducted. Infected colonies were distributed in aggregates, similar to what has been reported for cases of BBD infections of *Favia* in the Gulf of Aqaba (Zvuloni et al. 2009), a yellow band disease outbreak of *Orbicella* in the Gulf of Mexico (Foley et al. 2005), and *Aspergillosis* infections in sea fans in the Florida Keys (Jolles et al. 2002). The mode of transmission of the diseases is likely a result of a water-borne infection. BBD cases in *Montipora* exhibited significant variation in signs, even between colonies found within the same reef site. The rate of progression of the black band front varied from 1.1 to 5.5 mm per day in 2010 ($n = 15$) and 1.2–3.8 mm per day in 2011 ($n = 13$), with variation seen between colonies of the same reef. The band on each colony also changed in coloration and morphology as the diseased progressed.

Asoh (2008) reported multifocal pattern of circular or irregular areas of tissue loss on the branches of *Porites cylin-*

**Fig. 4.4** Coral diseases identified by Weil et al. (2012) in 2010 at Kerama Islands and the Ginowan region in Okinawa Prefecture. (**a** and **b**) Colonies of *Porites lutea* showing signs of *Porites* ulcerative white spots (*PUWS*). (**c** and **d**) Pigmentation responses expressed as *pink spot* and *pink lines* on *Porites* colonies. (**e**) Brown band-like disease signs on *Acropora nobilis*. (**f**) Growth anomalies (*GAs*) on *A. cytherea*. (**g**) Black band disease (*BBD*) on *Pachyseris speciosa*. (**h**) White syndrome (*WS*) on *A. florida*. (**i**) Crustose coralline white syndrome on unidentified coralline algae (Photos by E. Weil)

*drica* in Shiraho reef, Ishigaki Island, Okinawa, in August and September of 2007. Areas of skeleton exposure were characterized by thin tissue with a white opaque border. Line transect surveys were also conducted in September of 2007 and January of 2008 in order to determine the mean proportion of affected *P. cylindrica* branches. As the result, 0.52 ± 0.16% and 0.08 ± 0.06% of branches exhibited signs in the two survey months, respectively. Lesion size also varied between the 2 months, with colonies exhibiting larger lesions in higher proportions of the branches (33/34), compared to January of 2008 (5/39), suggesting that severity of the infection increases during the summer months. While the cause of the lesions is unknown, the signs were similar to those

**Fig. 4.5** Coral diseases and bleaching identified by Wada et al. (in press) between 2010 and 2011 in Kerama Islands and the region of Okinawa Prefecture. (**a**) Black band disease (*BBD*) on *Montipora* sp. (**b**) Skeletal eroding band (*SEB*) on *Montipora* sp. (**c**) Growth anomalies (*GAs*) on *Montipora* sp. (**d**) White syndrome (*WS*) on *Acropora* sp. (**e**) Bleaching on *Pocillopora* sp. An enlarged view of the box is provided in the corresponding plate labeled (***i***) (Wada et al. in press)

recently reported for cases of WS in *P. cylindrica* in Guam (Lozada-Misa et al. 2015).

In 2008, colonies of *Turbinaria peltata* in the temperate waters around Miyazaki Prefecture were found to have bleached spots approximately 1 cm in diameter on the coenosac (Yamashiro and Fukuda 2009; Fig. 4.7). Named white spot syndrome (WSS), the disease was characterized by gradually increasing spots found covering the colony. A greater than 2/3 reduction in *Symbiodinium* densities was also observed, with the disease eventually causing necrosis of the tissue. Of the 100 colonies surveyed, WSS was confirmed for 1/3 of the coral colonies. Furthermore, colonies of *T. peltata* in Kushimoto, Honshu, were also found to show sign of WSS (Sect. 4.2.1 National Government-Based Surveys), suggesting the potential for future spread of the disease.

In 2011, a survey for the presence of GAs in tabulate acroporids was conducted in Amuro Island of the Kerama Islands. GAs were confirmed for 11 acroporid species (*A. cytherea*, *A. latistella*, *A. hyacinthus*, *A. clathrata*, *A. florida*, *A. valenciennesi*, *A. palifera*, *A. abrotanoides*, *A. samoensis*, *A. digitifera*, and *A. secale*), with *A. cytherea* showing the highest frequency of infection (70/273) in colonies (Irikawa et al. 2011). The researchers found relatively large sizes of *A. cytherea*, with GAs appearing and frequency of occurrence of the disease increasing toward the center of the colony.

### 4.3.2 Pathological Studies

Pathological studies of coral diseases of Japan have been conducted for GAs, cyanobacteria blooms in gorgonian coral, and BBD.

GAs have been found to affect a variety of coral species from the Caribbean and the Indian Ocean. GAs are tumor-like diseases marked by abnormal skeletal growth (Peters et al. 1997; Sutherland et al. 2004). Since the initial report of tumor-like growth in Japanese corals in 2000 (Yamashiro et al. 2000), several additional cases of the disease have been reported (Yamashiro et al. 2001; Irikawa et al. 2011; Yasuda et al. 2012; Yasuda and Hidaka 2012). Abnormal growth lesions found on colonies of encrusting *Montipora informis* in Sesoko Island, Okinawa showed a reduction in the number of polyps and *Symbiodinium* density in the tissues, as well as finer skeletal structure than normal (Yamashiro et al. 2000). A significant reduction in fecundity was also associated with infected colonies. A threefold reduction in lipid storage relative to dry weight was also seen between healthy (32.2%) and diseased (10.6%) tissue (Yamashiro et al. 2001), as a direct result of reduction in stored wax and triacylglycerol. Furthermore, Yasuda and Hidaka (2012) also reported a lower number of apoptotic nuclei within the oral gastrodermis and calicodermis in disease lesions compared to healthy tissue in *M. informis* and

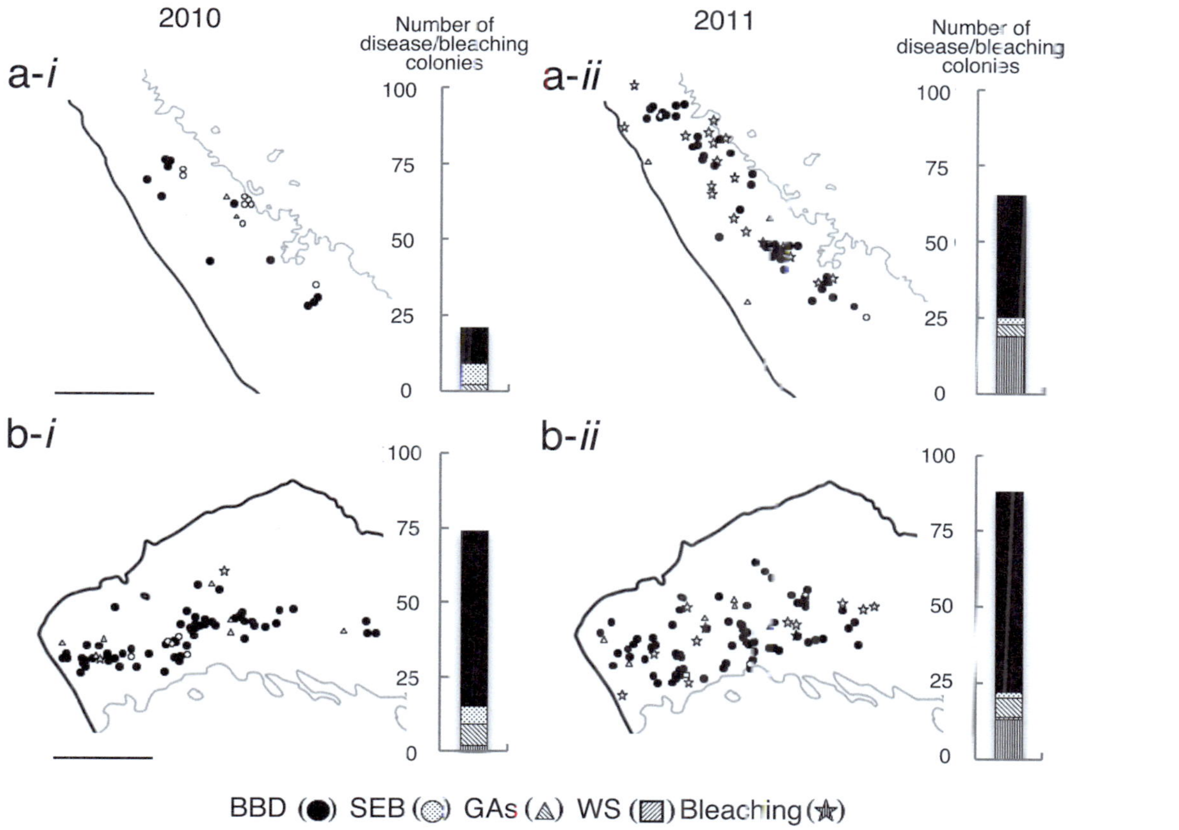


**Fig. 4.6** Distribution and prevalence of coral diseases and bleaching identified by Wada et al. (in press) between 2010 and 2011 in Kerama Islands and the region of Okinawa Prefecture. Distribution of colonies showing disease/bleaching signs of black band disease (*BBD*), skeletal eroding band (*SEB*), growth anomalies (*GAs*), white syndrome (*WS*), and bleaching at (**a**) Nishihama and (**b**) Mazyanohama at Kerama Islands, Okinawa in (*i*) 2010 and (*ii*) 2011. *Histograms* indicate the number of diseased and bleached colonies at each site. *Scale bars* indicate 100 m (Wada et al. in press)

**Fig. 4.7** White spot syndrome (WSS) identified by Yamashiro and Fukuda (2009) in 2008 in the waters around Miyazaki Prefecture. (**a**) White spots on the colony of *Turbinaria peltata*. (**b**) Enlarged view of the spots. (**c**) Colony fully covered by WSS

**Fig. 4.8** (**a–c**) Growth anomalies (*GAs*) identified by Yasuda et al. (2012) from 2003 to 2009. (**a**) GAs on *Porites australiensis* surrounded by healthy tissue. (**b**) GA from plate A observed 4 years later (May 2007) with tissue death occurring from the center outward. (**c**) GAs located at the margin of a colony (April 2009) with tissue death beginning from the margin. (**d–g**) Laboratory experiment in which colonies were placed in contact with: (**d** and **e**) healthy tissue before (**d**) and 6 months later (**e**, **f**, and **g**) diseased tissue before (**f**) and 6 months later (**g**). *Squares* and *scale bars* indicate 10 × 10 cm and 1 cm, respectively

*Porites australiensis*. An increase in BrDU-labeled proliferative cells within the epidermis, gastrodermis, and calicodermis of lesion areas was also seen for both *M. informis* and *P. australiensis*. These results suggest changes to cellular kinetics of disease lesions relative to healthy tissue. Interestingly, disease lesions of *A. cytherea* from Amuro Island, Kerama Islands, and Okinawa showed greater rates of progression in lesions smaller than 15 cm. Lesions greater than 10 cm were characterized by necrosis at the center (Irikawa et al. 2011). Long-term monitoring of *P. australiensis* colonies found in the back reef at Kayo, Okinawa, between 2003 and 2009 showed variable progression of the GAs depending on the location of the lesion (Yasuda et al. 2012). In cases where the disease lesions were located near the colony center, progression into necrosis also began at proximal to the center. However, for lesions toward the colony margins, the tissue necrosis began from the margin of the lesions (Yasuda et al. 2012; Fig. 4.8a–c). The mode of transmission of GA is currently a topic of contention. Field transplant experiments in which healthy *A. cytherea* fragments were placed in contact with disease lesions did not result in transmission of the disease to healthy fragments, even after 2 weeks of contact (Irikawa et al. 2011). Aquarium experiments on *P. australiensis* by Yasuda et al. (2012) also showed GAs to be noncontagious by contact, as healthy fragments in contact with diseased lesions continued healthy growth. Furthermore, the corals continued to show normal growth even after 6 months and fused completely with the healthy and diseased fragments (Fig. 4.8d–g). On the other hand, Kaczmarsky and Richardson (2007) reported contrasting results as successful transmission was established for *Porites* corals.

Aspergillosis, the most commonly reported disease in gorgonian corals (sea fans), has been previously reported

**Fig. 4.9** Cyanobacteria bloom on the gorgonian coral *Annella reticulata* in Kerama Islands identified by Yamashiro et al. (2014). (**a**) Colonies of the gorgonian coral *Annella reticulata*. (**b**) Colonies overgrown by the cyanobacterium *Moorea bouillonii*. (**c**) Enlarged view of *M. bouillonii*. (**d**) Close up of coral branches entangled by cyanobacterial mats. (**e**) Removal of cyanobacterial mats revealed white dead branch areas and collapsed branch leaving only a central axis (*arrow*). (**f**) *M. bouillonii* penetrating into host coral tissue. (**g**) Terminal end of *M. bouillonii* showing a swollen structure composed of a multilayer sheath

from the Caribbean in *Gorgonia ventalina* and is caused by the fungal pathogen *Aspergillus sydowii* (Geiser et al. 1998). While a disease has been reported for gorgonians in Japan, the causative agent appears to be cyanobacteria. In March of 2009, branches of *Annella reticulata* colonies, which found at 20 m depth off the coast of Kerama Islands in Okinawa, were covered by a filamentous cyanobacterial mat, which appeared to cause tissue necrosis (Yamashiro et al. 2014; Fig. 4.9). Determination of the bacterial mats by 16S sequencing revealed the biofilm mostly comprised of a microbe closely related to common benthic cyanobacterium *Moorea bouillonii* (Yamashiro et al. 2014). The authors pointed out the anoxic environment at the boundary between tissue and the cyanobacterial mat as the cause of necrosis but also suggested the involvement of cytotoxic macrolides and peptides produced by *M. bouillonii* (Matthew et al. 2010). Observations by microscopy revealed tissue penetration by the cyanobacteria lending strong support to the infection process and resultant necrosis (Yamashiro et al. 2014).

First reported in 1973 in the Caribbean waters off the coast of Belize, BBD has been subsequently identified in corals worldwide (Sutherland et al. 2004). An aggregation of cyanobacteria, sulfate-reducing bacteria, sulfide-oxidizing bacteria, and a multitude of heterotrophic bacteria often characterize the distinguishable black band associated with the disease. Progression of the disease usually leads to tissue necrosis (Richardson 2004; Bourne et al. 2009). Transmission has been demonstrated in experiments in which healthy colonies are placed in contact with those that are diseased, suggesting the disease to be highly contagious (Kuta and Richardson 1996). Since its confirmation in 2003 at Sekisei Lagoon, found between Ishigaki Island and Iriomote Island (Sect.

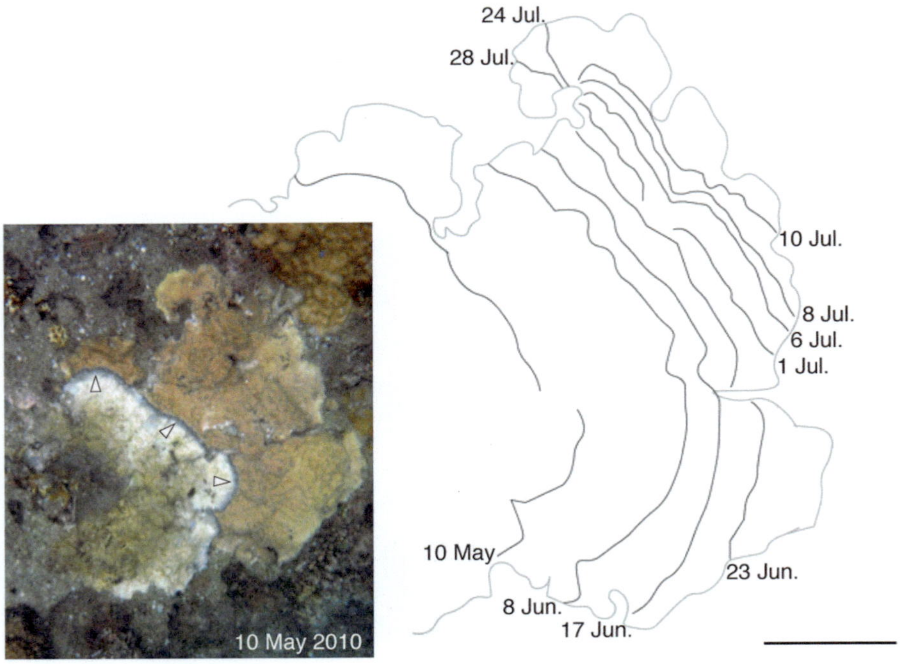


**Fig. 4.10** Lesion front (*arrow heads*) of black band disease (*BBD*) observed by Wada et al. (in press) in 2010 at Kerama Island, Okinawa. Illustration shows the progression front of the BBD lesion from May 10th to July 28th in 2010. *Scale bar* indicates 10 cm (Wada et al. in press)

4.2.1 National Government-Based Surveys), additional reports have been made in areas such as Kerama Islands (Wada et al. in press; Weil et al. 2012) and Okinawa (Yamashiro 2014) of Japan. Based on the initial observations of variation in lesion progression of BBD in encrusting *Montipora* spp., a comprehensive survey of *Montipora* colonies off of Akajima Island, Kerama Islands was conducted over a span of 3 months (Wada et al. in press). While the band forming lesion front was found to radiate from a centralized area, as reported by other studies (Fig. 4.10), observations also revealed band coloration to vary within a colony, ranging from grey to white in some areas. Stereomicroscopic observations of samples collected at the end of the study revealed a filamentous cyanobacterium to be highly abundant between the boundary of healthy and diseased tissue (Fig. 4.11a). Additionally, a white filamentous microbe was also found to be associated with areas directly trailing or above the lesion front (Fig. 4.11b). A microbe with two conspicuous peristomial wing, in addition to the ciliate previously identified with SEB, *Halofolliculina corallasia* (Antonius and Lipscomb 2000), were also found to be present in areas of skeleton exposure (Fig. 4.11c). Microscopic observations via decalcification and paraffin sectioning revealed an approximately 3 µm cyanobacterium to dominate around the boundary between healthy and diseased tissue (Fig. 4.11d–e). However, a larger cyanobacterium (approximately 10 µm) (Fig. 4.11e) as well as ciliates (Fig. 4.11f) were also present in some specimens. Furthermore, the increase in sulfate-reducing bacteria with subsequent increase in production of sulfides in sites lesion formation may be related to formation of the black band (Bourne et al. 2011; Glas et al. 2012; Sato et al. 2016). The cyanobacteria associated with BBD is also known to produce the cyanotoxin microcystin (Richardson et al. 2007), which may also play a role in initiating the disease. However, spatial localization of each bacterial taxa associated with the lesion is still unknown, and the mode of progression of the lesion has yet to be determined. Efforts are currently underway to determine the bacterial composition, localization within the lesion, and mode of progression of BBD using techniques that will be discussed in the next section.

## 4.4 Future Directions

Throughout the archipelago of Japan, corals totaling 415 species have been reported from as far south as the Yaeyama Islands of the Ryukyu archipelago and as far north as Boso Peninsula (Nishihira 2004). Increased sea surface temperature due to anthropogenic climate change and elevated $CO_2$ levels leading to ocean acidification is major concerns to these coral reef ecosystems (Hoegh-Guldberg 2011). Coral bleaching and diseases caused by high sea surface temperatures are considered major threats to coral reefs (Harvell et al. 1999, 2002; Hoegh-Guldberg 2011). A survey conducted from 1995 to 2009 across 47 reefs in the Great Barrier Reef showed coral bleaching and disease to be the cause of 5.6% and 6.5% reduction in coral cover over the 15 years, respectively (Osborne et al. 2011). The major bleaching event of 1998 caused significant damage to Japanese reefs (Loya et al. 2001; Kayane et al. 2002). The reefs around Okinawa Island and Ishigaki Island experienced two major bleaching events between 1998 and 2009. As a result, major shifts in coral species composition

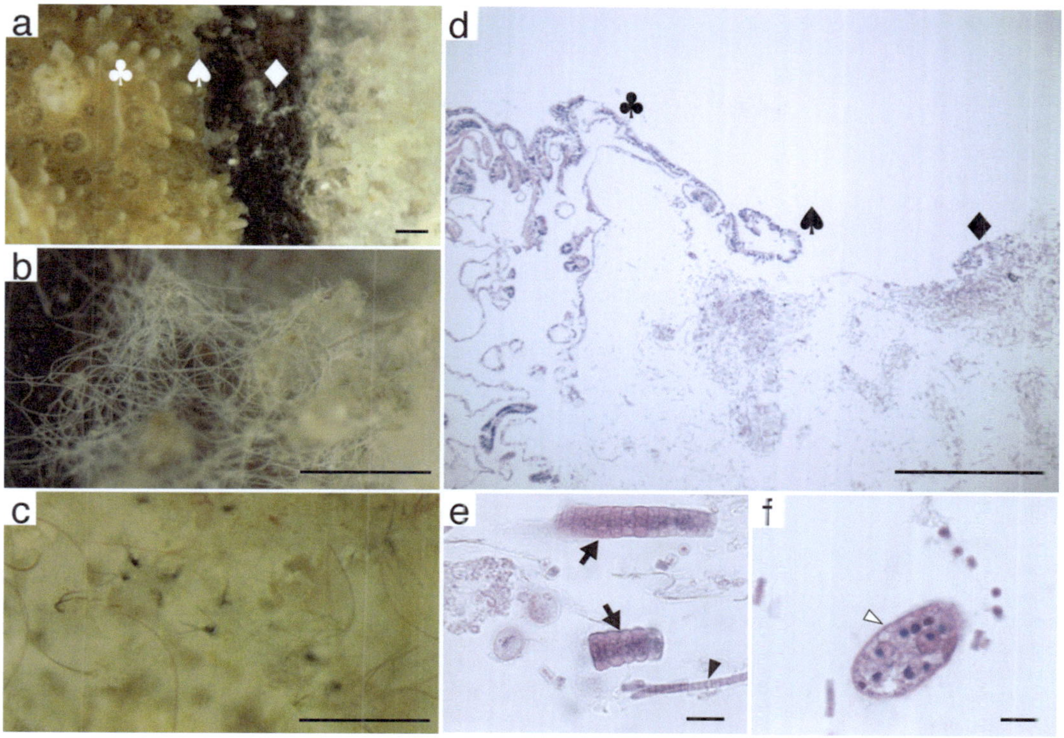

**Fig. 4.11** *Montipora* sp. infected with black band disease (BBD) from Kerama Island. (**a**) Overall image of the lesion boundary displaying a black band between tissue and exposed white skeleton. (**b**) White filamentous organisms above the black band surface. (**c**) The ciliate-like *Halofolliculina corallasia* on the skeleton surface. (**d**) Histological section of BBD showed the cyanobacterial mat penetrated into the tissue from right to left. (**e**) Thick (*arrows*) and thin (*arrowhead*) types of the cyanobacteria found in the mats. (**f**) The ciliates containing dinoflagellates found in the mats. The *symbols* indicate the orientation of lesions showing apparently intact coral tissue (♣), lesion front (♠), and cyanobacterial mat (♦). *Scale bars* represent 1 mm (**a**–**c**), 500 μm (**d**) and 10 μm (**e** and **f**).

within the reefs occurred (Hongo and Yamano 2013; Harii et al. 2014). However, little is known on the diversity and impact of coral disease on Japanese reefs. Due to the increasing deterioration of coral reef by infectious diseases worldwide, gathering data concerning types, occurrence, and impact of the coral diseases will be imperative in developing effective conservation strategies in Japan. Efforts toward understanding the causes of these diseases are also necessary.

While there are over 40 different diseases currently identified worldwide (Bruckner 2015), the cause is only known for a handful of infectious diseases, mainly bacteria (Bourne et al. 2009; Pollock et al. 2011). While fulfilling Koch's postulate (Koch 1884; Grimes 2006) has often confounded identification of the causative agent of coral diseases, these difficulties stem from several factors:

(i) The majority of marine bacteria are unculturable (Kogure et al. 1980; Ferguson et al. 1984; Eguchi and Ishida 1990).
(ii) Disease pathogens can be masked by opportunistic infections (Ainsworth et al. 2007).
(iii) There are difficulties in replicating natural conditions in a controlled or laboratory setting (Lesser et al. 2007).
(iv) Some coral diseases are a synergistic effect of multiple causative agents (e.g., BBD is considered a polymicrobial disease) (Sato et al. 2016).

Thus, molecular genomics (Pollock et al. 2011), immunological tools (Palmer et al. 2008), as well as histopathology (Pollock et al. 2011, Work and Meteyer 2014) are important approaches in enhancing our understanding of coral disease. Although a trained eye is required for histopathology work, only a light or fluorescence microscope is necessary to detect the spatial presence of lesions associated with microbes (fungi, protozoa, and bacteria), making it a common diagnostic tool in coral pathology (Work and Meteyer 2014). Conventional histological preparation for coral studies requires removal of skeleton consisting calcium carbonate (i.e., aragonite, Lowenstam and Weiner 1989). However, the specimens treated with a decalcifying agent potentially leave the interactive orientation between tissue, skeleton canals of gastrovascular canal network uncertain (Gladfelter 1983), owing to the vacant spaces on histological sections resulting from skeleton removal. Additionally, disruption of already fragile tissue can result in loss of tissue or associated microbes can be problematic for histopathology (Bourne et al. 2014). In order to circumvent this issue, the authors utilized a non-decalcifying sectioning method for observations of diseased coral tissue (Wada et al. 2016a: Fig. 4.12). While a cryostat and expensive sectioning knives compared to conventional methods are necessary, this method can dramatically reduce pro-

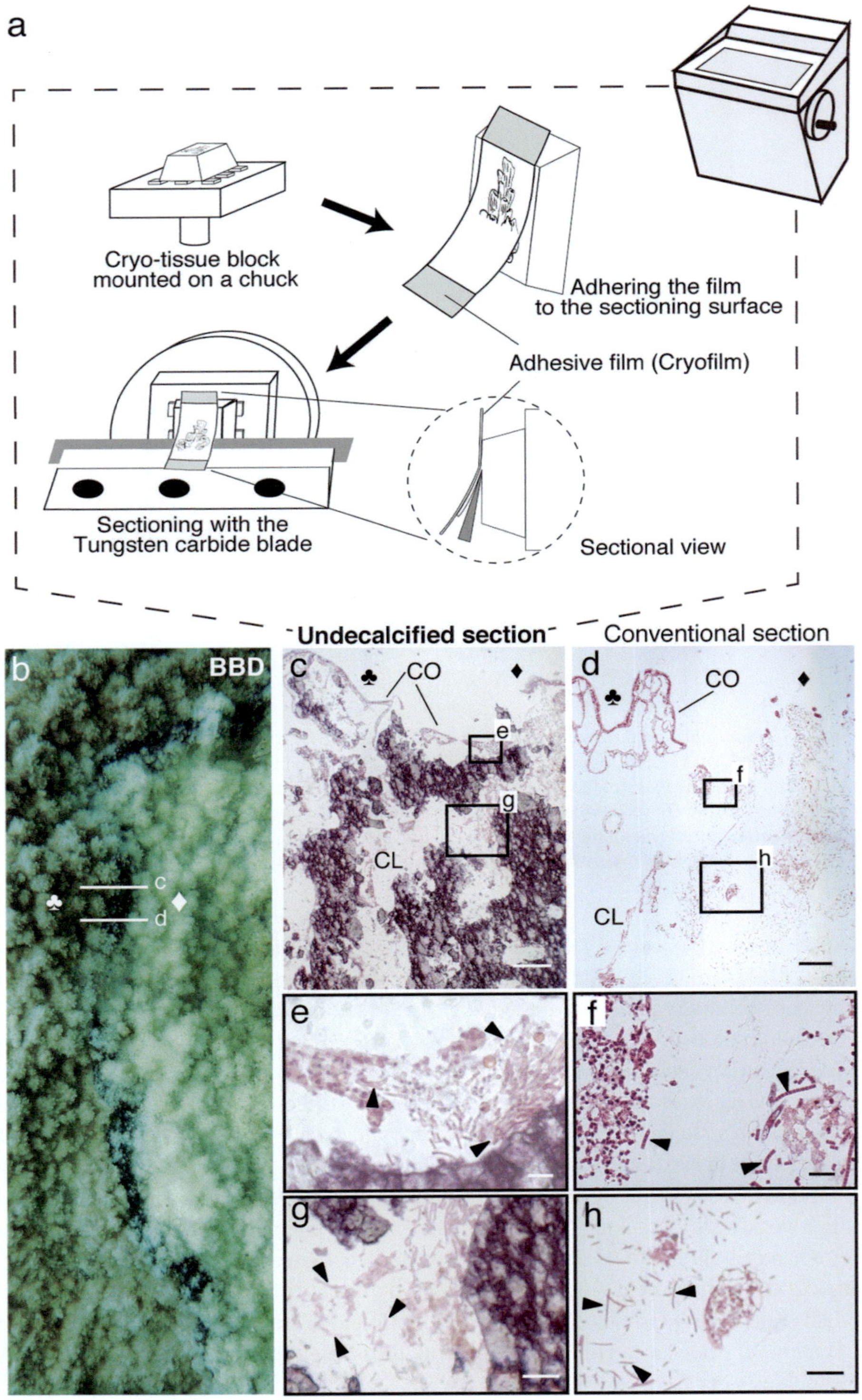


**Fig. 4.12** (**a**) Simplified illustration of the un-decalcified sectioning method shown by Wada et al. (2016a). (**b**) *Montipora* sp. infected with black band disease (*BBD*). (**c**, **e** and **g**) *Montipora* sp. disease lesion sectioned via the un-decalcified sectioning method. Dense aggregation of cyanobacterial filaments on the lesion surface as well as between skeletal parts (*black arrowheads*) can be seen clearly. (**d**, **f** and **h**) *Montipora* sp. disease lesion sectioned via the conventional decalcified sectioning method. Cyanobacterial filaments appear loosely associated with BBD-affected tissue in decalcified sections. The *symbols* indicate the orientation of lesions showing apparently intact coral tissue (♣) and exposed skeleton (♦). *Scale bars* represent 200 μm (**c** and **d**), 20 μm (**e** and **f**), and 50 μm (**g** and **h**)

**Fig. 4.13** (**a**) Aquarium at a coral aqua culture facility. (**b**) Rapid tissue necrosis (*RTN*; Borneman 2001, Luna et al. 2007) affecting *Acropora* sp. in an aquaculture aquarium. (**c**) Enlarged image of RTN affecting *Acropora* sp. (Photos by R. Nakamura)

cessing time as well as leave tissue near its original state. Research that a part of authors is currently being done on BBD utilizing this technique, in conjunction with identification of bacterial localization using FISH (Wada et al. 2016b), as well as next generation sequencing of the lesion microbiome. This new approach to coral histopathology should become an important tool in determining causative agents for coral disease.

Currently, there have been reported cases of coral diseases in aquariums and coral aquaculture facilities (Sweet et al. 2012). Rapid tissue necrosis (RTN: Borneman 2001, Luna et al. 2007), where sloughing off or death of tissue leading to colony mortality has been common. Cases of RTN have also been confirmed in Japanese aquariums (Wada et al. 2012) and aquaculture facilities (Wada et al. 2015) (Fig. 4.13). Researching diseases affecting cultured corals will also be an important as a holistic approach to coral conservation.

## References


Ainsworth TD, Kramasky-Winter E, Loya Y et al (2007) Coral disease diagnostics: what's between a plague and a band? Appl Environ Microbiol 73:981–992. https://doi.org/10.1128/AEM.02172-06

Antonius A (1973) New observations on coral destruction in reefs. 10th Meeting of the Association of Island Marine Laboratories of the Caribbean, University of Puerto Rico (Mayaguez)

Antonius A (1977) Coral mortality in reefs: a problem for science and management. Miami, pp 617–623

Antonius A (1999) *Halofolliculina corallasia*, a new coral-killing ciliate on Indo-Pacific reefs. Coral Reefs 18:300–300. https://doi.org/10.1007/s003380050199

Antonius AA, Lipscomb D (2000) First protozoan coral-killer identified in the Indo-Pacific. Atoll Res Bull 481:1–21

Aronson RB, Precht WF (2001) White-band disease and the changing face of Caribbean coral reefs. In: Porter JW (ed) The ecology and etiology of newly emerging marine diseases. Springer, Netherlands, pp 25–38

Asoh K (2008) Loss of live tissue and exposure of coral skeleton in *Porites cylindrica* at Shiraho, Ishigaki Island, Japan. Galaxea J Coral Reef Stud 10 45–45. https://doi.org/10.3755/galaxea.10.45

Asoh K (2009) Recovery of *Porites cylindrica* from the 2007-summer tissue loss at Shiraho, Ishigaki Island, Japan. Galaxea J Coral Reef Stud 11:27–32. https://doi.org/10.3755/galaxea.11.27

Bak RPM (1983) Neoplasia, regeneration and growth in the reef-building coral Acropora palmata. Mar Biol 77:221–227. https://doi.org/10.1007/BF00395810

Bellwood DR, Hughes TP, Folke C, Nyström M (2004) Confronting the coral reef crisis. Nature 429:827–833. https://doi.org/10.1038/nature02691

Ben-Haim Y, Rosenberg E (2002) A novel *Vibrio* sp. pathogen of the coral *Pocillopora damicornis*. Mar Biol 141:47–55. https://doi.org/10.1007/s00227-002-0797-6

Ben-Haim Y, Zicherman-Keren M (2003) Temperature-regulated bleaching and lysis of the coral *Pocillopora damicornis* by the novel pathogen *Vibrio coralliilyticus*. Appl Environ Microbiol 69:4236–4242. https://doi.org/10.1128/AEM.69.7.4236

Biodiversity Center of Japan (2003) The annual report of the "Monitoring Sites 1000" in Japan: coral reefs. Biodiversity Center of Japan, Ministry of the Environment of Japan, Japan. (in Japanese)

Biodiversity Center of Japan (2004) The annual report of the "Monitoring Sites 1000" in Japan: coral reefs. Biodiversity Center of Japan, Ministry of the Environment of Japan, Japan. (in Japanese)

Biodiversity Center of Japan (2005) The annual report of the "Monitoring Sites 1000" in Japan: coral reefs. Biodiversity Center of Japan, Ministry of the Environment of Japan, Japan. (in Japanese)
Biodiversity Center of Japan (2006) The annual report of the "Monitoring Sites 1000" in Japan: coral reefs. Biodiversity Center of Japan, Ministry of the Environment of Japan, Japan. (in Japanese)
Biodiversity Center of Japan (2007) The annual report of the "Monitoring Sites 1000" in Japan: coral reefs. Biodiversity Center of Japan, Ministry of the Environment of Japan, Japan. (in Japanese)
Biodiversity Center of Japan (2008) The annual report of the "Monitoring Sites 1000" in Japan: coral reefs. Biodiversity Center of Japan, Ministry of the Environment of Japan, Japan. (in Japanese)
Biodiversity Center of Japan (2009) The annual report of the "Monitoring Sites 1000" in Japan: coral reefs. Biodiversity Center of Japan, Ministry of the Environment of Japan, Japan. (in Japanese)
Biodiversity Center of Japan (2010) The annual report of the "Monitoring Sites 1000" in Japan: coral reefs. Biodiversity Center of Japan, Ministry of the Environment of Japan, Japan. (in Japanese)
Biodiversity Center of Japan (2011) The annual report of the "Monitoring Sites 1000" in Japan: coral reefs. Biodiversity Center of Japan, Ministry of the Environment of Japan, Japan. (in Japanese)
Biodiversity Center of Japan (2012) The annual report of the "Monitoring Sites 1000" in Japan: coral reefs. Biodiversity Center of Japan, Ministry of the Environment of Japan, Japan. (in Japanese)
Biodiversity Center of Japan (2013) The annual report of the "Monitoring Sites 1000" in Japan: coral reefs. Biodiversity Center of Japan, Ministry of the Environment of Japan, Japan. (in Japanese)
Biodiversity Center of Japan (2014) The annual report of the "Monitoring Sites 1000" in Japan: coral reefs. Biodiversity Center of Japan, Ministry of the Environment of Japan, Japan. (in Japanese)
Borneman E (2001) Aquarium corals: selection, husbandry, and natural history. Microcosm Limited, Neptune City
Bourne DG, Garren M, Work TM et al (2009) Microbial disease and the coral holobiont. Trends Microbiol 17:554–562. https://doi.org/10.1016/j.tim.2009.09.004
Bourne DG, Muirhead A, Sato Y (2011) Changes in sulfate-reducing bacterial populations during the onset of black band disease. ISME J 5:559–564. https://doi.org/10.1038/ismej.2010.143
Bourne DG, Ainsworth TD, Pollock FJ, Willis BL (2014) Towards a better understanding of white syndromes and their causes on Indo-Pacific coral reefs. Coral Reefs 34:233–242. https://doi.org/10.1007/s00338-014-1239-x
Bruckner AW (2015) History of coral disease research. In: Woodley CM, Downs CA, Bruckner AW et al (eds) Diseases of coral. Wiley, Hoboken, pp 52–84
Bruno JF, Selig ER (2007) Regional decline of coral cover in the Indo-Pacific: timing, extent, and subregional comparisons. PLoS One 2:e711–e711. https://doi.org/10.1371/journal.pone.0000711
Carpenter KE, Abrar M, Aeby G et al (2008) One-third of reef-building corals face elevated extinction risk from climate change and local impacts. Science 321:560–563. https://doi.org/10.1126/science.1159196
Denner EBM, Smith GW, Busse HJ et al (2003) *Aurantimonas coralicida* gen. nov., sp. nov., the causative agent of white plague type II on Caribbean scleractinian corals. Int J Syst Evol Microbiol 53:1115–1122. https://doi.org/10.1099/ijs.0.02359-0
Dustan P (1977) Vitality of reef coral populations off key largo, Florida: recruitment and mortality. Environ Geol 2:51–58. https://doi.org/10.1007/BF02430665
Dustan P (1999) Coral reefs under stress: sources of mortality in the Florida keys. Nat Resour Forum 23:147–155
Eguchi M, Ishida Y (1990) Oligotrophic properties of heterotrophic bacteria and in situ heterotrophic activity in pelagic seawater. FEMS Microbiol Ecol 6:23–30
English S, Baker V (1994) Survey manual for tropical marine resources. Townsv Aust Inst Mar Sci 390:390–390
Ferguson RL, Buckley E, Palumbo A (1984) Response of marine bacterioplankton to differential filtration and confinement. Appl Environ Microbiol 47:49–55
Foley JE, Sokolow SH, Girvetz E et al (2005) Spatial epidemiology of Caribbean yellow band syndrome in *Montastrea* spp. coral in the Eastern Yucatan, Mexico. Hydrobiologia 548:33–40. https://doi.org/10.1007/s10750-005-3619-2
Garzón-Ferreira J, Gil-Agudelo DL, Barrios LM, Zea S (2001) Stony coral diseases observed in southwestern Caribbean reefs. In: Porter JW (ed) The ecology and etiology of newly emerging marine diseases. Springer, Netherlands, pp 65–69
Geiser DM, Taylor JW, Ritchie KB, Smith GW (1998) Cause of sea fan death in the West Indies. Nature 394:137–138. https://doi.org/10.1038/28079
Gladfelter WB (1982) White-band disease in *Acropora palmata*: implications for the structure and growth of shallow reefs. Bull Mar Sci 32:639–643
Gladfelter EH (1983) Circulation of fluids in the gastrovascular system of the reef coral *Acropora cervicornis*. Biol Bull 165:619–636. https://doi.org/10.2307/1541469
Glas MS, Sato Y, Ulstrup KE, Bourne DG (2012) Biogeochemical conditions determine virulence of black band disease in corals. ISME J 6:1526–1534. https://doi.org/10.1038/ismej.2012.2
Grimes DJ (2006) Koch's postulates-then and now. Microbe-Am Soc Microbiol 1:223
Harii S, Hongo C, Ishihara M et al (2014) Impacts of multiple disturbances on coral communities at Ishigaki Island, Okinawa, Japan, during a 15-year survey. Mar Ecol Prog Ser 509:171–180
Harvell CD, Kim K, Burkholder JM et al (1999) Emerging marine diseases – climate links and anthropogenic factors. Science 285:1505–1510. https://doi.org/10.1126/science.285.5433.1505
Harvell CD, Mitchell CE, Ward JR et al (2002) Climate warming and disease risks for terrestrial and marine biota. Science 296:2158–2162. https://doi.org/10.1126/science.1063699
Harvell D, Jordán-Dahlgren E, Merkel S et al (2007) Coral disease, environmental drivers, and the balance between coral and microbial associates. Oceanography 20:172–195
Hoegh-Guldberg O (2011) Coral reef ecosystems and anthropogenic climate change. Reg Environ Chang 11:215–227. https://doi.org/10.1007/s10113-010-0189-2
Holden C (1996) Coral disease hot spot in the Florida keys. Science:5295–5295
Hongo C, Yamano H (2013) Species-specific responses of corals to bleaching events on anthropogenically turbid reefs on Okinawa Island, Japan, over a 15-year period (1995–2009). PLoS One 8:e60952. https://doi.org/10.1371/journal.pone.0060952
Hughes TP, Baird AH, Bellwood DR et al (2003) Climate change, human impacts, and the resilience of coral reefs. Science 301:929–933. https://doi.org/10.1126/science.1085046
Irikawa A, Casareto BE, Suzuki Y et al (2011) Growth anomalies on *Acropora cytherea* corals. Mar Pollut Bull 62:1702–1707. https://doi.org/10.1016/j.marpolbul.2011.05.033
Jolles AE, Sullivan P, Alker AP, Harvell CD (2002) Disease transmission of aspergillosis in sea fans: inferring process from spatial pattern. Ecology 83:2373–2378. https://doi.org/10.1890/0012-9658(2002)083[2373:DTOAIS]2.0.CO;2
Kaczmarsky L, Richardson LL (2007) Transmission of growth anomalies between Indo-Pacific Porites corals. J Invertebr Pathol 94:218–221. https://doi.org/10.1016/j.jip.2006.11.007
Kayanne H, Harii S, Ide Y, Akimoto F (2002) Recovery of coral populations after the 1998 bleaching on Shiraho Reef, in the southern Ryukyus, NW Pacific. Mar Ecol Prog Ser 239:93–103

Koch R (1884) Die Aetiologie der Tuberkulose. Klin Wochenschr 2:1–88. https://doi.org/10.1007/BF01765224
Kogure K, Simidu U, Taga N (1980) Distribution of viable marine bacteria in neritic seawater around Japan. Can J Microbiol 26:318–323
Korrubel JL, Riegl B (1998) A new coral disease from the southern Arabian Gulf. Coral Reefs 17:22–22. https://doi.org/10.1007/s003380050088
Kushmaro A, Rosenberg E, Fine M, Loya Y (1997) Bleaching of the coral *Oculina patagonica* by Vibrio AK-1. Mar Ecol Prog Ser 147:159–165. https://doi.org/10.3354/meps147159
Kushmaro A, Banin E, Loya Y et al (2001) *Vibrio shiloi* sp. nov., the causative agent of bleaching of the coral *Oculina patagonica*. Int J Syst Evol Microbiol 51:1383–1388
Kuta K, Richardson L (1996) Abundance and distribution of black band disease on coral reefs in the northern Florida Keys. Coral Reefs:219–223
Lesser MP, Bythell JC, Gates RD et al (2007) Are infectious diseases really killing corals? Alternative interpretations of the experimental and ecological data. J Exp Mar Biol Ecol 346:36–44. https://doi.org/10.1016/j.jembe.2007.02.015
Lowenstam HA, Weiner S (1989) On biomineralization. Oxford University Press, Oxford, p 324
Loya Y, Sakai K, Nakano Y, Woesik RV (2001) Coral bleaching: the winners and the losers. Ecol Lett 4:122–131. https://doi.org/10.1046/j.1461-0248.2001.00203.x
Lozada-Misa P, Kerr A, Raymundo L (2015) Contrasting lesion dynamics of white syndrome among the scleractinian corals *Porites* spp. PLoS One 10:e0129841. https://doi.org/10.1371/journal.pone.0129841
Luna GM, Biavasco F, Danovaro R (2007) Bacteria associated with the rapid tissue necrosis of stony corals. Environ Microbiol 9:1851–1857. https://doi.org/10.1111/j.1462-2920.2007.01287.x
Matthew S, Salvador LA, Schupp PJ et al (2010) Cytotoxic halogenated macrolides and modified peptides from the apratoxin-producing marine cyanobacterium *Lyngbya bouillonii* from Guam. J Nat Prod 73:1544–1552. https://doi.org/10.1021/np1004032
Maynard J, van Hooidonk R, Eakin CM et al (2015) Projections of climate conditions that increase coral disease susceptibility and pathogen abundance and virulence. Nat Clim Chang. https://doi.org/10.1038/nclimate2625
Nishihira M (2004) Hermatypic corals of Japan. In: Tsuchiya M, Nadaoka K, Kayanne H, Yamano H (eds) Coral reefs of Japan. Ministry of the Environment and Japanese Coral Reef Society, Tokyo
Nomura K (2004) Appendix 3 spot check method for coral reef monitoring. In: Tsuchiya M, Nadaoka K, Kayanne H, Yamano H (eds) Coral reefs of Japan. Ministry of the Environment and Japanese Coral Reef Society, Tokyo
Okinawa Prefectural Government (2009) Report on the coral reef resources information preparation program. Nature Conservation Division, Department of Environmental Affairs, Okinawa Prefuctural Government, Japan. (in Japanese)
Okinawa Prefectural Government (2010) Report on the coral reef resources information preparation program. Nature Conservation Division, Department of Environmental Affairs, Okinawa Prefectural Government, Japan. (in Japanese)
Okinawa Prefectural Government (2011) Report on the coral reef resources information preparation program. Nature Conservation Division, Department of Environmental Affairs, Okinawa Prefectural Government, Japan. (in Japanese)
Osborne K, Dolman AM, Burgess SC, Johns KA (2011) Disturbance and the dynamics of coral cover on the great barrier reef (1995–2009). PLoS One. https://doi.org/10.1371/journal.pone.0017516
Palmer CV, Mydlarz LD, Willis BL (2008) Evidence of an inflammatory-like response in non-normally pigmented tissues of two scleractinian corals. Proc Biol Sci 275:2687–2693. https://doi.org/10.1098/rspb.2008.0335
Patterson KL, Porter JW, Ritchie KB et al (2002) The etiology of white pox, a lethal disease of the Caribbean elkhorn coral, *Acropora palmata*. Proc Natl Acad Sci U S A 99:8725–8730. https://doi.org/10.1073/pnas.092260099
Peters EC, Gassman NJ, Firman JC et al (1997) Ecotoxicology of tropical marine ecosystems. Environ Toxicol Chem 16:12–40. https://doi.org/10.1002/etc.5620160103
Pollock FJ, Morris PJ, Willis BL, Bourne DG (2011) The urgent need for robust coral disease diagnostics. PLoS Pathog 7:e1002183. https://doi.org/10.1371/journal.ppat.1002183
Porter J, Dustan P, Jaap W, Patterson K (2001) Patterns of spread of coral disease in the Florida keys. Hydrobiologia:1–24
Ravindran J, Raghukumar C (2002) Pink line syndrome (PLS) in the scleractinian coral Porites lutea. Coral Reefs 21:252–252
Raymundo LJH, Harvell CD, Reynolds TL (2003) Porites ulcerative white spot disease: description, prevalence, and host range of a new coral disease affecting Indo-Pacific reefs. Dis Aquat Org 55:95–104. https://doi.org/10.3354/dao056095
Raymundo LJ, Couch CS, Bruckner AW, Harvell CD (2008) Coral disease handbook guidelines for assessment. Coral reef targeted research and capacity building for management program
Reeves L (1994) Newly discovered: yellow band disease strikes keys reefs. Underw USA 11:16
Richardson LL (2004) Black band disease. In: Rosenberg PE, Loya PY (eds) Coral health and disease. Springer Berlin, Heidelberg, pp 325–336
Richardson LL, Goldberg WM, Carlton RG, Halas JC (1998) Coral disease outbreak in the Florida Keys: Plague Type II. Rev Biol Trop 46:187–198
Richardson LL, Smith GW, Ritchie KB, Carlton RG (2001) Integrating microbiological, microsensor, molecular, and physiologic techniques in the study of coral disease pathogenesis. Methods 460:71–89. https://doi.org/10.2307/176940
Richardson LL, Sekar R, Myers JL et al (2007) The presence of the cyanobacterial toxin microcystin in black band disease of corals. FEMS Microbiol Lett 272:182–187. https://doi.org/10.1111/j.1574-6968.2007.00751.x
Ritchie KB, Smith GW (1998) Type II white-band disease. Rev Biol Trop 46:199–203
Santavy DL, Peters EC (1997) Microbial pests: coral disease research in the western Atlantic. In: 8th int. coral reef symp., pp. 607–601
Sato Y, Civiello M, Bell SC et al (2016) Integrated approach to understanding the onset and pathogenesis of black band disease in corals. Environ Microbiol 18:752–765. https://doi.org/10.1111/1462-2920.13122
Sussman M, Bourne DG, Willis BL (2006) A single cyanobacterial ribotype is associated with both red and black bands on diseased corals from Palau. Dis Aquat Org 69:111–118
Sussman M, Willis BL, Victor S, Bourne DG (2008) Coral pathogens identified for white syndrome (WS) epizootics in the indo-Pacific. PLoS One 3:e2393. https://doi.org/10.1371/journal.pone.0002393
Sutherland K, Porter J, Torres C (2004) Disease and immunity in Caribbean and Indo-Pacific zooxanthellate corals. Mar Ecol Prog Ser 266:273–302. https://doi.org/10.3354/meps266273
Sweet M, Jones R, Bythell J (2012) Coral diseases in aquaria and in nature. J Mar Biol Assoc U K 92:791–801
Thompson FL, Barash Y, Sawabe T et al (2006) Thalassomonas loyana sp. nov., a causative agent of the white plague-like disease of cor-

als on the Eilat coral reef. Int J Syst Evol Microbiol 56:365–368. https://doi.org/10.1099/ijs.0.63800-0

Wada M, Nishihara G, Isomura N (2012) Probing stony coral rapid tissue necrosis with optical oxygen sensors. In: 12th international coral reef symposium, Cairns, Queensland, Australia

Wada N, Shouhei I, Shinpei S et al (2015) Vibrio spp. isolated from scleractinian coral *Acropora muricata* with rapid tissue necrosis (RTN) and reared in aquarium. BioMicroWorld 2015, Barcelona

Wada N, Kawamoto T, Sato Y, Mano N (2016a) A novel application of a cryosectioning technique to undecalcified coral specimens. Mar Biol 163:117. https://doi.org/10.1007/s00227-016-2895-x

Wada N, Pollock FJ, Willis BL et al (2016b) In situ visualization of bacterial populations in coral tissues: pitfalls and solutions. PeerJ 4:e2424. https://doi.org/10.7717/peerj.2424

Wada N, Mano N, Yanagisawa Y, Mori T (in press) Occurrence of coral diseases in Akajima, Okinawa, Japan in 2010 and 2011. GALAXEA, Jpn Coral Reef Soc

Weil E, Irikawa A, Casareto B, Suzuki Y (2012) Extended geographic distribution of several Indo-Pacific coral reef diseases. Dis Aquat Org 98:163–170. https://doi.org/10.3354/dao02433

Wilkinson C (2008) Status of the coral reefs of the world: 2008. Global Coral Reef Monitoring Network, Townsville

Willis BL, Page CA, Dinsdale EA (2004) Coral disease on the great barrier reef. In: Rosenberg PE, Loya PY (eds) Coral health and disease. Springer Berlin, Heidelberg, pp 69–104

Woodley CM, Bruckner AW, McLenon AL et al (2008) Field manual for investigating coral disease outbreaks. National Oceanic and Atmospheric Administration, Silver Spring

Work T, Meteyer C (2014) To understand coral disease, look at coral cells. EcoHealth 11:610–618. https://doi.org/10.1007/s10393-014-0931-1

Yamashiro H (2014) Black band disease in Sesoko Island, Japan. The 3rd Asia-Pacific Coral Reef Symposium, Kenting, Pingtung, Taiwan

Yamashiro H, Fukuda M (2009) White spot syndrome of *Turbinaria peltata* in the temperate region of Japan. Coral Reefs 28:893–893. https://doi.org/10.1007/s00338-009-0514-8

Yamashiro H, Yamamoto M, van Woesik R (2000) Tumor formation on the coral *Montipora informis*. Dis Aquat Org 41:211–217. https://doi.org/10.3354/dao041211

Yamashiro H, Oku H, Onaga K et al (2001) Coral tumors store reduced level of lipids. J Exp Mar Biol Ecol 265:171–179. https://doi.org/10.1016/S0022-0981(01)00333-1

Yamashiro H, Isomura N, Sakai K (2014) Bloom of the cyanobacterium *Moorea bouillonii* on the gorgonian coral *Annella reticulata* in Japan. Sci Rep. https://doi.org/10.1038/srep06032

Yasuda N, Hidaka M (2012) Cellular kinetics in growth anomalies of the scleractinian corals Porites australiensis and *Montipora informis*. Dis Aquat Org 102:1–11. https://doi.org/10.3354/dao02530

Yasuda N, Nakano Y, Yamashiro H, Hidaka M (2012) Skeletal structure and progression of growth anomalies in *Porites australiensis* in Okinawa, Japan. Dis Aquat Org 97:237–247. https://doi.org/10.3354/dao02408

Zvuloni A, Artzy-Randrup Y, Stone L et al (2009) Spatio-temporal transmission patterns of black-band disease in a coral community. PLoS One 4:e4993–e4993. https://doi.org/10.1371/journal.pone.0004993

# Ocean Acidification Studies in Coral Reefs of Japan

5

Shoji Yamamoto

**Abstract**

Increasing anthropogenic $CO_2$ emissions cause progressive ocean acidification, reducing the calcium carbonate saturation state and coral reef calcification rate. The future uptake of $CO_2$ by the world ocean is predicted to reduce seawater pH by 0.3–0.5 units over the next few decades, which corresponds to a rate 100 times faster than that observed at any time during the last 20 million years. In this chapter, we discuss the effects of ocean acidification on coral reefs, which have been initially probed by culture experiments at several decreased pH conditions, being subsequently investigated by multiple stress factor experiments and field observations of acidified sites. By considering previous studies, we propose that the evaluation and prediction of future ecosystem dynamics require the development of convenient and inexpensive carbonate chemistry-related field measurement techniques such as pH logging, additionally highlighting the importance of studying two naturally acidified sites in Japan, namely, the Iwotorishima and Shikine Islands.

**Keywords**

Ocean acidification • Coral • Foraminifera • Ecosystem • Culture experiment • Japan

## 5.1 Overview of Ocean Acidification

The world ocean is a large carbon reservoir that absorbs atmospheric $CO_2$ to convert it into bicarbonate ($HCO_3^-$) and carbonate ($CO_3^{2-}$) ions. More than 30 % of anthropogenic $CO_2$ is taken up by the ocean, lowering the pH of surface water and decreasing the saturation states of carbonate minerals (Sabine et al. 2004; Kleypas et al. 2006). Over the next few decades, the ongoing uptake of $CO_2$ is predicted to reduce seawater pH by 0.3–0.5 units (Caldeira and Wickett 2003), which is known as ocean acidification.

Zeebe and Wolf-Gladrow (2003) and Millero (2006) described the fundamental seawater chemistry of inorganic carbon, revealing that it is present mainly as free aqueous carbon dioxide ($CO_2$(aq)), bicarbonate ($HCO_3^-$), and carbonate ($CO_3^{2-}$) ions, which are interconverted by the following equilibrium:

$$CO_2 + H_2O \leftrightarrow HCO_3^- + H^+ \leftrightarrow CO_3^{2-} + 2H^+ \quad (5.1)$$

The sum of dissolved $CO_2$, $HCO_3^-$, and $CO_3^{2-}$ concentrations is termed total dissolved inorganic carbon ($C_T$):

$$C_T = [CO_2] + [HCO_3^-] + [CO_3^{2-}], \quad (5.2)$$

and total alkalinity ($A_T$) describes the capacity of water to neutralize acids:

S. Yamamoto (✉)
Department of Earth and Planetary Science, The University of Tokyo, Hongo 7-3-1-Sci1-737, Bunkyo-ku, Tokyo 113-0033, Japan
e-mail: yamamoto-s-@eps.s.u-tokyo.ac.jp


A. Iguchi, C. Hongo (eds.), *Coral Reef Studies of Japan*, Coral Reefs of the World 13,
https://doi.org/10.1007/978-981-10-6473-9_5

$$A_T = [HCO_3^-] + 2[CO_3^{2-}] + [B(OH)_4^-] + [OH^-] - [H^+] + \text{minor components}$$
$$([HPO_4^{2-}] + 2[PO_4^{3-}] + [H_3SiO_4^-] + [NH_3] + [HS^-] - [HSO_4^-] - [HF] - [H_3PO_4]) \quad (5.3)$$

Ocean acidity can be defined in terms of pH (pH = $-\log_{10}[H^+]$), which can be reported using the National Bureau of Standards ($pH_{NBS}$), seawater ($pH_{SWS}$), free ($pH_F$), and total ($pH_T$) scales. Herein, the total scale is used, as recommended by Dickson (2010).

Future changes in calcification and carbonate dissolution induced by ocean acidification can be drastic and sufficiently large, making coral reef communities and carbonate ecosystems in general subject to a net loss of $CaCO_3$ (Kleypas et al. 1999; Andersson et al. 2009; Gattuso and Hansson 2011).

For a given mineral, seawater acidity is defined in terms of saturation state, which is a measure of the thermodynamic potential of this mineral to precipitate or dissolve. For example, the saturation state of calcium carbonate (Ω) is defined as

$$\Omega = [Ca^{2+}] \times [CO_3^{2-}] / K_{sp}^*, \quad (5.4)$$

where $K_{sp}^*$ is the solubility product of a given $CaCO_3$ modification (e.g., aragonite, calcite, or Mg-calcite).

Calcium carbonate formation is thermodynamically favorable at $\Omega > 1.0$. As mentioned above, oceanic $CO_2$ uptake decreases seawater pH and thus the carbonate ion concentration $[CO_3^{2-}]$, leading to a decrease in Ω.

Carbonate dissolution, described as

$$CaCO_3 + CO_2 + H_2O \rightarrow Ca^{2+} + 2HCO_3^-, \quad (5.5)$$

increases $A_T$, $C_T$, and Ω values. The carbonate-bicarbonate system buffers atmospheric $CO_2$ levels and ocean acidification, implying that carbonate dissolution increases the buffering capacity of seawater (i.e., its total alkalinity). Thus, an accurate understanding of the carbonate buffer system is needed to correctly predict future changes of carbon partitioning between the atmosphere and the ocean and the resulting degree of ocean acidification.

Hoegh-Guldberg et al. (2007) modeled changes in the saturation state of aragonite ($\Omega_a$) in response to increasing atmospheric $CO_2$ concentrations. Currently, $\Omega_a$ in high-latitude regions ranges from 1 to 2 and equals 3–4.5 in tropical regions. Since tropical $\Omega_a$ is predicted to be higher than one even at $pCO_2$ = 650 ppm, no chemical dissolution of biogenic carbonate (e.g., that in coral reefs) at low latitudes is expected. Previous studies on coral reefs have mainly dealt with the effects of elevated $pCO_2$ on the net photosynthesis and calcification of marine organisms in laboratory experiments.

Ries et al. (2009) performed 60-day laboratory experiments to investigate the effects of $CO_2$-induced ocean acidification on calcification in 18 benthic marine organisms, showing that 10 of 18 studied species exhibited reduced net calcification rates and observing net dissolution at elevated $pCO_2$ in some cases. However, seven species (crab, shrimp, lobster, limpet, purple urchin, coralline red alga, and *Halimeda*) showed increased net calcification at intermediate and/or maximum $pCO_2$ levels, and one species (blue mussel) showed no calcification change. Doney et al. (2009) and Kroeker et al. (2013) summarized the biological responses to increasing $pCO_2$, revealing decreases of survival rate, calcification, growth, development, and abundance for a pool of marine organisms. However, the magnitude of these responses varies among taxonomic groups, suggesting predictable trait-based variations in sensitivity. Furthermore, these results suggest that other factors such as nutritional status or source population could cause substantial response variations.

## 5.2 Publication Bias for Ocean Acidification Research

Figure 5.1 shows the temporal change of the number of publications categorized by topic words, using data provided by the Web of Science. Although this method does not reflect all performed research, it is used in the following discussion, indicating that bleaching investigations have gradually gained volume since the global bleaching event of 1997–1998, with almost 150 papers per year being currently published on this topic. On the other hand, the effect of ocean acidification on coral reefs is a relatively new topic, with the corresponding exploration dating back to 2005. The number of related studies has dramatically increased in the last 10 years, with more than 200 papers being annually published at present. The corresponding Japanese studies also started at the same time, accounting for 75 out of 1292 papers (5.8 %) published since 2005. In 2016, the Japanese accounted for 4.2 % of the International Coral Reef Symposium participants, which is a value similar to the one above.

Initially, the responses of marine organisms to increasing $CO_2$ levels were observed at various growth stages. Subsequently, laboratory experiments became more complex, accounting not only for increased $CO_2$ levels but also

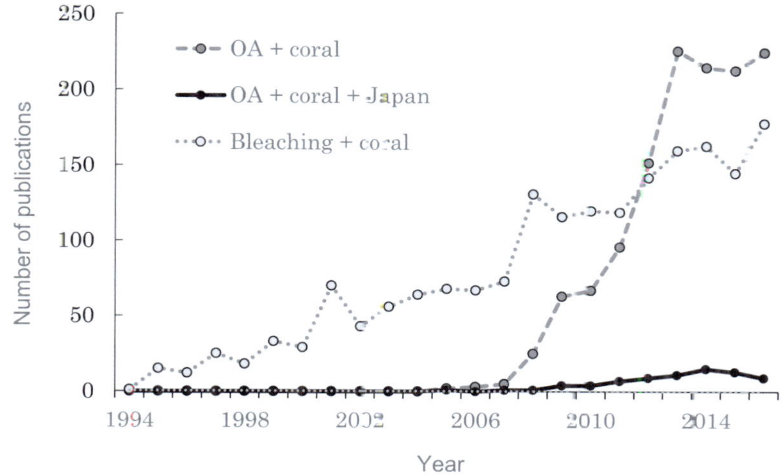


**Fig. 5.1** Number of annual publications returned by Web of Science searches for "ocean acidification" and "coral" (*gray dashed line*); "ocean acidification," "coral," and "Japan" (*black line*); and "bleaching" and "coral" (*gray dotted line*) keywords

for changes of temperature and nutrient supply. Furthermore, the research focus shifted to forecasting the dynamics of reef ecosystems. The following section summarizes the results of laboratory culture experiments in Japan, followed by those of other investigations, with future directions discussed in the last section.

### 5.2.1 Ocean Acidification Research in Japan (Culture Experiment)

A large number of experimental studies reflecting ocean acidification have been conducted in Japan, pioneered by Kurihara et al. (2008). Table 5.1 summarizes the above experimental studies, which investigate the impact of ocean acidification on coral reefs using culture experiments. Initially, both $CO_2$ gas bubbling and acid treatment were used for changing seawater pH, with most recent studies mainly utilizing only $CO_2$ gas bubbling to achieve more precise simulations. These experiments were conducted at 27–32 °C, mainly under artificial light illumination (except for Ref. 5 and 14 in Table 5.1). The impact of ocean acidification has mostly been studied for *Acropora digitifera* (Ref. 1–9 in Table 5.1), as opposed to foraminifera and coralline algae. In general, ocean acidification was demonstrated to favor photosynthesis and disfavor calcification, although some investigations showed that increasing $CO_2$ levels have no effect (Ref. 2, 4, 7–9, 11, 12, and 14 in Table 5.1). In addition to the influence of $CO_2$, recent studies have also evaluated the effects of other factors, such as temperature and nutrient supply (Ref. 8 and 14 in Table 5.1). However, the experimental conditions of laboratory experiments, e.g., light intensity and water flow, are not realistic compared to environmental factors in the field. Thus, direct observation of naturally acidified environments would enable us to predict how reef organisms respond to future ocean acidification under more realistic conditions.

### 5.2.2 Ocean Acidification Research in Japan: $CO_2$ Vent, Dissolution, Modeling, and New Techniques

In this section, we describe characteristic research methods other than culture experiments. The important investigation of Inoue et al. (2013) showed the possibility of community shift from reef-building hard corals to non-reef-building soft corals due to ocean acidification. Most projected community changes caused by ocean acidification describe transitions from hard corals to non-calcifying macroalgal communities. However, related studies show that the spatial distributions of both hard and soft coral communities in volcanically acidified environments (Iwotorishima Island (N 27° 53′ E 128° 13′), Japan) were related to $pCO_2$. Moreover, Agostini et al. (2015) found a $CO_2$ vent on the Shikine Island (N 34° 19′ E 139° 13′), which provides an opportunity to predict the effect of future ocean acidification on reef organisms in Japan (Fig. 5.2).

Research focusing on dissolution promotion (especially that of Mg-calcite) as opposed to calcification inhibition has also gained ground in recent years. As seawater $CO_2$ levels continuously increase the saturation states of carbonate minerals concomitantly decrease. Consequently, Mg-calcite phases are the first ones to experience unsaturation and dissolve, being "the canary in a coal mine" of ocean acidification (Andersson and Mackenzie 2012). Although changes of $CaCO_3$ dissolution behavior due to ocean acidification are potentially more important to the future accretion and maintenance of coral reef ecosystems than changes in calcification, the corresponding studies are underrepresented (Eyre et al. 2014). Yamamoto et al. (2012) experimentally determined that Mg-calcite derived from foraminifera and coralline algae dissolves at $\Omega_a$ values between 3.0 and 3.2, subsequently estimating the dissolution rate in a sandy area of the Shiraho coral reef at the Ishigaki Island in Japan under natural hydrodynamic conditions using eddy covariance and sedimentary $A_T$ profiles (Yamamoto et al. 2015).

**Table 5.1** Summary of Japanese experimental studies investigating the impact of ocean acidification on coral reefs

| Species and communities | | Growth stage | Temperature | Light | Acidification method | Effect for increasing $pCO_2$ | Result | Remarks | References |
|---|---|---|---|---|---|---|---|---|---|
| Coral | *A. digitifera* | Adult | 26.8 | Artificial | $CO_2$ gas | Sperm flagellar motility | ↘ | | 1 |
| | *A. digitifera* | Planula | 26.8 | Artificial | $CO_2$ gas | Mean survival rate | NC | | 2 |
| | *A. digitifera* | Primary polyp | 27 | Artificial | $CO_2$ gas | Polyp growth | ↘ | | |
| | *A. digitifera* | Primary polyp | 27 | Artificial | $CO_2$ gas | Polyp growth | ↘ | | 3 |
| | *A. digitifera* | Planula | 26–27 | Artificial | $CO_2$ gas | Oxygen consumption rate | NC | | 4 |
| | *A. digitifera* | Planula | 26–27 | Artificial | $CO_2$ gas | Survival rate | NC | | |
| | *A. digitifera* | Planula | 26–27 | Artificial | $CO_2$ gas | Metamorphosis rate | ↘ | | |
| | *A. digitifera* | Adult | 27.1 | Artificial | $CO_2$ gas | Sperm flagellar motility | ↘ | | 5 |
| | *A. digitifera* | Larvae | 27 | Natural | $CO_2$ gas | Hsp70, Hsp90, and Hsf1 | NC | | |
| | *A. digitifera* | Polyp with symbionts | 27 | Artificial | $CO_2$ gas | Polyp weight | ↘ | | 6 |
| | *A. digitifera* | Polyp without symbionts | 27 | Artificial | $CO_2$ gas | Polyp weight | ↘ | | |
| | *A. digitifera* | Adult fragments with symbionts | 27 | Artificial | $CO_2$ gas | Polyp weight | ↘ | | |
| | *A. digitifera* | Coral nubbins | 29.3 | Natural | $CO_2$ gas | Calcification rate | NC | | 7 |
| | *A. digitifera* | Coral nubbins | 29.3 | Natural | $CO_2$ gas | Respiration rate | NC | | |
| | *A. digitifera* | Coral nubbins | 29.3 | Natural | $CO_2$ gas | Photosynthesis rate | NC | | |
| | *A. digitifera* | Coral nubbins | 29.3 | Natural | $CO_2$ gas | Zooxanthellae density | NC | | |
| | *A. digitifera* | Post-settlement juvenile | 27.2 | Artificial | $CO_2$ gas | Calcification rate | ↘ | With low nutrient | 8 |
| | *A. digitifera* | Post-settlement juvenile | 27.2 | Artificial | $CO_2$ gas | Calcification rate | NC | With high nutrient | |
| | *A. digitifera* | Post-settlement juvenile | 27.2 | Artificial | $CO_2$ gas | Chl a content normalized to the endosymbiont cell | NC | With low nutrient | |
| | *A. digitifera* | Post-settlement juvenile | 27.2 | Artificial | $CO_2$ gas | Chl a content normalized to the endosymbiont cell | ↘ | With high nutrient | |
| | *A. digitifera* | Post-settlement juvenile | 27.2 | Artificial | $CO_2$ gas | Chl a content normalized to the skeletal weight | NC | With low nutrient | |
| | *A. digitifera* | Post-settlement juvenile | 27.2 | Artificial | $CO_2$ gas | Normalized to the skeletal weight | NC | With high nutrient | |
| | *A. digitifera* | Post-settlement juvenile | 27.2 | Artificial | $CO_2$ gas | Endosymbiont density normalized to the skeletal weight | NC | With low nutrient | |

| | | | | | | | | | |
|---|---|---|---|---|---|---|---|---|---|
| | *A. digitifera* | Post-settlement juvenile | 27.2 | Artificial | $CO_2$ gas | Endosymbiont density normalized to the skeletal weight | ↘ | With high nutrient | |
| | *A. digitifera* | Post-settlement juvenile | 27.2 | Artificial | $CO_2$ gas | $O_2$ production rate | ↗ | With low nutrient | |
| | *A. digitifera* | Post-settlement juvenile | 27.2 | Artificial | $CO_2$ gas | $O_2$ production rate | ↗ | With high nutrient | |
| | *A. digitifera* | Sperm and egg | 27/31 | – | $CO_2$ gas | Fertilization rate | NC | | 9 |
| | *A. tenuis* | Larvae | 26.8 | – | $CO_2$ gas | Polyp weight | ↘ | | 10 |
| | *A. tenuis* | Planula | 26.8 | Artificial | $CO_2$ gas | Mean survival rate | ↗ | | 2 |
| | *G. fascicularis* | Polyp | 27.2/32.0 | Artificial | $CO_2$ gas | Net photosynthesis rate | NC | | 11 |
| | *G. fascicularis* | Polyp | 27.2/32.0 | Artificial | $CO_2$ gas | Net respiration rate | NC | | |
| | *G. fascicularis* | Polyp | 27.2/32.0 | Artificial | $CO_2$ gas | Host ATP content | NC | | |
| | *P. australiensis* | Coral nubbins | 27 | Artificial | $CO_2$ gas | Calcification rate | ↘ | | 12 |
| | *P. australiensis* | Coral nubbins | 27 | Artificial | $CO_2$ gas | Symbiotic algae density | NC | | |
| | *P. australiensis* | Coral nubbins | 27 | Artificial | $CO_2$ gas | Chlorophyll content | NC | | |
| | *P. australiensis* | Coral nubbins | 27 | Artificial | $CO_2$ gas | Fluorescence yield | ↘ | | |
| | *P. australiensis* | Coral nubbins | 27 | Artificial | $CO_2$ gas | Calcification rate | ↘ | | 13 |
| | *I. palifera* | Coral nubbins | 27 | Artificial | $CO_2$ gas | Calcification rate | ↘ | | |
| | *F. fungites* | Larvae | 28/31/34 | Natural | $CO_2$ gas | Survival rate | NC | | 14 |
| | *L. repanda* | Larvae | 28/31/34 | Natural | $CO_2$ gas | Survival rate | NC | | |
| Foraminifera | *M. kudakajimensis* | Early life stage | 25 | Artificial | HCl, NaOH | Calcification rate | ↘ | | 15 |
| | *B. sphaerulata* | Adult individuals | 27.5 | Artificial | $CO_2$ gas | Calcification rate | ↗↘ | First increase, then decrease | 16 |
| | *C. gandichaudii* | Adult individuals | 27.5 | Artificial | $CO_2$ gas | Calcification rate | ↗↘ | | |
| | *A. hemprichii* | Adult individuals | 27.5 | Artificial | $CO_2$ gas | Calcification rate | ↘ | | |
| | *A. kudakajimensis* | Early life stage | 27.1 | Artificial | $CO_2$ gas | Calcification rate | ↘ | | 17 |
| | *C. gandichaudii* | Early life stage | 27.1 | Artificial | $CO_2$ gas | Calcification rate | ↗ | | |
| Coralline algae | *L. Kotschyanum* | Adult | 27.3 | Artificial | $CO_2$ gas | Calcification rate | ↘ | | 18 |
| | *H. Samoense* | Adult | 27.3 | Artificial | $CO_2$ gas | Calcification rate | ↘ | | |

1: Morita et al. (2009); 2: Suwa et al. (2010); 3: Inoue et al. (2011); 4: Nakamura et al. (2011); 5: Nakamura et al. (2012); 6: Ohki et al. (2013); 7: Takahashi and Kurihara (2013); 8: Tanaka et al. (2014); 9: Iguchi et al. (2015); 10: Kurihara et al. (2008); 11: Agostini et al. (2013); 12: Iguchi et al. (2012); 13: Iguchi et al. (2014); 14: Baria et al. (2015); 15: Kuroyanagi et al. (2009); 16: Fujita et al. (2011); 17: Hikami et al. (2011); 18: Kato et al. (2014)

*NC* means no change

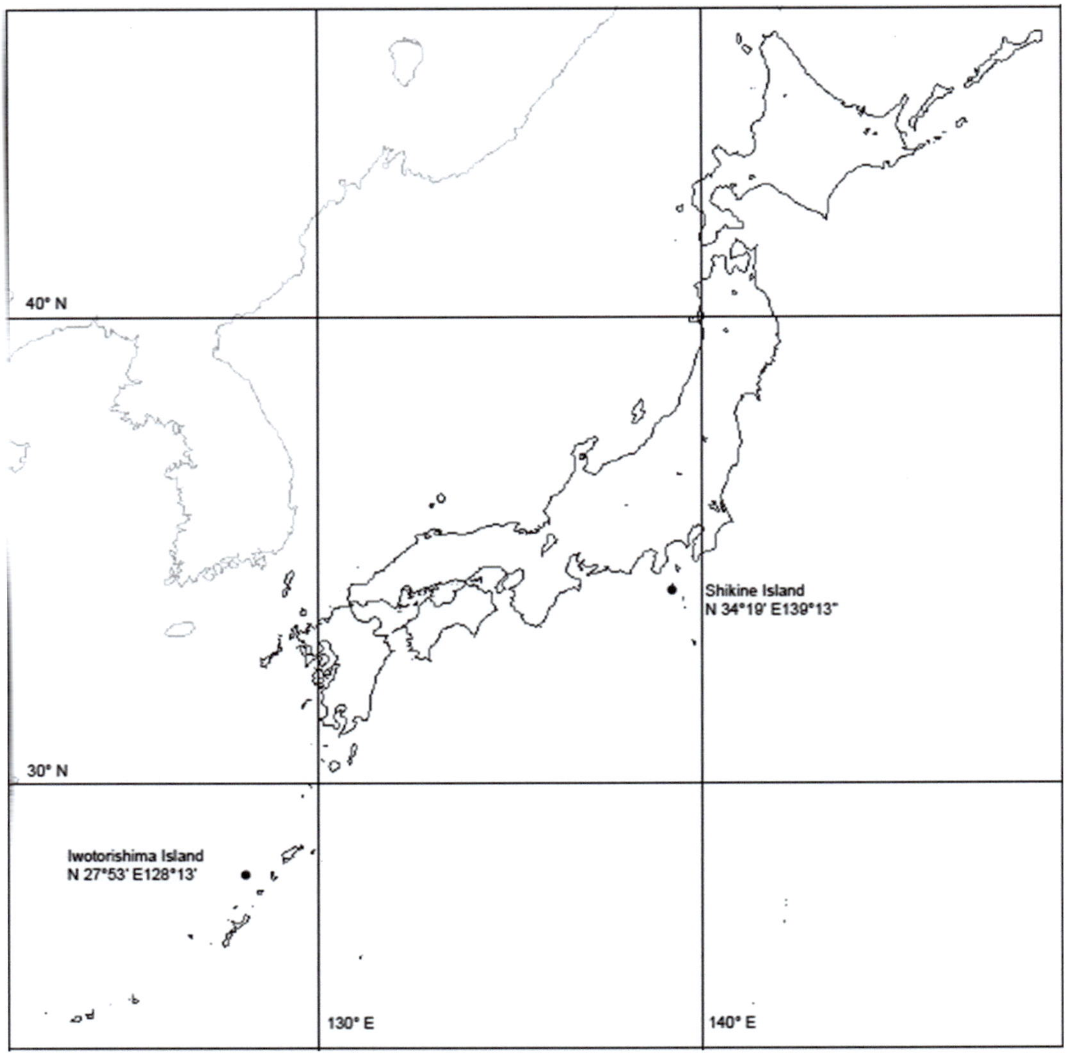


**Fig. 5.2** Map of Iwotorishima and Shikine Islands

pH is an important ocean acidification index that can be backtracked in time by analyzing the boron isotope ratio of coral deposits. Shinjo et al. (2013) reported an annually resolved time series of a pH proxy record for 1940–1999 using a modern massive of *Porites* coral from the Guam Island. Kubota et al. (2014) reported seawater pH and $pCO_2$ variations based on analysis of fossil *Porites* corals from a location offshore of the Tahiti Island. Kubota et al. (2017) also showed that ocean acidification affected the pH of the calcification fluid in *Porites* corals within the western North Pacific Subtropical Gyre at two separate locations, Chichijima Island (Ogasawara Archipelago) and Kikaijima Island based on boron isotopic ratio measurements.

Yara et al. (2012, 2016) and Takao et al. (2015) estimated the effects of future global warming and ocean acidification on potential habitats of tropical/subtropical and temperate coral communities in the seas around Japan using numerical simulation models. Indices of the currently observed regional ranges for temperature and $\Omega_a$ were used for classification, revealing that coral habitats will be narrowed between the northern region, where $\Omega_a$ decreases, and the southern region, where coral bleaching occurs. Another simulation work was performed by Nakamura et al. (2013), who constructed a numerical model describing coral polyp photosynthesis, respiration, and calcification. The above model comprised three components (ambient seawater, coelenteron, and calcifying fluid), predicting that ocean acidification will decrease coral calcification but increase the net photosynthesis rate.

Ohno et al. (2017) performed direct pH imaging at calcification sites (subcalicoblastic medium, SCM) to visualize active pH upregulation in live aposymbiotic primary coral polyps treated with HCl-acidified seawater. Toyofuku et al. (2017) also investigated the effect of external pH changes on the calcification of perforate foraminifera *Ammonia* sp. for a range of $pCO_2$ values. HPTS was used as a pH indicator in both investigations. Such live imaging techniques are a promising approach to understanding how ocean acidification affects calcification processes in reef organisms.

## 5.3 Future Studies

Ocean acidification reflects the ongoing decrease of seawater pH due to increasing atmospheric levels of anthropogenic $CO_2$, affecting marine ecosystems and species sensitive to changes in seawater carbonate chemistry. The report of COP 19 (19th Conference of the Parties to the United Nations Convention on Climate Change) stated that the current seawater acidification rate is 100 times faster than that observed at any time during the last 20 million years. Moreover, the rising seawater temperature, deoxygenation, and food chain changes act as co-stressors, since both their causes and effects are often mechanistically linked to acidification, making the corresponding studies even more complicated.

Coral reefs are the world's most biodiverse ecosystems, being the habitat of more than 25 % of marine species despite occupying less than 1 % of the ocean area. This ecosystem is not only highly resilient to changing environmental conditions but is also highly productive, providing millions of people with food, medicine, protection from storms, and revenue from fishing and tourism. Therefore, in the future, the impact of ocean acidification on coral reefs needs to be estimated based on a deep understanding of the current and forecast ocean chemistry.

First of all, we have to understand what is now happening to coral reefs, which are experiencing a high fluctuation of $pCO_2$ levels (typically between 200 and 600 ppm) due to the metabolic activity of marine organisms. Moreover, $pCO_2$ values are also subject to significant intra-reef variation. Nevertheless, studies of seawater carbonate chemistry are scarce in view of the high cost of automatic measurement systems suitable for field use and the difficulty of operating them in coral reefs. Thus, convenient and inexpensive measurement systems should be developed, and a basic characterization of seawater carbonate content coupled with long-term monitoring is required. Currently, pH and $pCO_2$ are continuously monitored by discrete sample measurements of total alkalinity and total inorganic carbon. However, the combined effects of pH and $pCO_2$ lead to rather large errors in determining carbonate chemistry, thus requiring continuous monitoring of total alkalinity and/or total inorganic carbon.

Subsequently, we need to evaluate and predict how and when ocean acidification will affect coral reefs. A large number of laboratory investigations conducted under controlled conditions reveal the physiological responses of different species to ocean acidification. However, the influence of other environmental stressors such as temperature, light intensity, salinity, water flow, reef history, and surrounding habitat conditions results in gaps between these studies and in situ ecosystem responses.

Moreover, laboratory experiments cannot always evaluate species adaptability, competition among species, and the resilience of the whole coral reef ecosystem. To project the future response of coral reefs to ocean acidification, its impact should be evaluated on an ecosystem scale, including various heterotrophs and inorganic sediment. Fortunately or unfortunately, two acidified sites are located in Japan, namely, Iwotorishima (tropical area) and Shikine (temperate area) Islands, being well suited for such natural experiments. Japanese coral reefs are located along the peripheral edge of reef distribution, being situated at a location most sensitive to ocean acidification and global warming, necessitating the examination and monitoring of their resilience threshold.

**Acknowledgment** I thank all people who supported my work on ocean acidification and two reviewers whose comments improved its quality. This work was supported by JST Core Research for Evolutional Science and Technology (CREST) Grant Number JPMJCR13A1, Japan.

## References


Agostini S, Fujimura H, Higuchi T, Yuyama I, Casareto BE, Suzuki Y, Nakano Y (2013) The effects of thermal and high-$CO_2$ stresses on the metabolism and surrounding microenvironment of the coral *Galaxea fascicularis*. C R Biologies 336(8):384–391. https://doi.org/10.1016/j.crvi.2013.07.003

Agostini S, Wada S, Kon K, Omori A, Kohtsuka H, Fujimura H, Tsuchiya Y, Sato T, Shinagawa H, Yamada Y, Inaba K (2015) Geochemistry of two shallow $CO_2$ seeps in Shikine Island (Japan) and their potential for ocean acidification research. Reg Stud Mar Sci 2:45–53. https://doi.org/10.1016/j.rsma.2015.07.004

Andersson AJ, Mackenzie FT (2012) Revisiting four scientific debates in ocean acidification research. Biogeosciences 9:893–905

Andersson AJ, Kuffner IB, Mackenzie FT, Jokiel PL, Rodgers KS, Tan AA (2009) Net loss of $CaCO_3$ from a subtropical calcifying community due to seawater acidification: mesocosm-scale experimental evidence. Biogeosciences 6:1811–1823

Baria MVB, Kurihara H, Harii S (2015) Tolerance to elevated temperature and ocean acidification of the larvae of the solitary corals *Fungia fungites* (Linnaues, 1758) and *Lithophyllon repanda* (Dana, 1846). Zool Sci 32(5):447–454. https://doi.org/10.2108/zs150036

Caldeira K, Wickett ME (2003) Anthropogenic carbon and ocean pH. Nature 425:365

Dickson AG (2010) The carbon dioxide system in seawater: equilibrium chemistry and measurements. In: Riebesell U, Fabry VJ, Hansson L, Gattuso JP (eds) Guide to best practices in ocean acidification research and data reporting. Office for Official Publications of the European Communities, Luxembourg, pp 17–40

Doney S, Balch WM, Fabry VJ, Feely RA (2009) Ocean acidification a critical emerging problem for the ocean sciences. Oceanography 22:16–25

Eyre B, Andersson AJ, Cyronak T (2014) Benthic coral reef calcium carbonate dissolution in an acidifying ocean. Nat Clim Change 4:969–976. https://doi.org/10.1038/nclimate2380

Fujita K, Hikami M, Suzuki A, Kuroyanagi A, Sakai K, Kawahata H, Nojiri Y (2011) Effects of ocean acidification on calcification of symbiont-bearing reef foraminifers. Biogeosciences 8:2089–2098. https://doi.org/10.5194/bg-8-2089-2011

Gattuso JP, Hansson L (2011) Ocean acidification. Oxford University Press, Oxford. 311pp

Hikami M, Ushie H, Irie T, Fujita K, Kuroyanagi A, Sakai K, Nojiri Y, Suzuki A, Kawahata H (2011) Contrasting calcification responses to ocean acidification between two reef foraminifers harboring different algal symbionts. Geophys Res Lett 38:L19601. https://doi.org/10.1029/2011GL048501

Hoegh-Guldberg O, Mumby PJ, Hooten AJ, Steneck RS, Greenfield P et al (2007) Coral reefs under rapid climate change and ocean acidification. Science 318:1737–1742

Iguchi A, Ozaki S, Nakamura T, Inoue M, Tanaka Y, Suzuki A, Kawahata H, Sakai K (2012) Effects of acidified seawater on coral calcification and symbiotic algae on the massive coral *Porites australiensis*. Mar Environ Res 73:32–36. https://doi.org/10.1016/j.marenvres.2011.10.008

Iguchi A, Kumagai NH, Nakamura T, Suzuki A, Sakai K, Nojiri Y (2014) Responses of calcification of massive and encrusting corals to past, present, and near-future ocean carbon dioxide concentrations. Mar Pollut Bull 89:348–355. https://doi.org/10.1016/j.marpolbul.2014.09.037

Iguchi A, Suzuki A, Sakai K, Nojiri Y (2015) Comparison of the effects of thermal stress and $CO_2$-driven acidified seawater on fertilization in coral *Acropora digitifera*. Zygote 23:631–634. https://doi.org/10.1017/S0967199414000185

Inoue M, Suwa R, Suzuki A, Sakai K, Kawahata H (2011) Effects of seawater pH on growth and skeletal U/Ca ratios of *Acropora digitifera* coral polyps. Geophys Res Lett 38:L12809. https://doi.org/10.1029/2011GL047786

Inoue S, Kayanne H, Yamamoto S, Kurihara H (2013) Spatial community shift from hard to soft corals in acidified water. Nat Clim Chang 3(7):683–687. https://doi.org/10.1038/NCLIMATE1855

Kato A, Hikami M, Kumagai NH, Suzuki A, Nojiri Y, Sakai K (2014) Negative effects of ocean acidification on two crustose coralline species using genetically homogeneous samples. Mar Environ Res 94:1–6. https://doi.org/10.1016/j.marenvres.2013.10.010

Kleypas JA, Buddemeier RW, Archer D, Gattuso JP, Langdon C, Opdyke BN (1999) Geochemical consequences of increased atmospheric carbon dioxide on coral reefs. Science 284:118–120

Kleypas JA, Feely RA, Fabry VJ, Langdon C, Sabine CL, Robbins LL (2006) Impacts of ocean acidification on coral reefs and other marine calcifiers: a guide for future research, report of a workshop held 18–20 April 2005. St. Petersburg, FL, sponsored by NSF, NOAA, and the US Geological Survey

Kroeker KS, Kordas RL, Crim R, Hendriks IE, Ramajo L, Singh GS, Duarte CM, Gattuso JP (2013) Impacts of ocean acidification on marine organisms: quantifying sensitivities and interaction with warming. Glob Chang Biol 19:1884–1896. https://doi.org/10.1111/gcb.12179

Kubota K, Yokoyama Y, Ishikawa T, Obrochta S, Suzuki A (2014) Larger $CO_2$ source at the equatorial Pacific during the last deglaciation. Sci Rep 4:5261. https://doi.org/10.1038/srep05261

Kubota K, Yokoyama Y, Ishikawa T, Suzuki A, Ishii M (2017) Rapid decline in pH of coral calcification fluid due to incorporation of anthropogenic CO2. Sci Rep 7(1):7694

Kurihara H (2008) Effects of $CO_2$-driven ocean acidification on the early developmental stages of invertebrates. Mar Ecol Prog Ser 373:275–284. https://doi.org/10.3354/meps07802

Kuroyanagi A, Kawahata H, Suzuki A, Fujita K, Irie T (2009) Impacts of ocean acidification on large benthic foraminifers: results from laboratory experiments. Mar Micropaleontol 73:190–195. https://doi.org/10.1016/j.marmicro.2009.09.003

Millero FJ (2006) Chemical oceanography. CRC/Taylor and Francis, Boca Raton

Morita M, Suwa R, Iguchi A, Nakamura M, Shimada K, Sakai K, Suzuki A (2009) Ocean acidification reduces sperm flagellar motility in broadcast spawning reef invertebrates. Zygote 18:103–107. https://doi.org/10.1017/S0967199409990177

Nakamura M, Morita M (2012) Sperm motility of the scleractinian coral *Acropora digitifera* under preindustrial, current, and predicted ocean acidification regimes. Aquat Biol 15:299–302. https://doi.org/10.3354/ab00436

Nakamura M, Ohki S, Suzuki A, Sakai K (2011) Coral larvae under ocean acidification: survival, metabolism, and metamorphosis. PLoS One 6(1):e14521. https://doi.org/10.1371/journal.pone.0014521

Nakamura T, Nadaoka K, Watanabe A (2013) A coral polyp model of photosynthesis, respiration and calcification incorporating a transcellular ion transport mechanism. Coral Reefs 32(3):779–794. https://doi.org/10.1007/s00338-013-1032-2

Ohno Y, Iguchi A, Shinzato C, Inoue M, Suzuki A, Sakai K, Nakamura T (2017) An aposymbiotic primary coral polyp counteracts acidification by active pH regulation. Sci Rep 7:40324. https://doi.org/10.1038/srep40324

Ohki S, Irie T, Inoue M, Shinmen K, Kawahata H, Nakamura T, Kato A, Nojiri Y, Suzuki A, Sakai K, Woesik R (2013) Calcification responses of symbiotic and aposymbiotic corals to near-future levels of ocean acidification. Biogeosciences 10:6807–6814. https://doi.org/10.5194/bg-10-6807-2013

Ries JB, Cohen AL, McCorkle DC (2009) Marine calcifiers exhibit mixed responses to $CO_2$-induced ocean acidification. Geology 37:1131–1134. https://doi.org/10.1130/G30210A.1

Sabine CL, Feely RA, Gruber N, Key RM, Lee K et al (2004) The oceanic sink for anthropogenic $CO_2$. Science 305:367–371

Shinjo R, Asami R, Huang KF, You CF, Iryu Y (2013) Ocean acidification trend in the tropical North Pacific since the mid-20th century reconstructed from a coral archive. Mar Geol 342:58–64. https://doi.org/10.1016/j.margeo.2013.06.002

Suwa R, Nakamura M, Morita M, Shimada K, Iguchi A, Sakai K, Suzuki A (2010) Effects of acidified seawater on early life stages of scleractinian corals (genus *Acropora*). Fish Sci 76:93–99. https://doi.org/10.1007/s12562-009-0189-7

Takahashi A, Kurihara H (2013) Ocean acidification does not affect the physiology of the tropical coral *Acropora digitifera* during a 5-week experiment. Coral Reefs 32:305–314. https://doi.org/10.1007/s00338-012-0979-8

Takao S, Yamano H, Sugihara K, Kumagai NH, Fujii M, Yamanaka Y (2015) An improved estimation of the poleward expansion of coral habitats based on the inter-annual variation of sea surface temperatures. Coral Reefs 34:1125–1137. https://doi.org/10.1007/s00338-015-1347-2

Tanaka Y, Iguchi A, Nishida K, Inoue M, Nakamura T, Suzuki A, Sakai K (2014) Nutrient availability affects the response of juvenile corals and the endosymbionts to ocean acidification. Limnol Oceanogr 59(5):1468–1476. https://doi.org/10.4319/lo.2014.59.5.1468

Toyofuku T, Matsuo MY, De Nooijer LJ, Nagai Y, Kawada S, Fujita K, Reichart GJ, Nomaki H, Tsuchiya M, Sakaguchi H, Kitazato H (2017) Proton pumping accompanies calcification in foraminifera. Nat Commun 8:14145. https://doi.org/10.1038/ncomms14145

Yamamoto S, Kayanne H, Terai M, Watanabe A, Kato K, Negishi A, Nozaki K (2012) Threshold of carbonate saturation state determined by $CO_2$ control experiment. Biogeosciences 9:1441–1450. https://doi.org/10.5194/bg-9-1441-2012

Yamamoto S, Kayanne H, Tokoro T, Kuwae T, Watanabe A (2015) Total alkalinity flux in coral reefs estimated from eddy covariance and sediment pore-water profiles. Limnol Oceanogr 60:229–241. https://doi.org/10.1002/lno.10018

Yara Y, Vogt M, Fujii M, Yamano H, Hauri C, Steinacher M, Gruber N, Yamanaka Y (2012) Ocean acidification limits temperature-induced poleward expansion of coral habitats around Japan. Biogeosciences 9:4955–4968. https://doi.org/10.5194/bg-9-4955-2012

Yara Y, Yamano H, Steinacher M, Fujii M, Vogt M, Gruber N, Yamanaka Y (2016) Potential future coral habitats around Japan depend strongly on anthropogenic $CO_2$ emissions, Aquatic Biodiversity Conservation and Ecosystem Services. Springer, Singapore, pp 41–56

Zeebe RE, Wolf-Gladrow D (2003) CO2 in seawater: equilibrium, kinetics, isotopes. Elsevier Science, New York

# Frontiers of Coral-Based Sclerochronological Studies in Japan

6

Kohki Sowa and Kentaro Tanaka

**Abstract**

Corals deposit calcium carbonate ($CaCO_3$: aragonite) skeletons with paired high- and low-density bands that reflect past sea surface conditions during their growth. In addition, the elemental and isotopic compositions of a coral skeleton are affected by environmental factors during deposition, such as sea surface temperature (SST), sea surface salinity (SSS), nutrients, turbidity, pH, and dissolved inorganic carbon (DIC). Therefore, physical and chemical analyses of coral skeletons can provide a record of paleoclimatic and oceanographic changes and coral physiological responses. Because the southwestern part of Japan is at the northern habitable limit of reef-building corals, Japanese corals have distinct biological characteristics, and the area's climatic and oceanographic setting is unique (e.g., East Asian Monsoon, Pacific Decadal Oscillation, and Kuroshio Current). Studying corals in marginal regions can facilitate understanding of the mechanisms of climatic and oceanographic changes and coral physiological responses and contribute to predicting future changes. This chapter will introduce the use of physical and chemical analyses of coral skeletons as chronological indices, highlight previous coral studies in Japan, and propose future directions for paleoenvironmental research using corals.

**Keywords**

Coral growth • Trace elements • Isotopic composition • Quaternary • EAM • PDO • Kuroshio Current

## 6.1 Introduction: Coral Skeletons as Archives

Coral communities and coral reefs in Japan have unique characteristics because of the local oceanographic and climatic setting in the subtropical and temperate regions. The Kuroshio Current is a major current that flows from the southwestern part of Japan to the north along the Pacific side of the Japanese islands, bringing warm and salty water masses from the tropics (Fig. 6.1a, c). Therefore, modern high-latitude coral communities are found in Tateyama (Fig. 6.1b, d), with the northern limit of modern coral reefs at Iki (Fig. 6.1b) (Yamano et al. 2001, 2011; Halfar et al. 2005). Substantial outcrops of mid-Holocene fossil corals (5–6 kyr BP; 1 kyr = 1000 years) are also found in Tateyama, representing a time when SSTs were ~3 °C warmer than today at Tokyo Bay (Veron 1992; Schöne et al. 2004). The East Asian Monsoon (EAM) is a major climatic system that affects Japan (e.g., Ding and Chan 2005). The EAM has two seasonal components, the East Asian Winter Monsoon (EAWM) and the East Asian Summer Monsoon (EASM) (Fig. 6.1) (e.g., Wen et al. 2000; Wu and Wang 2002; Kosaka and Nakamura 2010; Kurita et al. 2015). In general, EAWM is characterized by strong and dry northwesterly winds

K. Sowa (✉)
Department of Chemistry, Faculty of Science, Toho University, Funabashi, Chiba, Japan
e-mail: sowa@frontier.hokudai.ac.jp

K. Tanaka
Atmosphere and Ocean Research Institute, The University of Tokyo, Kashiwa, Chiba, Japan


A. Iguchi, C. Hongo (eds.), *Coral Reef Studies of Japan*, Coral Reefs of the World 13,
https://doi.org/10.1007/978-981-10-6473-9_6

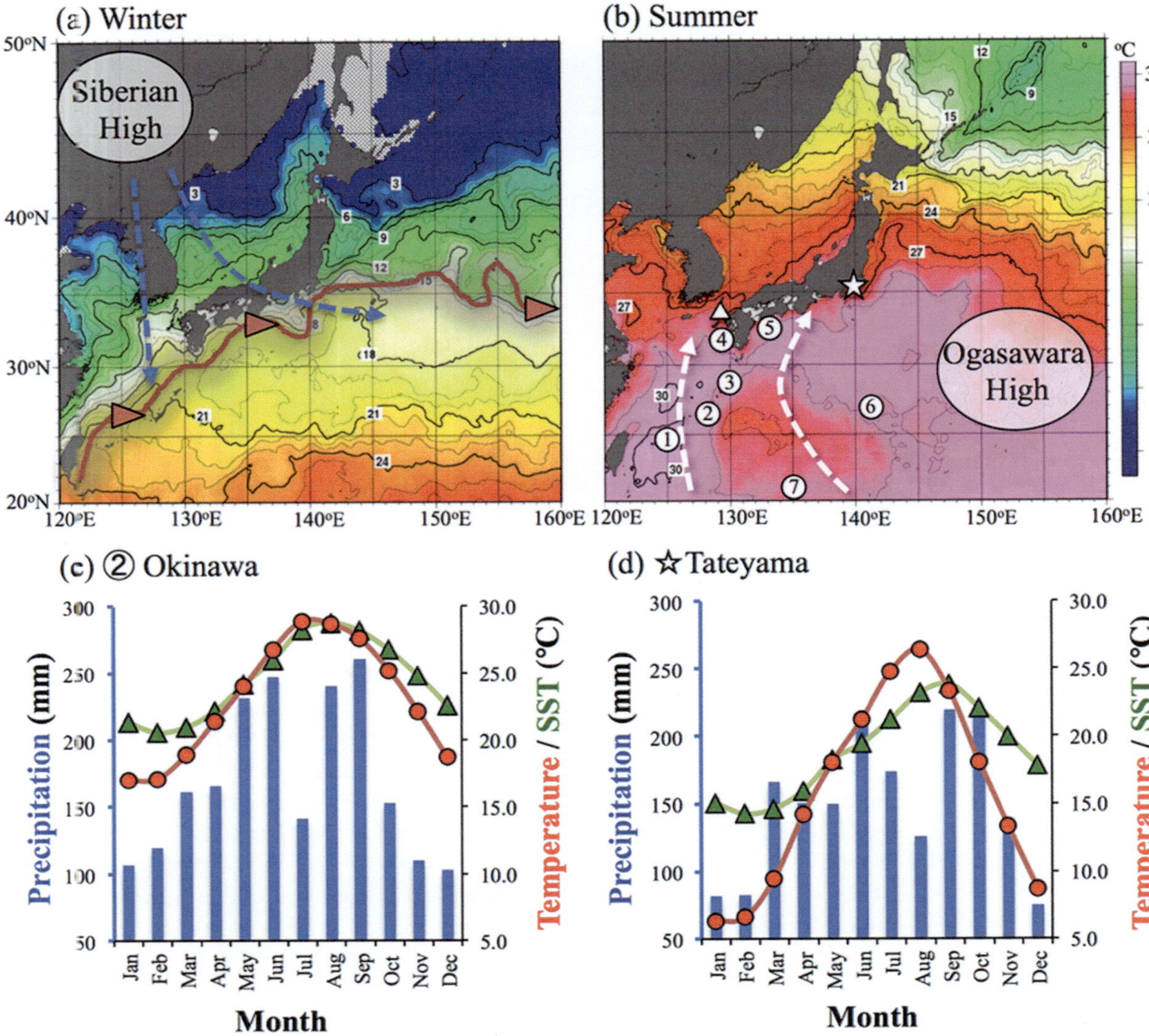


**Fig. 6.1** Maps showing typical SST isotherms around Japan during the (**a**) winter and (**b**) summer; SSTs are indicated by numbers in small *white boxes*. Circled numbers in (**b**) show coral sampling sites from previous studies of coral skeletons (see also Table 6.1). (**c**) Monthly average SSTs (2002–2009) at Motobu and temperature and precipitation (1981–2010) at Naha, Okinawa. (**d**) Monthly average SSTs (2002–2008 except for 2005–2006) at Chikura, which is located 10 km east of Tateyama, and temperature and precipitation (1981–2010) at Tateyama, Chiba. The white *triangle* and *star* in (**b**) indicate the positions of the highest-latitude coral reef and the highest-latitude outcrops of mid-Holocene fossil corals, respectively. The *red arrow* in (**a**) shows the Kuroshio Current flowing northward along the Pacific coast of Japan; the *blue* and *white arrows* in (**a**) and (**b**) represent the prevailing directions of the monsoon winds (SST and temperature/precipitation data are from the Japan Meteorological Agency (http://www.jma.go.jp/jma/indexe.html) and the Japan Oceanographic Data Center (http://www.jodc.go.jp/index.html)

at the surface near Honshu and northerlies and northeasterlies at Okinawa because of the cold Siberian High (SH), which appears over the Asian continent during the winter (Okinawa Meteorological Observatory, Japan Meteorological Agency; http://www.jma-net.go.jp/okinawa/index.html). In contrast, the EASM is characterized by humid southerlies at the surface because of the warm Ogasawara (Bonin) High.

Recent environmental changes at the global, including global warming and ocean acidification (OA), and local, including increased sedimentation and nutrient loading, levels have the potential to positively and negatively impact Japanese coral communities and coral reefs (e.g., Fabricius 2005, 2011; Hoegh-Guldberg et al. 2007; Cooper et al. 2012; McCulloch et al. 2012a). Recent ocean warming has promoted a poleward range shift for Japanese corals of up to 14 km/year based on an 80-year coral occurrence record (Yamano et al. 2011), although ocean warming triggered mass coral bleaching at Okinawa during 1998 (Fujioka 1999; Kayanne et al. 1999; Taniguchi et al. 1999; Yamazato 1999; Loya et al. 2001). In addition, anthropogenic pollution related to terrestrial development, such as red soil loading onto coral reefs, has reduced species diversity (Hongo and Yamano 2013). Further, the dominant coral taxa have also shifted because of disturbances such as warmer SSTs, strong typhoons, and sediment loading (Harii et al. 2014). Thus, coral communities and reefs in Japan are affected in multiple ways by both global and local environmental changes.

Understanding the response of coral communities and coral reefs in Japan to environmental changes over time will help predict their future behaviors and facilitate establishing effective conservation plans to promote their protection and recovery at the northern habitable limit of reef-building corals. Thus, long-term and high-frequency observational records of seawater environmental factors and coral responses are valuable; however, these records are temporally and spatially restricted (e.g., Brown et al. 2011; Morrongiello et al. 2012; Hongo and Yamano 2013; Stocker et al. 2013; Harii et al. 2014). The skeletons of marine organisms have been used as archives of environmental parameters. In particular, calcifying organisms, such as corals, bivalves, foraminifers, and sclerosponges, can offer paleoenvironmental information spanning long periods of time (>several decades). In the case of hermatypic corals, their skeleton, which consists of aragonite, have density bands that correspond to annual growth cycles. The variations in coral skeletal growth rate are related to coral reef environments and physiological processes (e.g., Lough and Cooper 2011; Lough and Cantin 2014; Pratchett et al. 2015). In addition, the skeletal composition also reflects environmental conditions, such as SST and SSS (e.g., de Villiers et al. 1995; Gagan 1998; Marion et al. 2005; Pelejero et al. 2005; Corrège 2006; Holcomb et al. 2016). Therefore, skeletal growth rate and composition of hermatypic corals archive the past environments and compensate for the lack of observational records.

This review focuses on recent physical and chemical analyses of coral skeleton and its application to paleoenvironmental studies in Japan. Section 2 summarizes the analytical procedures for measuring coral skeletal growth using massive *Porites* spp. coral skeletons; this species of reef-building coral is common throughout the Pacific Ocean (Hongo 2012). Section 3 summarizes the major geochemical proxies derived from coral skeletons (SST, SSS, nutrients, sediment, pH, and DIC). Section 4 describes studies investigating paleoclimatology and coral physiology in Japan. Section 5 proposes future directions for coral research in Japan.

## 6.2 Measurements of Past Coral Skeletal Growth

Massive coral skeletons serve as long-term records of the many factors that affect their growth. To date, several massive coral species, including *Porites* (in the Indo-Pacific region), *Orbicella* (formerly *Montastraea*, in the Atlantic/Caribbean region), *Siderastrea*, and *Diploastrea* (in the Atlantic/Caribbean region), have been used as paleoenvironmental proxies (e.g., Lough 2010; Lough and Cooper 2011; Lough and Cantin 2014).

Coral skeletal growth can be described in terms of the calcification rate, linear extension rate, and skeletal density. A typical linear extension rate for massive *Porites* corals is 1–2 cm/year (Pratchett et al. 2015). In general, the skeletal density of most Indo-Pacific massive *Porites* corals averages 0.92–1.96 g/cm$^3$ (Lough and Barnes 2000), which is lower than the density of aragonite (2.98 g/cm$^3$). The skeletal density of corals is low because of variations in the skeletal architecture (e.g., dissepiment, septa, wall, and theca columella) that lead to high porosity; therefore, low skeletal density is not related to the chemical composition (Barnes and Devereux 1988). The calcification rate is calculated as the product of the linear extension rate and the skeletal density. Based on the typical linear extension rate and skeletal density of massive *Porites* corals, the calcification rate is typically 1–4 g/cm$^2$/year. In this section, we describe the general model of coral skeletal growth, sampling procedures and analytical methods.

The following function is used to model coral skeletal growth:

$$G_t = f\left(E_t^{\mathrm{sw}} + E_t^{\mathrm{bio}} + I_t^{\mathrm{age}} + I_t^{\mathrm{bio}} + C_{t-1} + e_t\right)$$

where $G_t$ is the coral skeletal growth in the year $t$. Coral skeletal growth are affected by both extrinsic ($E$) and intrinsic ($I$) factors (see more detail in Morrongiello et al. 2012). $E_t^{\mathrm{SW}}$ is related to environmental factors, such as SST, SSS, nutrients, light attenuation, and water flow. $E_t^{\mathrm{bio}}$ is related to extrinsic biological interactions, such as competition and feeding. $I_t^{\mathrm{age}}$ is the age-related growth trend (sometimes called the "age effect") of a deceleration in growth with aging (which occurs, e.g., in bivalves, otoliths, and trees) (see more detail in Speer 2010; Morrongiello et al. 2012). $I_t^{\mathrm{bio}}$ encompasses individual-specific factors, such as genetic predisposition and the ontogenetic effect. $C_{t-1}$ is related to the effects of autocorrelation, in which the current year's growth is affected by the previous year's growth. $e_t$ includes all the errors that are not controlled for by the above factors. These parameters are treated as unknowns, except for $t$ (the "age" or "year") and $E_t^{\mathrm{SW}}$. Among these factors, $I_t^{\mathrm{age}}$ and $C_{t-1}$ can be statistically detrended. In biochronological analysis, the detrending of $I_t^{\mathrm{age}}$ is the first step in statistical analysis (e.g., Speer 2010; Morrongiello et al. 2012). However, detrending procedures have not been established for the age effect in coral research. We note that Lough (2008) reported that the linear extension rates and skeletal density are influenced by the age effect, whereas the calcification rate is not. However, age effects may be species specific; for example, no effect of age on the linear extension rate of *O. faveolata* in the Florida

Keys was noted (Helmle et al. 2011). Conversely, $C_{t-1}$ can be experimentally determined or included in statistical models (Lough and Barnes 1997; Castillo et al. 2011, 2012; Kwiatkowski et al. 2013; Sowa et al. 2014; Manzello et al. 2015b). Recent studies have used complex models, such as generalized linear models (GLMs), generalized additive models (GAMs), linear fixed-effect models, generalized additive mixed models (GAMMs), autoregressive moving average (ARMA) models, and generalized state space models (e.g., De'ath et al. 2009; Carilli et al. 2010; Castillo et al. 2011; Cooper et al. 2012; Kwiatkowski et al. 2013; Tanzil et al. 2013; Sowa et al. 2014).

To collect cores of the skeletons of large massive coral colonies, various hydraulic and pneumatic drilling systems have been developed (e.g., Macintyre 1978; Isdale and Daniel 1989; Adachi and Abe 2003). The core bit of a drill is placed against the surface of a coral colony (Fig. 6.2a); the length of the core barrel is typically 50–70 cm, and the diameter (φ) is generally less than 3 inch. The hole left by the collection of the coral cores is filled with a conical plug made of concrete, which prevents coral colony mortality by bioerosion and sedimentation (Fig. 6.2b). Matson (2011) showed that a *Porites* colony in the Great Barrier Reef covered such a plug with new growth within 30 months. Few studies have made recommendations regarding the appropriate number of samples to collect from a given site. However, it is recommended to collect two coral cores from the same colony or multiple cores from several colonies at one site, as in dendrochronological studies (Spencer et al. 2000; Pratchett et al. 2015).

Prior to coral growth measurements, coral slabs are prepared from the coral cores. The slabs (which are usually 3 to 10-mm thick) are prepared using a rock saw (Fig. 6.2c). The slab thickness is approximately equal to several coral polyps, allowing measurements that average the skeletal architecture described above to be obtained (Barnes and Devereux 1988; Pratchett et al. 2015). Prior to collecting the coral core powder samples for geochemical analysis, whole-core chemical treatments using NaClO are recommended (e.g., Nagtegaal et al. 2012). In general, analyzing the paths that cross the valleys located between corallite fans or disorganized corallite fans should be avoided because they may adversely affect the geochemical signature (e.g., Alibert and McCulloch 1997; Cohen and Hart 1997; DeLong et al. 2013).

Coral paleoenvironmental research requires well-preserved samples. Diagenesis can alter the skeletal density and chemical composition of both living and fossil corals through effects such as secondary infilling (precipitation of aragonite or calcite in the pores of the coral skeleton) and the dissolution and replacement of aragonite by calcite (e.g., Schroeder 1969; Bathurst 1975; Fairchild and Killawee 1995; Enmar et al. 2000; Müller et al. 2001, 2004; McGregor and Gagan 2003; Quinn and Taylor 2006; Allison et al. 2007; Hendy et al. 2007; Montaggioni and Braithwaite 2009; McLaren 2011; Perrin 2011). The Sr/Ca and $\delta^{18}O$ values of secondary aragonite (Enmar et al. 2000; Müller et al. 2001; Allison et al. 2007; Hendy et al. 2007) and infilling calcite (McGregor and Gagan 2003) have been shown to be higher and lower than those of primary coral skeletal material. Laboratory experiments have shown that seawater preferentially dissolves Ca in the coral skeleton relative to Sr (Schroeder 1969), resulting in higher Sr/Ca values (and, thus, a bias toward lower inferred SSTs). In contrast, freshwater leaching tends to dissolve more Sr than Ca from coral skeletons, resulting in lower Sr/Ca values (and, thus, a bias toward higher SSTs) (Fairchild and Killawee 1995). Therefore, prior to analyzing the coral growth parameters (this section) and chemical composition (Sect. 3), the degree of diagenesis should be determined. Several methods have been used to identify digenetic textures and the amount of calcite in coral skeletons: petrographic thin sections, X-radiography, scanning electron microscopy (SEM), X-ray diffraction (XRD), and Raman spectroscopy. Synchrotron-XRD and nondestructive 2D-XRD have also been applied (e.g., Sowa et al. 2008; Watanabe et al. 2011; Smodej et al. 2015).

Paired high- and low-skeletal-density bands that reflect seasonal cycles allow for the conversion of coral skeletal measurements into years. Knutson et al. (1972) used the signatures of nuclear weapon tests at Eniwetok Atoll as depositional markers in *Favia, Goniastrea, Porites, Psammocora, Platygyra*, and *Orbicella* skeletons. These tests increased the concentration of $^{90}Sr$ in the coral skeletons, and thus, the authors were able to demonstrate that a pair of density bands corresponded to 1 year of growth. Coral skeletal growth is currently measured using both indirect and direct methods. In coral cores and slabs, skeletal growth has been measured using two-dimensional X-radiography (Fig. 6.2c), computed tomography (CT; three-dimensional X-radiography), and γ-densitometry (Buddemeier 1974; Chalker et al. 1985; Chalker and Barnes 1990; Bosscher 1993; Cantin and Lough 2014; Lough and Cantin 2014). In general, X-radiography and CT are the most common methods used to measure growth and determine sampling transects (e.g., Bosscher 1993; Bessat and Buigues 2001; Cantin et al. 2010; Carilli et al. 2012; DeCarlo et al. 2015; Manzello et al. 2015a, b). Many studies of coral skeletal growth have used CT scanning instead of X-radiography because it does not require the preparation of coral slabs, which dramatically reduces the number of steps required to collect coral growth measurements, especially for long cores. Direct measurements of skeletal density via water or mercury displacement have also been performed to determine calcification rates during culture experiments and in coral cores (e.g., Dustan 1975; Brown et al. 1990; D'Olivo et al. 2013). It is important to determine the core chronology prior to using the displacement method

**Fig. 6.2** Images of coral sampling and coral slices. (**a**) Coral sampling using a pneumatic drill; (**b**) core plug (φ = 2.5 inch: *white arrow*) filling the hole made by collection of a coral core; and (**c**) from *left* to *right*, luminescent photograph under ultraviolet light, positive X-radiograph and optical light photograph of a *Porites* sp. coral collected from a temperate region of Japan (Provided by Dr. T Watanabe). *Gray arrow* indicates the growth direction of coral skeleton

for skeletal density. Most studies in Japan have used X-radiography for coral growth measurements (see references in Table 6.1).

Chronology development based on density bands can be problematic because there may be absolute chronological errors with ±2 to 3 years (e.g., Lough and Barnes 1997). Previous studies have reported several potential sources of error, including the deposition of multiple pairs of density bands in 1 year, the variable depositional timing of high-density bands because of sexual differences and different growth directions within the same colony (Brown et al. 1986; Hart and Cohen 1996; Carricart-Ganivet et al. 2013). These results suggest that simply counting density bands is insufficient to obtain a precise chronology.

Temporal fluctuations in SST proxies (Sect. 3) along the primary growth axis can support the creation of a precise annual- to seasonal-scale chronology. In addition, pairs of luminescent and nonluminescent bands (under ultraviolet light; Fig. 6.2c) have also been used to create a chronology when the density bands are unclear in X-radiography; this method is much simpler than chemical analysis (e.g., Scoffin et al. 1992; Tanzil et al. 2009; Grove et al. 2010). Recently, spectral luminescence scanner was developed (Grove et al. 2010). However, the underlying cause of these luminescent bands remains debated, and previous studies have suggested that they originate from humic acids or changes in the skeletal cavity, aragonite crystal size and packing, and chemical composition (Boto and Isdale 1985; Susic et al. 1991; Matthews et al. 1996; Barnes and Taylor 2001a, b, 2005). Therefore, variations in the chemical composition of the skeleton provide a more precise chronology than can be obtained by counting density or luminescent bands.

To understand the effects of environmental changes on coral growth in the geologic past, absolute ages must be determined for fossil coral skeletons. The most common dating method for corals less than 50 kyr old is $^{14}C$ dating (Fallon 2011). Interactions between $^{14}N$ atoms and cosmic rays produce $^{14}C$, which has a half-life of 5.73 kyr. Corals precipitate $^{14}C$ in their skeletons along with lighter C atoms. Currently, $^{14}C$ analysis is performed using accelerator mass spectrometry (AMS) Approximately 5–50 mg of coral powder is required for high-precision AMS measurements. The uncertainty associated with $^{14}C$ dating is approximately 20–40 $^{14}C$ years for ages less than 10 kyr BP (Nakamura et al. 2016). $^{14}C$ dating methods require the following information: (1) variations in the atmospheric concentration of $^{14}C$, (2) the carbon isotopic fractionation factor, and (3) the $^{14}C$ marine reservoir effect (Reimer et al. 2013).

From the late Pleistocene to the present (since approximately 500 kyr BP), uranium series dating using the

**Table 6.1** Summary of coral chronological studies in Japan

| Research topics | Modern/fossi coral | References | Major geochemical signals | Research sites |
|---|---|---|---|---|
| Climatology, oceanography, and environmental dynamics | | | | Site, [Fig. 6.1 index] |
| EAWM | Modern | Mishima et al. (2010) | $\delta^{18}O$, Sr/Ca | Ishigaki, [1] |
| PDO | Modern | Watanabe et al. (2014) | $\delta^{18}O$ | Koshiki, [4] |
| Seawater freshening | Modern | Felis et al. (2009) | $\delta^{18}O$, Sr/Ca, U/Ca | Ogasawara, [6] |
| SST regime shift | Modern | Tsunoda et al. (2006, 2008) | $\delta^{18}O$ | Ishigaki, [1] |
| Oceanic advection | Modern | Morimoto et al. (2004) | $\Delta^{14}C$ | Kikai, [3] |
| SST/biological response | Fossil | Abram et al. (2001) | $\delta^{18}O$ | Kikai, [3] |
| Evaporation/precipitation | Fossil | Suzuki et al. (2001) | $\delta^{18}O$ | Yonaguni, [1] |
| EAM | Fossil | Morimoto et al. (2007) | $\delta^{18}O$, Sr/Ca | Kikai, [3] |
| EAM | Fossil | Mishima et al. (2009) | $\delta^{18}O$, Sr/Ca | Miyako, [1] |
| EAWM | Fossil | Seki et al. (2012) | $\delta^{18}O$, Sr/Ca | Kume, [2] |
| Pacific lead pollution | Modern | Inoue et al. (2006); Inoue and Tanimizu (2008) | Pb/Ca, Pb isotope | Ogasawara, [6] |
| Anthropogenic inputs of nutrient | Modern | Yamazaki et al. (2015) | $\delta^{15}N$ | Ishigaki, [1] |
| Coral skeletal growth characteristics | | | | |
| Growth cessation in winter, up-welling | Modern | Fallon et al. (1999) | Mg/Ca, Sr/Ca, Ba/Ca, U/Ca | Komame, [5] |
| Impact of bleaching event | Modern | Suzuki et al. (2003) | $\delta^{18}O$, $\delta^{13}C$ | Ishigaki, [1] |
| Growth cessation in winter | Modern | Omata et al. (2006) | $\delta^{18}O$ | Ushibuka, [4] |
| Impact of bleaching event | Modern | Rosenfeld et al. (2006) | $\delta^{18}O$, $\delta^{13}C$ | Okinawa, [2] |
| Growth cessation | Fossil | Suzuki et al. (2008) | $\delta^{18}O$ | Ishigaki, [1] |
| Growth reduction in winter | Modern | Yamazaki et al. (2009) | $\delta^{18}O$ | Tatsukushi, [5] |
| Anthropogenic impact to coral growth | Modern/Fossil | Sowa et al. (2014) | Sr/Ca, Ba/Ca | Ishigaki, [1] |
| Omega in calcification fluid | Modern | Tanaka et al. (2015) | $\delta^{11}B$, Sr/Ca | Okinawa, [2] |
| Proxy development and evaluation | | | | |
| Natural reefs | | | | |
| New proxy development | Modern | Mitsuguchi et al. (1996) | Mg/Ca | Ishigaki, [1] |
| Proxy evaluation | Modern | Suzuki et al. (1999) | $\delta^{18}O$ | Ishigaki, [1] |
| Proxy evaluation | Modern | Shimamura et al. (2008) | $\delta^{18}O$ | Ishigaki, [1] |
| Proxy evaluation | Modern | Armid et al. (2011) | Sr/Ca, Mg/Ca, U/Ca | Okinawa, [2] |
| Proxy evaluation | Modern | Yamazaki et al. (2011b) | $\delta^{15}N$ | Ishigaki, [1] |
| Proxy evaluation | Modern | Yamazaki et al. (2011a) | $\delta^{18}O$, $\delta^{13}C$, $\delta^{15}N$, Ba/Ca | Okinotori, [7] |
| Proxy evaluation | Modern | Hirabayashi et al. (2013) | Sr/Ca | Ushibuka, [4] |
| Proxy evaluation | Modern | Seo et al. (2013) | Sr/Ca | Iki, [△] |
| Proxy evaluation | Modern | Hayashi et al. (2013) | $\delta^{18}O$, Sr/Ca | Okinawa, [2] |
| Proxy evaluation | Modern | Kawakubo et al. (2014) | Sr/Ca | Kikai, [3] |
| Proxy evaluation | Modern | Inoue et al. (2014) | Fe/Ca, Mn/Ca | Ishigaki, [1] |
| Proxy evaluation | Modern | Kubota et al. (2015) | $\delta^{11}B$ | Ogasawara, [6] |
| Culturing setting | | | | |
| Proxy evaluation | Modern | Hönisch et al. (2004) | $\delta^{11}B$ | Okinawa, [2] |
| Proxy evaluation | Modern | Inoue et al. (2011) | U/Ca | Okinawa, [2] |
| Proxy evaluation | Modern | Hayashi et al. (2013) | $\delta^{18}O$, Sr/Ca | Okinawa, [2] |
| Proxy evaluation | Modern | Nishida et al. (2014a, b) | $\delta^{18}O$ | Okinawa, [2] |

$^{238}U$-$^{234}U$-$^{230}Th$ decay chain has also been applied to carbonates. In coral skeletons, U-series dating typically has a precision ranging from 0.35 (modern) to 19 years (in samples up to 6.9 kyr old) (e.g., Zhao and Collins 2011; Shen et al. 2012; Clark et al. 2014). The half-lives of $^{238}U$, $^{234}U$, and $^{230}Th$ are $4.468 \times 10^9$ years, $24.55 \times 10^4$ years, and $7.54 \times 10^4$ years, respectively. Uranium is soluble in seawater, whereas Th is relatively insoluble. As a result, corals only incorporate U into their skeletons. Indeed, JCp-1 (a *Porites* standard used for the analysis of coral samples) contains a very small amount of Th (approximately 50 ppb) relative to U (approximately 2,900 ppb) (Okai et al. 2002; Watanabe and Nakai 2006). The U-Th isotopic compositions can be analyzed using various techniques, including α-counting, thermal ionization mass spectrometry (TIMS), inductively coupled plasma mass spectrometry (ICP-MS), and multicollector (MC)-ICP-MS (e.g. Barnes et al. 1956; Chen et al. 1986; Shaw and Francois 1991; Luo et al. 1997; Shen et al. 2012; Grothe et al. 2016; Obert et al. 2016). Approximately 10–200 mg of coral powder is required for high-precision MC-ICP-MS measurements (Shen et al. 2012).

**Table 6.2** Summary of typical coral chemical and isotopic signals used as paleoenvironmental proxies

| Reconstruction target | Major isotopic or geochemical tracer | Comments |
|---|---|---|
| Absolute dating | U-series | 20 year–500 kyr BP |
| | $^{14}C$ | < 50 kyr BP |
| Seasonal dating | Sr/Ca, $\delta^{18}O$ (SST proxy) | Weekly to annual |
| | Luminescence bands | Annual |
| | Density bands | Annual |
| SST | Sr/Ca, $\delta^{18}O$ | |
| Salinity | $\delta^{18}O$ | |
| Nutrients/sediment loading | $\delta^{15}N$, P/Ca, Ba/Ca | |
| pH and DIC | $\delta^{11}B$, B/Ca | |

## 6.3 The Chemical Composition of Coral Skeletons as a Proxy for Seawater Conditions

The isotopic values and elemental concentrations of coral skeletons reflect environmental parameters and coral physiological activity. Based on these chemical signals, a number of proxies for environmental parameters have been developed (Smith et al. 1979; Hemming and Hanson 1992; Min et al. 1995; Mitsuguchi et al. 1996; Montagna et al. 2014) and used in a wide range of studies on coral geochemistry (Tables 6.1 and 6.2). The physiological activity of symbiotic corals can also affect the chemical composition of coral skeletons, potentially biasing paleoenvironmental reconstructions. Coral reefs form in coastal areas of Japan despite low wintertime SSTs (approximately 13 °C), although the dominant reef-building coral species around Japan differ from those in tropical and subtropical regions (e.g., Yamano et al. 2001). The genus *Porites*, which is the most common species used in coral paleoenvironmental studies, is present in Japan (AIMS 2013), although few massive *Porites* corals occur at Iki Island, the northern limit of coral reefs (Fig. 6.1) (Yamano et al. 2001). In addition, corals from the family Merulinidae (e.g., *Platygyra, Favites, Goniastrea, Dipsastraea*, and *Cyphastrea*) can occur at latitudes as high as 35°N (Yamano et al. 2001; Sugihara and Yamano 2004; Yamano et al. 2004). The low SSTs during the winter in temperate regions reduce or halt skeletal growth of corals. Coral growth cessation has been reported under conditions of lower SSTs than 18 °C (e.g., Jacques et al. 1977; Lough and Barnes 2000). In Komame, Kochi, Japan (Fig. 6.1b [5]), Fallon et al. (1999) used multiple trace elements to show that temperate *Porites* corals stopped growing when SSTs fell below 18 °C. Similarly, Omata et al. (2006) reported that the $\delta^{18}O_{carbonate}$ values of *Porites* corals from Ushibuka, Kumamoto, Japan (Fig. 6.1) indicated growth cessation in the winter because of physiological stresses associated with low SSTs. This slowdown in coral skeletal growth complicates the interpretation of coral geochemical proxies because the relationship between coral proxies and seawater conditions can be variable. In addition, the response of coral proxies to reef environments can differ among coral species (see more detail in Sadler et al. 2014).

These complexities related to the coral growth rate and species can increase the difficulty of quantitative interpretations on climate and ocean environmental proxies of coral chemical composition. Before now, several proxies involving subtropical and temperate corals in Japan have been evaluated (Table 6.1).

### 6.3.1 SST and SSS

SST is a major parameter that reflects climatic variability at both global and local scales via the interaction between the atmosphere and the oceans. Various elemental concentrations (e.g., Sr/Ca, U/Ca, Mg/Ca, B/Ca and Li/Mg) and isotopic compositions (e.g., $\delta^{18}O$ and $\Delta 47$) in coral skeletons are correlated with the SSTs at the time and location of coral growth (Beck et al. 1992; Min et al. 1995; Mitsuguchi et al. 1996; Sinclair et al. 1998; Fallon et al. 1999; Alibert et al. 2003; Ghosh et al. 2006; Montagna et al. 2014). Among these, the Sr/Ca composition and the $\delta^{18}O$ value of coral skeletons are most commonly used paleo-SST proxies. The skeletal Sr/Ca ratio is inversely proportional to SST (e.g., Marshall and Clode 2004; Corrège 2006) (Fig. 6.3a).

The distribution coefficient of Sr in coral aragonite depends on temperature and the ionic ratio of $Sr^{2+}$ to $Ca^{2+}$ in seawater, which is nearly constant in modern open oceans (de Villiers et al. 1995; Shen et al. 1996). The rate of aragonite precipitation in a calcifying fluid can influence the aragonite's Sr/Ca value, assuming that coral biomineralization occurs in a nearly closed system (Gaetani and Cohen 2006). The sensitivity of the coral Sr/Ca thermometer in massive corals to seawater temperature is typically from −0.061 to −0.084 mmol/mol/°C (summarized in Corrège 2006; Gagan et al. 2012) (Fig. 6.3).

The stable oxygen isotopic ratio ($^{18}O/^{16}O$) in $CaCO_3$ is expressed relative to the Vienna Pee Dee Belemnite (VPDB). This ratio is conventionally written as $\delta^{18}O$ value of coral skeletons ($\delta^{18}O_{aragonite}$) and reported in per mil (‰), where $\delta^{18}O_{aragonite} = \{(^{18}O/^{16}O)_{aragonite}/(^{18}O/^{16}O)_{VPDB} - 1\} \times 10^3$. The equilibrium fractionation factor depends on the temperature (Urey 1947; McCrea 1950). Therefore, the$\delta^{18}O_{aragonite}$ is constrained by the $\delta^{18}O$ value of the seawater ($\delta^{18}O_{water}$) and the SST (Fig. 6.3b). The hydrologic cycle in the oceans (e.g., evaporation, precipitation, and freshwater inputs) controls the $\delta^{18}O_{water}$ value. Freshwater is enriched in $^{16}O$ (and, thus, has a lower $\delta^{18}O_{water}$ value) relative to seawater water, resulting in a salinity effect on the $\delta^{18}O_{water}$ value, generating a positive relationship between the two parameters. If variations in salinity are relatively small, the $\delta^{18}O_{aragonite}$ value of a coral skeleton will mainly reflect the SST. The sensitivity of massive coral $\delta^{18}O_{aragonite}$ values to seawater temperature typically ranges from −0.18 to −0.23 ‰/°C (summarized in Juliet-Leclerc and Schmidt 2001; Corrège 2006; Gagan et al. 2012).

The use of both Sr/Ca and $\delta^{18}O_{aragonite}$ proxies in the same coral skeleton can serve as a reliable indicator of past SST and SSS (e.g., McCulloch et al. 1994; Gagan et al. 2002; Asami et al. 2009). As mentioned above, Sr/Ca mainly reflects the SST, whereas the $\delta^{18}O_{aragonite}$ value is constrained by the SST and $\delta^{18}O_{water}$ value (or SSS). Therefore, the $\delta^{18}O_{water}$ value (as a proxy for SSS) can be calculated based on the difference between SST and the $\delta^{18}O_{aragonite}$ value. SST and SSS variability in Japan has been reconstructed using both Sr/Ca and $\delta^{18}O_{aragonite}$ proxies in *Porites* cores (more detail is given in Sect. 4).

Biases in these SST proxies have been evaluated by comparing the empirical relationships of chemical parameters between coral skeletons and inorganically precipitated aragonite. These biases have been recognized as (1) different relations (especially, offsets of intercepts) and (2) exponential function-like correlations between geochemical signals and the extension rate when it is lower than approximately 6 mm/year (McConnaughey 1989; Felis et al. 2003; Corrège 2006). Figure 6.3 highlights the large difference in intercepts between the temperature-Sr/Ca and temperature-$\delta^{18}O_{aragonite}$ relationships in coral skeletons versus inorganically precipitated carbonate. For example, the Sr/Ca values of aragonite

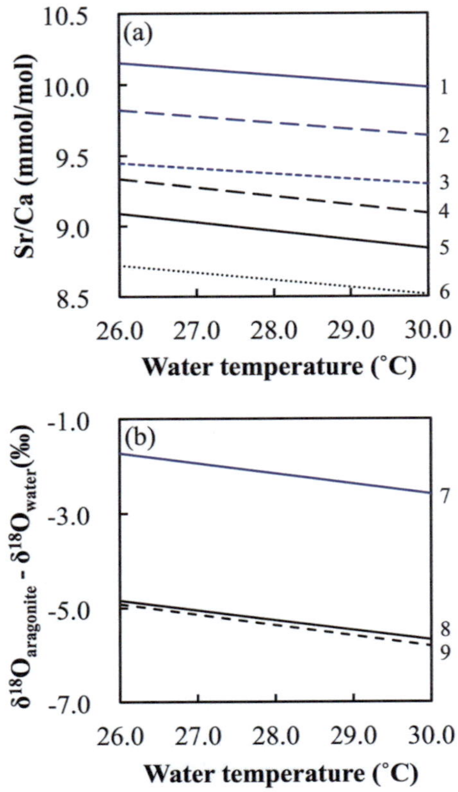


**Fig. 6.3** Relationship between water temperature and (**a**) Sr/Ca and (**b**) $\delta^{18}O_{aragonite}$-$\delta^{18}O_{water}$. The *blue line* indicates the relationship obtained from inorganic experiments; the *black lines* represent the relationships between local SSTs and coral skeletal proxies (*1* Dietzel et al. 2004, *2* Kinsman and Holland (1969), *3* Gaetani and Cohen (2006), *4* Wu et al. (2014), *5* McCulloch et al. (1999), *6* Quinn and Sampson (2002), *7* Kim et al. (2007), *8* McConnaughey (1989), *9* Wellington et al. (1996))

inorganically precipitated at a given temperature (20–28 °C, which is typical of coral reefs) are systematically higher than those of coral aragonites by approximately 1 mmol/mol (e.g., Kinsman and Holland 1969; Dietzel et al. 2004) (Fig. 6.3). The absence of symbiotic algae in coral also resulted in lower Sr/Ca values for coral aragonite relative to inorganic aragonite (Cohen et al. 2002). However, McConnaughey (1989) suggested that the offset of $\delta^{18}O_{aragonite}$ values remains constant over time.

The exponential function-like correlations for skeletal extension rates of lower than approximately 6 mm/year have been suggested to result from kinetic isotope disequilibrium (e.g., McConnaughey 1989). The growth effect on Sr/Ca and $\delta^{18}O_{aragonite}$ values has been evaluated in both natural and culture experiments in Japan. In a natural setting, Omata et al. (2006) and Hirabayashi et al. (2013) compared Sr/Ca and $\delta^{18}O_{aragonite}$ values to the growth rates (linear extension rates) of *Porites* corals collected from Ushibuka, Kumamoto, Japan (Fig. 6.1), where the SST varies between 13 and 27 °C (Fig. 6.3). These investigations showed that the growth effect on Sr/Ca was insignificant, although it became significant for

$\delta^{18}O_{aragonite}$ when the linear extension rate was lower than 6 mm/year. A similar trend was observed in culture experiments (Hayashi et al. 2013; Nishida et al. 2014b). Hayashi et al. (2013) examined the effects of growth rate on coral proxies in five *Porites* colonies cultured under natural SST conditions (approximately 19–30 °C) in Okinawa. The growth effects on the $\delta^{18}O_{aragonite}$ and Sr/Ca values were limited despite large inter-colony differences in the annual extension rate (2.6–11 mm/year). Armid et al. (2011) conducted a culture experiment under different SST conditions (22, 26 and 30 °C) using branching corals (*Porites cylindrica*), which grow more rapidly than massive *Porites* corals. The Sr/Ca ratio was significantly correlated with SST, as in massive *Porites* corals, and was independent of the extension rate. These results suggest that paleoenvironmental reconstructions using branching *Porites* corals may be possible.

Relationships between skeletal composition and seawater conditions have been also evaluated in several coral species other than *Porites*. Shimamura et al. (2008) demonstrated that the empirical relationship between the $\delta^{18}O_{aragonite}$ values of *Platygyra* corals from Ishigaki and Okinawa Islands and the SST yielded a slope of −0.219 ‰/°C, similar to the average slope reported for *Porites* spp. Seo et al. (2013) analyzed the coral Sr/Ca values of *Dipsastraea* from Iki Island (Fig. 6.1). The Sr/Ca ratio varied seasonally over most of the sampling transect, although low amplitudes of the Sr/Ca variations were occasionally observed when the growth rate was significantly decreased. Nishida et al. (2014b) reported that the $\delta^{18}O_{aragonite}$ values of juvenile *Acropora* corals depended on the calcification rate. Similarly, Nishida et al. (2014a) examined the $\delta^{18}O_{aragonite}$ values of *Isopora palifera* cultivated under five SST settings ranging from 21.1 to 29.5 °C in Okinawa. They found an insignificant relationship between calcification and the $\delta^{18}O_{aragonite}$ value of the corals and a significant correlation between the $\delta^{18}O_{aragonite}$ value and SST with a slope of −0.15 ‰/°C.

Coral SST proxies may be affected by variations in the growth rate and may differ between species. However, the biases associated with Sr/Ca values are minor or insignificant. For $\delta^{18}O_{aragonite}$ values, slow calcification rates (linear extension rates lower than ~6 mm/year) can result in higher $\delta^{18}O_{aragonite}$ values than expected from SSTs and seawater $\delta^{18}O$. These proxies have also been shown to be reliable in coral species other than *Porites*. Thus, with care, it is possible to use temperate corals in Japan as environmental archives.

### 6.3.2 Nutrients and Sediment Loading

Primary production in the surface oceans is affected by the dynamics of nutrients, including nitrate, phosphate, and silicate, as well as sediment loading. The nutrient concentrations in coral reefs are influenced by the inflow of freshwater from rivers and groundwater and by the upwelling of deeper ocean water (e.g., Devlin and Brodie 2005; D'Croz and O'Dea 2007). Recently, anthropogenic effects, such as eutrophication and land development, have also influenced the nutrient dynamics in coral reefs in some coastal areas. As with nutrient dynamics, sediment loading has also increased because of human activity (e.g., Fabricius 2005). The chemical composition of coral skeletons can reflect seasonal and/or abrupt changes in nutrient conditions and sediment loading.

The stable nitrogen isotopic ratio is conventionally represented as $\delta^{15}N$ ($\delta^{15}N = \{(^{15}N/^{14}N)/(^{15}N/^{14}N)_{air} - 1\} \times 10^3$). $\delta^{15}N$ of organic matter preserved in coral skeletons commonly record the changes in the nitrogen sources around coral reefs, including $N_2$ fixation, upwelling, terrestrial input, and anthropogenic pollutants (Marion et al. 2005; Yamazaki et al. 2009, 2011a, b, 2015; Erler et al. 2015, 2016; Wang et al. 2015). At Ishigaki Island, for example, the $\delta^{15}N$ value of intra-skeletal organic matter in *Porites* corals decreases with the increasing distance from shoreline close to the mouth of the Todoroki River, and correlates with the $\delta^{15}N$ value of nitrate ions (Yamazaki et al. 2011a, b) (Fig. 6.4).

Additionally, the phosphorus concentration of coral skeletons (P/Ca) may record variations in the phosphate concentration near coral reefs (Dodge et al. 1984; Shotyk et al. 1995; Kumarsingh et al. 1998; Alibert et al. 2003; LaVigne et al. 2008; LaVigne et al. 2010; Mallela et al. 2013). The Ba/Ca value of coral skeletons has been used as a proxy for upwelling and terrestrial runoff because the Ba concentrations in deep water and terrestrial runoff are higher than in the surface water near coral reefs (e.g., Shen and Boyle 1988; Lea and Boyle 1989; Guzmán and Jiménez 1992; Fallon

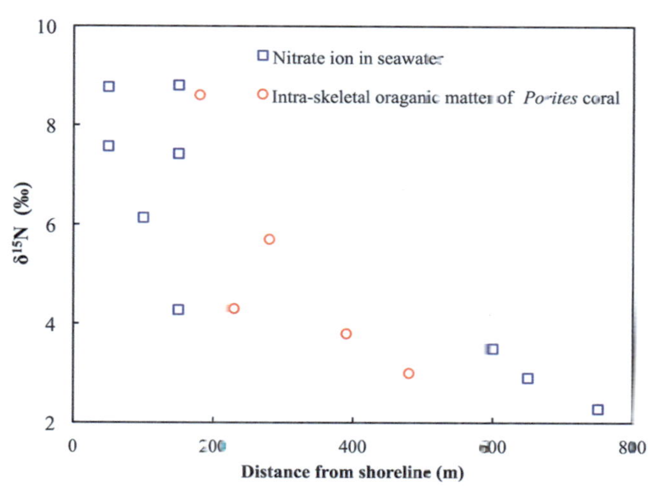


**Fig. 6.4** Relationship between $\delta^{15}N$ and the distance from the shoreline. *Red circles* and *blue squares* indicate the $\delta^{15}N$ values of intra-skeletal organic matters of *Porites* coral and seawater nitrate, respectively (Figure modified and redrawn from Yamazaki et al. 2015)

et al. 1999; McCulloch et al. 2003; Carilli et al. 2009; Prouty et al. 2010, 2014; Grove et al. 2012, 2013; Horta-Puga and Carriquiry 2012; LaVigne et al. 2016).

### 6.3.3 pH and DIC

Ocean carbonate chemistry has changed dramatically on geologic timescales (e.g., Zachos et al. 2005). More recently, anthropogenic $CO_2$ emissions since the Industrial Revolution have reduced the pH of the surface oceans by approximately 0.1 pH unit (Orr et al. 2005). Seawater pH in the western North Pacific has decreased by ~0.001–0.002 pH unit/year after 1984 (Midorikawa et al. 2010). Hence, a demand for pH proxies that can be used to understand the past carbon flux between the oceans and the atmosphere and, therefore, to predict the future changes has emerged. Hemming and Hanson (1992) proposed a theoretical relationship between the boron isotopic ratio ($\delta^{11}B$) and the pH of seawater ($pH_{SW}$). Boron in seawater exists as boric acid ($B(OH)_3$) and borate ions ($B(OH)_4^-$), and the proportions of these species depend on the pH. The boron isotope ratio is conventionally represented as $\delta^{11}B$ ($\delta^{11}B = \{(^{11}B/^{10}B)/(^{11}B/^{10}B)_{NIST951} - 1\} \times 10^3$). $\delta^{11}B$ of $B(OH)_4^-$ is lighter than that of $B(OH)_3$ by approximately 27 ‰ and the two values become lighter with decreasing pH (Fig. 6.5a). Assuming that only $B(OH)_4^-$ is incorporated into aragonite skeletons, the $\delta^{11}B$ value of coral skeletons has been used to estimate past changes in $pH_{SW}$ and carbon flux (Pelejero et al. 2005; Wei et al. 2009; Shinjo et al. 2013; Kubota et al. 2014). In addition, a number of studies have assessed the relationship between skeletal $\delta^{11}B$ values and $pH_{SW}$ in culture experiments and natural settings (Hönisch et al. 2004; Reynaud et al. 2004; Krief et al. 2010). Recent studies suggest that coral $\delta^{11}B$ values are more representative of the pH in the calcifying fluid ($pH_{CF}$) than $pH_{sw}$, although $pH_{CF}$ is moderately proportional to $pH_{sw}$ (see Fig. 6.5b, Trotter et al. 2011; McCulloch et al. 2012a, b). These studies demonstrated that the $pH_{CF}$ values derived from $\delta^{11}B$ measurements are higher than the ambient $pH_{SW}$ by 0.4–0.6, which is consistent with direct measurements of $pH_{CF}$ (Al-Horani et al. 2003; Venn et al. 2011, 2013). Currently, based on the relationships between $\delta^{11}B$ values, $pH_{CF}$, and $pH_{SW}$, the $\delta^{11}B$ value of coral skeletons is being used to understand various physiological aspects of symbiotic corals, such as bleaching events (Schoepf et al. 2014; Dishon et al. 2015) and the omega value in the calcifying fluid (Allison et al. 2014; Tanaka et al. 2015). However, the relationships between $\delta^{11}B$ values and $pH_{SW}$ derived from culture experiments suggest that the $\delta^{11}B$ values of *Porites* corals may not be suitable for calculating absolute value of $pH_{SW}$ directly. More realistic calibrations have been proposed based on field calibration techniques (Kubota et al. 2015).

Boron concentration in coral skeletons (B/Ca) also potentially records DIC species in seawater. Several parameters related with carbonate chemistry (e.g., concentration of borate ion, bicarbonate ion and carbonate ion) were proposed as factors controlling B/Ca in marine biogenic carbonate (e.g., Yu and Elderfield 2007; Douville et al. 2010; Allen et al. 2012). Recent studies conducted inorganic experiments to specify factors controlling B/Ca in calcium carbonate precipitated from seawater (Uchikawa et al. 2015; Holcomb et al. 2016). According to the inorganic experiments, B/Ca in aragonite represents the most strongest correlation with $[B(OH)_4^-]/[CO_3^{2-}]^{0.5}$ in seawater, where $[B(OH)_4^-]$ and $[CO_3^{2-}]$ are the concentration of borate and carbonate ion, respectively (Holcomb et al. 2016). Since $[B(OH)_4^-]$ is determined from pH and total boron concentration, which is proportional to salinity (Uppström 1974; Lee et al. 2010), B/Ca is possibly useful to estimate $[CO_3^{2-}]$ in seawater. Thus, the combination of two boron proxies ($\delta^{11}B$ and B/Ca) in the coral skeleton informs carbonate chemistry in the past oceans.

## 6.4 Previous Coral Paleoenvironmental Studies in Japan

### 6.4.1 Paleoenvironmental Reconstructions: Climatic and Oceanic Changes

A common approach for characterizing climatic or oceanographic changes is to detect its patterns, such as an oscillation or regime shift, which can be defined as "a transition from one climatic state to another within a period substantially shorter than the lengths of the individual epochs of each climate state" (Minobe 1997). The major climatic and oceanic modes that directly or indirectly influence environmental changes around Japan have timescales ranging from interannual to multidecadal and are associated with the El Niño-Southern Oscillation (ENSO), the Pacific Decadal Oscillation (PDO) or Interdecadal Pacific Oscillation (IPO), the Kuroshio Meander, and the regime shifts in wintertime SSTs (Mizuno and White 1983; Mantua et al. 1997; Minobe 1997; Zhang et al. 1997; Yasunaka and Hanawa 2002; Di Lorenzo et al. 2008; Urabe and Maeda 2014). SST and SSS are important parameters that can be used to detect these modes. Paleo-SST and SSS can highlight the patterns associated with these and other modes and to determine their causes. Thus, geochemical proxies in modern and fossil massive *Porites* from Japan can be used to reconstruct seasonal to interannual SST and SSS variations in the local subtropical to temperate oceans since the last interglacial period to today (Fig. 6.1 and Table 6.1).

Based on the observational records (such as atmospheric temperature and SST) in the twentieth century, regime shifts

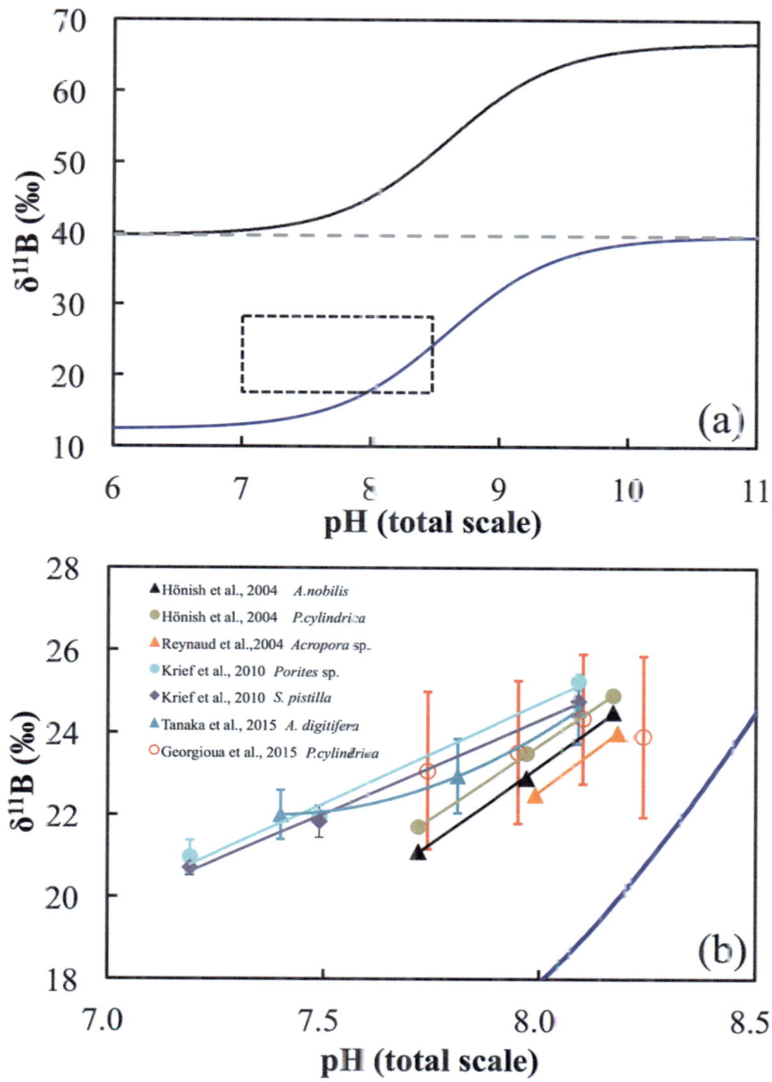


**Fig. 6.5** (**a**) Theoretical relationships between seawater pH and the boron isotopic compositions ($\delta^{11}B$) of borate (*black line*) and borate ions (*blue line*). The *black dotted rectangle* indicates the range shown in (**b**). (**b**) Empirical relationships between seawater pH and the $\delta^{11}B$ values of coral skeletons cultured under various seawater pH conditions; the theoretical curve is represented by the *blue line*

in wintertime SSTs have been detected in 1925/1926, 1945/1946, 1957/1958, 1970/1971, 1976/1977, and 1988/1989 in the Northern Hemisphere (Yasunaka and Hanawa 2002). Tsunoda et al. (2006, 2008) reported that a modern *Porites* coral record of wintertime $\delta^{18}O_{aragonite}$ values from Ishigaki Island (Fig. 6.1) spanning 30 years, from 1971 to 1998, included the 1988/1989 regime shift in wintertime SSTs. Instrumentally determined winter SST values also indicate SST warming (approximately 1 °C) in 1988/1989. The correlation between climate indices and coral winter $\delta^{18}O_{aragonite}$ values also changed in 1988/1989. From 1971 to 1988, the coral winter $\delta^{18}O_{aragonite}$ values and instrumental wintertime SSTs were correlated to only the Monsoon Index, which is defined as the difference in sea level pressure between Irkutsk and Nemuro and a proxy for the strength of the EAWM (Hanawa 1988). In contrast, from 1989 to 2000, coral winter $\delta^{18}O_{aragonite}$ values and instrumental winter SSTs were correlated with only the Southern Oscillation Index (SOI).

Other than the 1988/1989 regime shift, the five SST regime shifts were not reported based on the coral wintertime $\delta^{18}O_{aragonite}$ values in previous modern coral paleoenvironmental studies. However, a regime shift in SSS was reported in the early twentieth century. Felis et al. (2009) reconstructed $\delta^{18}O_{water}$ values from an Ogasawara coral and found that the $\delta^{18}O_{water}$ value abruptly decreased by 0.345 ‰ between 1905 and 1910, corresponding to a SSS decrease of 0.82 (Fig. 6.6). They noted that no significant change in precipitation occurred during this time in Okinawa (26.2°N) or Tokyo (35.8°N). They suggested that seawater freshening could have resulted from decreased evaporation or reduced water mass transport resulting from weakened westerly winds. Interestingly, during a similar time period (1900–1905), Mishima et al. (2010) detected decreasing wintertime SSTs at Ishigaki Island (Fig. 6.6). They analyzed a 165-year record of $\delta^{18}O_{aragonite}$ and Sr/Ca values in a subtropical *Porites* coral, showing that the $\delta^{18}O_{aragonite}$ and Sr/Ca values recorded an abrupt decrease in wintertime SSTs between 1900 and 1905 (Fig. 6.6). They inferred that the development of a very cold SH brought intensely cold air masses to East Asia and weakened the winds, resulting in cooler winters. In summary, Felis et al. (2009) and Mishima et al. (2010) suggested that during the early twentieth century, the development of a strong SH and weakened winds resulted in cold SSTs at Ishigaki Island and reduced the Kuroshio transport and evaporation over the ocean. In the North Pacific region, a maximum-salinity water mass called the North Pacific Tropical Water (NPTW) exists near the center of the North Pacific (Cannon 1966; Tsuchiya 1968). Studies have demonstrated relationships between wind stress forcing, Kuroshio transport, the subtropical gyre, and the westward expansion of NPTW (e.g., Shuto 1997; Miller et al. 1998; Deser et al. 1999; Suga et al. 2000; Li and Wang 2012). Therefore, reduced Kuroshio transport by weakening winds contract NPTW eastward, which reduced the salinity at Ogasawara Island. Seasonal to interannual changes in Kuroshio transport (advection) are evidenced by seasonal coral $\Delta^{14}C$ measurements from Kikai Island, which serve as a proxy for DIC in seawater (Morimoto et al. 2004).

The PDO is a prominent oceanic oscillation in the North Pacific and is defined as the first empirical orthogonal function of SST in the North Pacific (>20°N) (Mantua et al. 1997). Several studies have reported a correlation between coral geochemical signals and the PDO index. Felis et al. (2010) reanalyzed wintertime Sr/Ca and U/Ca values in a coral skeleton from Ogasawara that were first reported by Felis et al. (2009). A combined index of coral Sr/Ca and U/Ca values and the growth record from a geoduck clam from the northeastern Pacific by Strom et al. (2004) (approximately 48.12°N, 122.95°W) was shown to be significantly correlated with the instrumentally determined winter PDO index. Watanabe et al. (2014) generated a 106-year seasonal $\delta^{18}O_{aragonite}$ record (1902–2008) using a temperate *Porites* coral from Koshiki Island, Japan (Fig. 6.1). They found that summertime $\delta^{18}O_{aragonite}$ values were correlated with the wintertime PDO index with a 1- to 3-year time lag and suggested that the EASM could modulate wintertime PDO variability in this region.

In order to understand the climatic change through the Holocene, the variation in the temperature and intensity of the EAWM and EASM have been reconstructed from stalagmites and planktonic foraminifera (e.g., Wang et al. 2005; Marcott et al. 2013; Sagawa et al. 2014) (Fig. 6.7). In addition, fossil coral skeletons from Japan provide records of interannual to decadal changes in paleo-SST and SSS in the

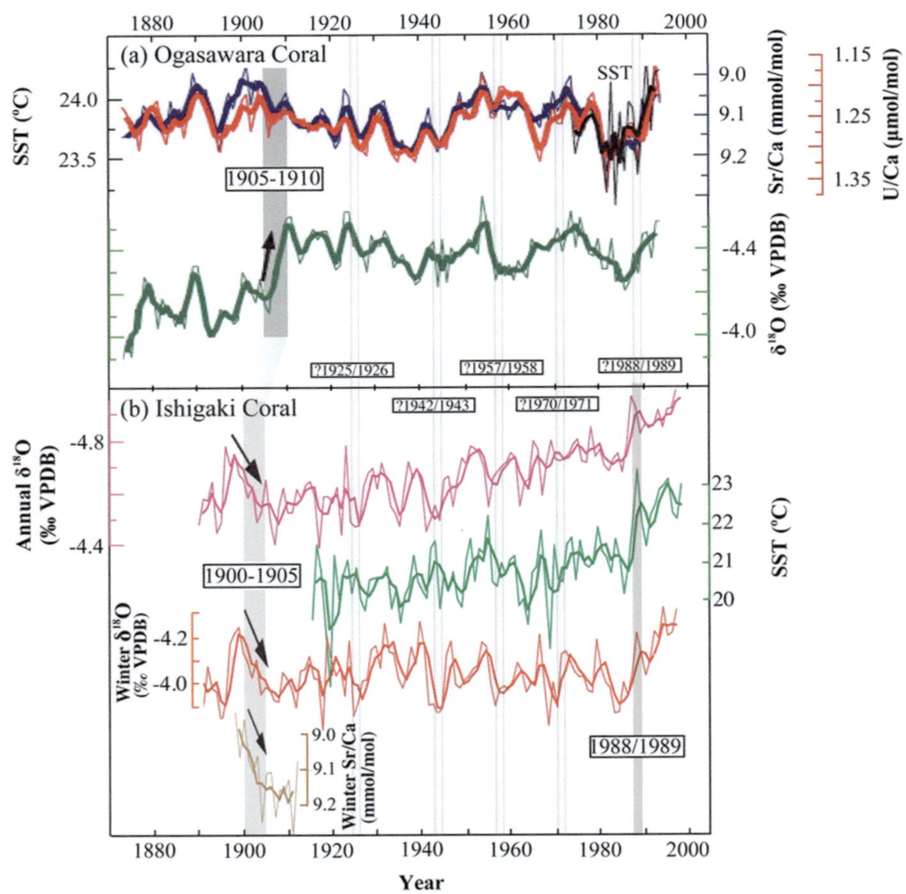


**Fig. 6.6** Comparison of *Porites* coral geochemical records and instrumentally determined SSTs between Ogasawara (Fig. 6.1b [6]) and Ishigaki Island (Fig. 6.1b [1]). (**a**) Ogasawara coral $\delta^{18}O$, Sr/Ca, U/Ca values and instrumentally determined SSTs. (**b**) Ishigaki Island coral $\delta^{18}O$ and Sr/Ca values and instrumentally determined SSTs. *Thin* and *bold lines* indicate annual and 3-year running means, respectively. *Black arrows* and the *dark gray vertical bar* indicate regime shifts for each parameter. The *light gray bar* indicates the expected regime shift in wintertime SSTs suggested by Yasunaka and Hanawa (2002) (Figure modified and redrawn from Tsunoda et al. 2006, 2008, Felis et al. 2009, and Mishima et al. 2010)

Holocene (Fig. 6.7). For example, Holocene SST and SSS were reconstructed at Kikai Island (Fig. 6.1). Kikai Island has unique geological characteristics: it is located on the eastern side of Amami Island and is near the Ryukyu Trench (Fig. 6.1b [3]). The island is surrounded by four Holocene reef terraces raised by episodic uplift (e.g., Sugihara et al. 2003; Hongo 2010). Several studies (Kiyama et al. 2000; Abram et al. 2001; Morimoto et al. 2007) attempted to reconstruct the paleo-SST and SSS from Kikai Island at 1.9, 3.4, 3.8, 4.2, 6.2, 7.0, and 9.2 kyr BP using the $\delta^{18}O_{aragonite}$ and Sr/Ca values of fossil *Porites* coral skeletons (Fig. 6.7). Abram et al. (2001) reported paleo-SSTs at 1.9, 3.4, 3.8, and 4.2 kyr BP from fossil coral $\delta^{18}O_{aragonite}$ values. The mean $\delta^{18}O_{aragonite}$ values for all these time periods were higher (cooler) than modern ones. In particular, the wintertime SSTs reconstructed from $\delta^{18}O_{aragonite}$ values at 3.4 and 3.8 kyr

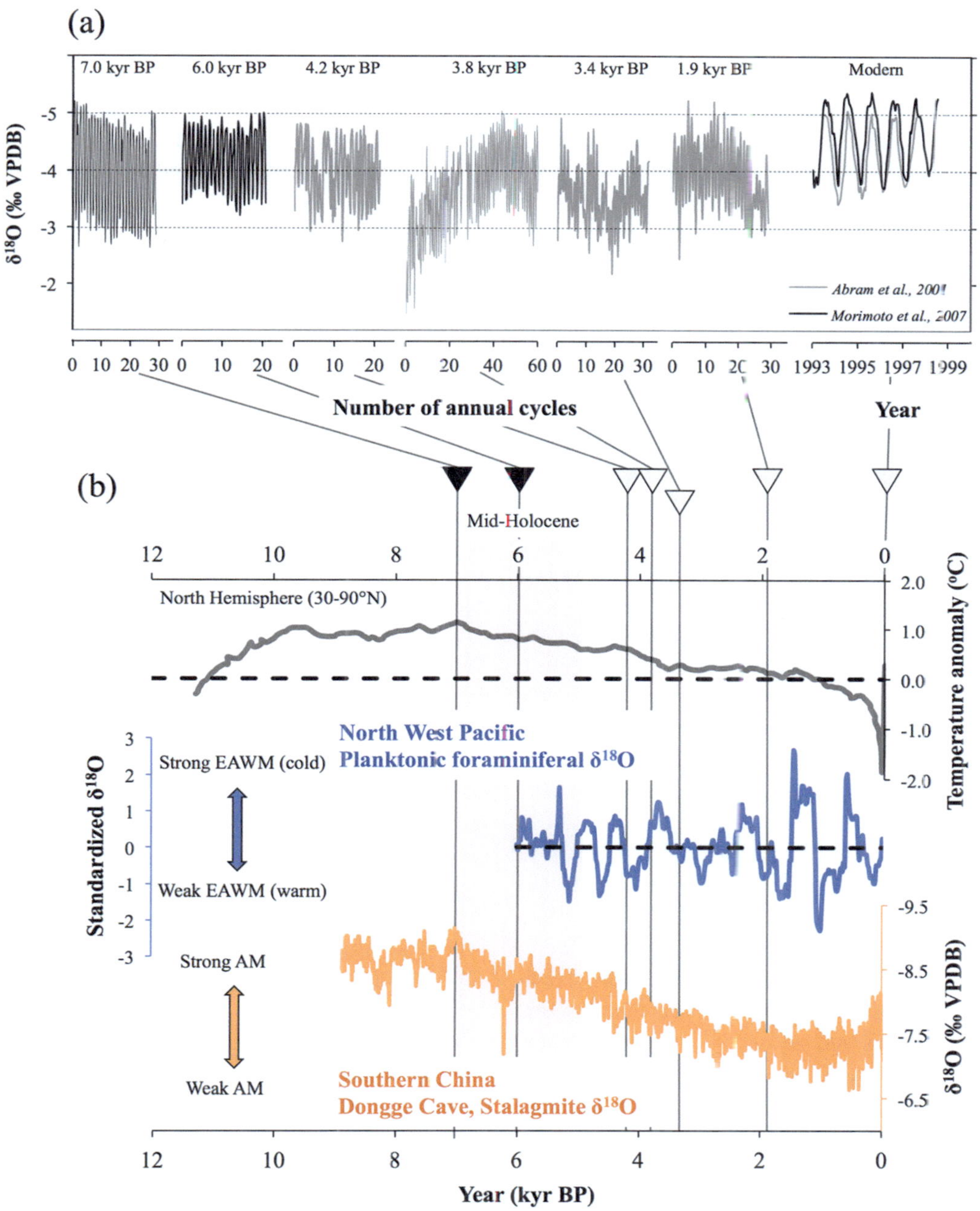


**Fig. 6.7** (**a**) Fossil coral $\delta^{18}O$ values in six time windows during the Holocene and (**b**) past temperature anomalies (Marcott et al. 2013) and intensity of the EAWM (Sagawa et al. 2014) inferred from planktonic foraminiferal $\delta^{18}O$ values and the Asia Monsoon (*AM*) intensity inferred from Dongge Cave stalagmite $\delta^{18}O$ values (Wang et al. 2005). The $\delta^{18}O$ values in (**a**) are from fossil and modern *Porites* corals from Kikai Island (Fig. 6.1b [3]). The *pink bar* indicates the term of mid-Holocene. (**a**) was modified and redrawn from Abram et al. 2001, Morimoto et al. 2007, and Kiyama et al. 2000. (**b**) was modified and redrawn from Marcott et al. 2013, Sagawa et al. 2014, and Wang et al. 2005)

BP were cooler than 18 °C, which is generally considered the minimum SST at which coral reefs can grow (Kleypas et al. 1999 and references therein). Ecological surveys of fossil corals in the terraces corresponding to 3.4 and 3.8 kyr BP (including measurements of total coral abundance, colony size, individual coral coverage, genera diversity, and evenness) suggested that coral communities were stressed by cooler SSTs but were able to continue growing (Abram et al. 2001). In addition, Morimoto et al. (2007) reported that the $\delta^{18}O_{water}$ values at 6.2 and 7.0 kyr BP derived from coral $\delta^{18}O_{aragonite}$ and Sr/Ca values were higher than the present values by 0.2–0.8 ‰, corresponding to an increase in the SSS value of 0.6–1.9. The SSTs at 6.2 kyr BP derived from coral Sr/Ca values were similar to modern ones, but those at 7.0 kyr BP were lower (−0.8 and −0.6 °C during the summer and winter, respectively). These values suggest the existence of more intense EAM effects on the evaporation-precipitation balance around Kikai Island during this period. We note that Kiyama et al. (2000) concluded that coral $\delta^{18}O_{aragonite}$ values at 9.2 kyr BP did not give a precise record of SST because of the vital effect, which is corresponding to the biases in Sect. 3.1.

Subtropical Holocene fossil corals can also be found in the Ryukyu Islands. Seki et al. (2012) examined the Sr/Ca values of fossil corals collected from Kume and Ishigaki Islands to estimate the SSTs at 3.8 and 4.5 kyr BP (Fig. 6.1b [1, 2]). They also calibrated observational SSTs and coral Sr/Ca values using a modern coral from Sesoko Island, Okinawa (Fig. 6.1b [2]). The low SSTs estimated using coral proxies during this period (−2.2 °C in summer and −0.5 °C in winter in 3.8 kyr comparing to pre industrial era) contradict those derived from $TEX_{86}$ and foraminifera Mg/Ca values collected from sediment cores along the Okinawa Trough (Jian et al. 2000; Seki et al. 2012, see references there in).

A few studies have reconstructed Pleistocene paleoclimate using Japanese fossil corals. Mishima et al. (2009) reported coral $\delta^{18}O_{aragonite}$, $\delta^{13}C$ and Sr/Ca values from *Dipsastraea* corals (approximately 16 kyr ago) collected from a sediment core from near Miyako Island. These data suggest that annual mean SST and $\delta^{18}O_{water}$ values were, respectively, 5 °C lower and 0.2 ‰ higher (more saline) than today. The authors proposed that intense winter monsoons could have lowered SSTs and enhanced evaporation (thus, higher SSS). Suzuki et al. (2001) analyzed the $\delta^{18}O$ and $\delta^{13}C$ values of *Porites* corals (from 127 ± 6 kyr BP) collected from Yonaguni Island. The fossil coral $\delta^{18}O_{aragonite}$ values were 1.1 ‰ higher than those of present-day corals. The authors suggested that the higher evaporation and flux volume in the Kuroshio Current could have caused higher $\delta^{18}O$ values during the last interglacial period. In summary, quaternary coral archives around Japan well documented the change in SST and SSS in association with climatic/oceanographic modes in the western North Pacific.

### 6.4.2 Reconstructing Anthropogenic Pollution in Local Coral Reefs

When anthropogenic pollutants from terrestrial inputs and airborne particles flow into the water near coral reefs, coral skeletons commonly record chemical signals of these inputs (e.g., heavy metals, such as Pb and Mn). Anthropogenic pollution in coral reefs can be assessed using these proxies. In the Western Pacific, Pb originates from Chinese loess and the combustion of leaded gasoline before its use was banned (e.g., Tatsumoto and Patterson 1963; Shen and Boyle 1988; Shotyk et al. 1998). Inoue et al. (2006) analyzed the Pb content of *Porites* coral skeletons collected from several sites in the Pacific (Amakusa, Ishigaki Island, Ogasawara, Hong Kong, Hainan Island, Jakarta, and Pohnpei Island) (Fig. 6.1). The skeletal Pb content decreased from the coasts of East Asian continent to the open ocean. Subsequently, Inoue and Tanimizu (2008) demonstrated that Pb isotopes ($^{206}Pb/^{207}Pb$ and $^{208}Pb/^{207}Pb$) had undergone temporal changes in Pacific coral skeletons from 1896 to 2001. They suggested that the increasing Pb content and the change in the Pb isotopic signature after the 1980s were derived from the input of Chinese aerosols to the surface waters of the Western Pacific.

Agriculture and land development can increase nutrient and sediment loading in coastal areas, and high inputs of nutrients and sediment can disturb coral reef ecosystems (e.g., Fabricius 2005, 2011; Fabricius et al. 2013). Yamazaki et al. (2015) demonstrated that the $\delta^{15}N$ value of intra-skeletal organic matter in *Porites* corals at the mouth of the Todoroki River on Ishigaki Island had changed from 1958 to 2010. The average $\delta^{15}N$ value decreased by 1 ‰ after 1981 compared to that prior to 1980. The authors suggested that this shift may have resulted from a land use change from paddy fields (manure contains high $\delta^{15}N$ values) to sugarcane fields. Sowa et al. (2014) assessed coral Ba/Ca and skeletal growth in modern, 1.2- and 3.5-kyr-BP corals from Ishigaki Island (Table 6.1 and Fig. 6.1). Although the modern coral Ba/Ca values were correlated with the calcification rate, those of the fossil corals did not. The authors inferred that the anthropogenic loading of sediments and nutrients was likely responsible for this phenomenon in the modern corals.

Past coral bleaching events can also be inferred from the $\delta^{18}O$ and $\delta^{13}C$ values of coral skeletons. Suzuki et al. (2003) and Rosenfeld et al. (2006) measured $\delta^{18}O$ and $\delta^{13}C$ values at a monthly resolution in coral skeletons collected from Ishigaki and Sesoko Islands in 1998 (Table 6.1, Fig. 6.1), when mass coral bleaching occurred in Okinawa (e.g., Fujioka 1999; Hasegawa et al. 1999; Yamazato 1999). The summertime $\delta^{18}O$ values abruptly decreased in 1998 at Ishigaki Island ($n = 1$), and the variation of summertime $\delta^{18}O$ values in 1998 were higher in comparison with 1997 and 1999 at Sesoko Island ($n = 6$). The authors proposed that a disturbance in skeletal growth was caused by coral bleaching.

In addition, they found lower skeletal $\delta^{13}C$ values or no negative $\delta^{13}C$ peaks during the summer of 1998, possibly attributable to a reduction in photosynthetic activity resulting from the bleaching event (Suzuki et al. 2003; Rosenfeld et al. 2006). These results suggest that together, skeletal growth and $\delta^{18}O$ and $\delta^{13}C$ values in coral skeletons could facilitate reconstructing past coral bleaching events.

## 6.5 Future Directions for Coral Paleoenvironmental Studies

Key areas of focus for coral paleoenvironmental research at the northern margins of coral habitats can be divided into (1) assessing and detecting paleoclimatic and oceanographic patterns on a geologic timescale and (2) assessing the combined impacts of global and local environmental changes on coral growth in subtropical to temperate regions.

Massive corals in the northern limits of their habitat can serve as paleoenvironmental proxies, promoting a better understanding of the impacts of climatic and oceanic variability in these regions. Information regarding regime shifts along the Pacific Rim will help characterize paleoclimatic changes over the Pacific Ocean and the East Asia. For example, wintertime shifts in SST regimes occurred in 1925/1926, 1945/1946, 1957/1958, 1970/1971, and 1976/1977 on a global scale and also did in 1988/1989 in the Northern Hemisphere (Tachibana et al. 1996; Yasunaka and Hanawa 2002, 2005). Minobe (1997) identified a 50–70-year climatic oscillation in the Northern Hemisphere (the North Pacific and North America). Several regime shifts and oscillations are apparent in coral geochemical signals (Sect. 4), but many regimes are not. To characterize the climatic or oceanic patterns of variability, future studies should aim to achieve greater spatial coverage during coral sampling and to create records longer than 100 years.

Fossil corals can reflect climatic and oceanic changes during the late quaternary. Fossil corals collected from reef terraces at Kikai and Okinawa Islands that were uplifted during the Holocene serve as "snapshots" of the environmental parameters at that time (e.g., Abram et al. 2001; Kawana and Kan 2002; Morimoto et al. 2007). In addition, numerous *Porites* coral boulders scattered along the coastlines of the fringing reefs in the southern Ryukyu Islands were transported to the present-day locations by tsunamis or storm-related waves (Kawana and Nakata 1994; Suzuki et al. 2008; Goto et al. 2009, 2010a, b, c, 2011; Araoka et al. 2010, 2013). Several boulders consist of massive *Porites* corals (Goto et al. 2010b). Suzuki et al. (2008) analyzed several years of seasonal $\delta^{18}O_{coral}$ values from seven *Porites* corals collected from boulders in the fringing coral reefs in Ishigaki Island (Fig. 6.1). They reported that the corals ceased growing during the summer and autumn, which correspond to the active typhoon season. Therefore, they suggested that the *Porites* coral boulders were transported by high-energy waves associated with typhoons. Araoka et al. (2010, 2013) noted that not all the *Porites* boulders were necessarily transported by tsunamis and the U/Th and the $^{14}C$ dating of *Porites* boulders ranged from 0.238 to 5.38 kyr BP. These results imply that geochemical analysis of the *Porites* boulders could produce a record of paleoclimatic change over the past 5 kyr in the Ryukyu Islands. We note that significant discrepancies exist between the SSTs derived from corals and other proxies in the region (Sect. 4). Resolving this problem could facilitate obtaining more precise estimates of SST and SSS, which would help characterize past climatic and oceanographic patterns.

Proxy development and evaluation remain an important research area. The coral species living in temperate areas differ from those in subtropical and tropical areas. The coral growth depressed by low SSTs (e.g., <14.5 °C) and the reliability of the Sr/Ca and $\delta^{18}O$ thermometers are genus and species specific (Omata et al. 2006; Shimamura et al. 2008; Mishima et al. 2009; Hirabayashi et al. 2013; Seo et al. 2013; Nishida et al. 2014a) Therefore, the development of new coral genus/species and chemical composition proxies is highly desirable. In coral reconstructions of SST in temperate areas, Li/Mg could serve as an alternative SST proxy (Montagna et al. 2014). The relationship between Li/Mg and seawater temperature has been observed in a wide range of coral species living in temperatures from 1 to 28 °C, although the calcification mechanisms and growth rates differ between species. The slope of the Li/Mg and seawater temperature relationship steepens as SST decreases, suggesting that Li/Mg could be a relatively sensitive thermometer in temperate regions in Japan.

Another interesting topic is the combined effects of various environmental changes, including ocean warming, OA and anthropogenic pollution, on coral growth. The response of coral growth to global environmental changes has been shown to vary by coral reef and genus (e.g., Cooper et al. 2012; Anderson et al. 2015; Sowa et al. 2014 and references there in). For example, ocean warming generally enhances coral growth, whereas OA possibly reduces it (e.g., Lough and Barnes 2000; Anthony et al. 2008, 2011). The controversial projection of coral skeletal growth to rising SSTs and decreasing pH has been reported (Kleypas 1999; McNeil et al. 2004; McCulloch et al. 2012a). Around Japan, rising SSTs have promoted the poleward expansion of tropical corals (Yamano et al. 2011). The projected decrease in pH will be greater in temperate oceans than in tropical ones (Midorikawa et al. 2010). Therefore, high-latitude corals near Japan are good candidates for addressing this issue in a natural setting. Paleo $pH_{SW}$ reconstructed using $\delta^{11}B$ values can be used to assess the degree of OA at mid-latitudes compared to those in the tropical and subtropical oceans.

Currently, the northern limit of *Porites* coral distribution is at Kikai, Tatsukushi, and Koshiki Islands (Fig. 6.1b [3–5]) (Yamazaki et al. 2009; Kawakubo et al. 2014; Watanabe et al. 2014). In addition, only a few datasets include long-term coral growth analyses of multiple coral cores (Omata et al. 2002). Therefore, combining geochemical analyses with measurements of coral skeletal growth will help characterize coral responses to environmental changes; to date, only a few studies have subjected the same core to both analyses (e.g., Bessat and Buigues 2001; Guzman et al. 2008; Storz and Gischler 2011; Sowa et al. 2014).

As mentioned above, fossil corals can be found at several sites in Japan. Analyses of Holocene corals can indicate the responses of coral skeletal growth to past climatic and oceanographic changes. Fossil *Porites* microatolls were formed at Ishigaki and Okinawa Islands from approximately 3–4 kyr BP (Sowa et al. 2014; Fujita et al. 2015). Fujita et al. (2015) found a *Porites* microatoll (>1 m in diameter) at the southeastern part of Okinawa Island. They estimated that the microatoll was one of approximately 1000 colonies that covered an area of 50,000 $m^2$. Analyzing additional microatolls could provide information on past coral responses to paleoenvironmental changes.

Although coral skeletons can record the paleoenvironmental changes that affect their growth, multiple uncertainties remain regarding coral growth and geochemical composition, especially in temperate regions, such as Japan. Recently, the Coral Core Center (http://ccc.kikaireefs.org/index.html) and the Kikai Institute for Coral Reef Sciences (http://kikaireefs.org) were established in Japan. These institutes provide field and analytical equipment and archive networks of coral cores and will greatly enhance Japanese coral reef researches. Further studies of corals will contribute to predicting climate change and its effects on coral growth in the future.

**Acknowledgments** We are grateful to Tsuyoshi Watanabe at Hokkaido University for providing photos of coral slabs. We also thank two editors of Akira Iguchi and Chuki Hongo, and three referees of Jens Zinke, Ryuji Asami, and Tsuyoshi Watanabe for constructive reviews that contributed to a substantially improved final manuscript. This chapter is supported by Kurita Water and Environment Foundation Grant (15B037) to KS.

## References

Abram N, Webster J, Davies P, Dullo W (2001) Biological response of coral reefs to sea surface temperature variation: evidence from the raised Holocene reefs of Kikai-jima (Ryukyu Islands, Japan). Coral Reefs 20(3):221–234

Adachi H, Abe O (2003) Air drill for submerged massive coral drilling. Mar Technol Soc J 37(2):31–36

AIMS (2013) Australian Institute of Marine Science (AIMS) Coral Fact Sheets – *Agaricia agaricites*. Viewed 01 Jan 2013 http://www.coral.aims.gov.au/factsheet.jsp?speciesCode=0561

Al-Horani F, Al-Moghrabi S, de Beer D (2003) The mechanism of calcification and its relation to photosynthesis and respiration in the scleractinian coral *Galaxea fascicularis*. Mar Biol 142(3):419–426

Alibert C, McCulloch MT (1997) Strontium/calcium ratios in modern *Porites* corals from the Great Barrier Reef as a proxy for sea surface temperature: calibration of the thermometer and monitoring of ENSO. Paleoceanography 12(3):345–363

Alibert C, Kinsley L, Fallon SJ, McCulloch MT, Berkelmans R, McAllister F (2003) Source of trace element variability in Great Barrier Reef corals affected by the Burdekin flood plumes. Geochim Cosmochim Acta 67(2):231–246

Allen KA, Hönisch B, Eggins SM, Rosenthal Y (2012) Environmental controls on B/Ca in calcite tests of the tropical planktic foraminifer species *Globigerinoides ruber* and *Globigerinoides sacculifer*. Earth Planet Sci Lett 351:270–280

Allison N, Finch AA, Webster JM, Clague DA (2007) Palaeoenvironmental records from fossil corals: the effects of submarine diagenesis on temperature and climate estimates. Geochim Cosmochim Acta 71(19):4693–4703

Allison N, Cohen I, Finch AA, Erez J, Tudhope AW (2014) Corals concentrate dissolved inorganic carbon to facilitate calcification. Nat Commun 5:5741

Anderson KD, Heron SF, Pratchett MS (2015) Species-specific declines in the linear extension of branching corals at a subtropical reef, Lord Howe Island. Coral Reefs 34(2):479–490

Anthony KRN, Kline DI, Diaz-Pulido G, Dove S, Hoegh-Guldberg O (2008) Ocean acidification causes bleaching and productivity loss in coral reef builders. Proc Natl Acad Sci U S A 105(45):17442–17446

Anthony KRN, Maynard JA, Diaz-Pulido G, Mumby PJ, Marshall PA, Cao L, Hoegh-Guldberg O (2011) Ocean acidification and warming will lower coral reef resilience. Glob Chang Biol 17(5):1798–1808

Araoka D, Inoue M, Suzuki A, Yokoyama Y, Edwards RL, Cheng H, Matsuzaki H, Kan H, Shikazono N, Kawahata H (2010) Historic 1771 Meiwa tsunami confirmed by high-resolution U/Th dating of massive *Porites* coral boulders at Ishigaki Island in the Ryukyus, Japan. Geochem Geophys Geosyst 11(6):Q06014

Araoka D, Yokoyama Y, Suzuki A, Goto K, Miyagi K, Miyazawa K, Matsuzaki H, Kawahata H (2013) Tsunami recurrence revealed by *Porites* coral boulders in the southern Ryukyu Islands, Japan. Geology 41(8):919–922

Armid A, Asami R, Fahmiati T, Sheikh MA, Fujimura H, Higuchi T, Taira E, Shinjo R, Oomori T (2011) Seawater temperature proxies based on $D_{Sr}$, $D_{Mg}$, and $D_U$ from culture experiments using the branching coral *Porites cylindrica*. Geochim Cosmochim Acta 75(15):4273–4285

Asami R, Felis T, Deschamps P, Hanawa K, Iryu Y, Bard E, Durand N, Murayama M (2009) Evidence for tropical South Pacific climate change during the younger Dryas and the Bølling–Allerød from geochemical records of fossil Tahiti corals. Earth Planet Sci Lett 288(1):96–107

Barnes DJ, Devereux MJ (1988) Variations in skeletal architecture associated with density banding in the hard coral *Porites*. J Exp Mar Biol Ecol 121(1):37–54

Barnes DJ, Taylor RB (2001a) Natural and artificial luminescence in a skeletal slice of *Porites*. Coral Reefs 19(3):270

Barnes DJ, Taylor RB (2001b) On the nature and causes of luminescent lines and bands in coral skeletons. Coral Reefs 19(3):221–230

Barnes DJ, Taylor RB (2005) On the nature and causes of luminescent lines and bands in coral skeletons: II. Contribution of skeletal crystals. J Exp Mar Biol Ecol 322(2):135–142

Barnes JW, Lang EJ, Potratz HA (1956) Ratio of ionium to uranium in coral limestone. Science 124(27):175–176

Bathurst RG (1975) Carbonate sediments and their diagenesis, Development in sedimentology. Elsevier, Amsterdam

Beck JW, Edwards RL, Ito E, Taylor FW, Recy J, Rougerie F, Joannot P, Henin C (1992) Sea-surface temperature from coral skeletal strontium/calcium ratios. Science 257(5070):644–647

Bessat F, Buigues D (2001) Two centuries of variation in coral growth in a massive *Porites* colony from Moorea (French Polynesia): a response of ocean-atmosphere variability from south central Pacific. Palaeogeogr Palaeoclimatol Palaeoecol 175(1–4):381–392

Bosscher H (1993) Computerized-tomography and skeletal density of coral skeletons. Coral Reefs 12(2):97–103

Boto K, Isdale P (1985) Fluorescent bands in massive corals result from terrestrial fulvic acid inputs to nearshore zone. Nature 315(6018):396–397

Brown B, Tissier M, Howard L, Charuchinda M, Jackson J (1986) Asynchronous deposition of dense skeletal bands in *Porites lutea*. Mar Biol 93(1):83–89

Brown BE, Letissier MDA, Scoffin TP, Tudhope AW (1990) Evaluation of the environmental-impact of dredging on intertidal coral reefs at Ko-Phuket, Thailand, using ecological and physiological-parameters. Mar Ecol Prog Ser 65(3):273–281

Brown CJ, Schoeman DS, Sydeman WJ, Brander K, Buckley LB, Burrows M, Duarte CM, Moore PJ, Pandolfi JM, Poloczanska E (2011) Quantitative approaches in climate change ecology. Glob Chang Biol 17(12):3697–3713

Buddemeier RW (1974) Environmental controls over annual and lunar monthly cycles in hermatypic coral calcification. In: Proc 2th Int Coral Reef Symp, pp 259–267

Cannon GA (1966) Tropical waters in the western Pacific Ocean, August–September 1957. In: Deep sea research and oceanographic abstracts, vol 6. Elsevier, pp 1139–1148

Cantin N, Lough J (2014) Surviving coral bleaching events: porites growth anomalies on the Great Barrier Reef. PLoS One 9(2):e88720

Cantin NE, Cohen AL, Karnauskas KB, Tarrant AM, McCorkle DC (2010) Ocean warming slows coral growth in the central Red Sea. Science 329(5989):322–325

Carilli JE, Prouty NG, Hughen KA, Norris RD (2009) Century-scale records of land-based activities recorded in Mesoamerican coral cores. Mar Pollut Bull 58(12):1835–1842

Carilli JE, Norris RD, Black B, Walsh SM, McField M (2010) Century-scale records of coral growth rates indicate that local stressors reduce coral thermal tolerance threshold. Glob Chang Biol 16(4):1247–1257

Carilli J, Donner SD, Hartmann AC (2012) Historical temperature variability affects coral response to heat stress. PLoS One 7(3):e34418

Carricart-Ganivet JP, Vásquez-Bedoya LF, Cabanillas-Terán N, Blanchon P (2013) Gender-related differences in the apparent timing of skeletal density bands in the reef-building coral *Siderastrea siderea*. Coral Reefs 32(3):769–777

Castillo KD, Ries JB, Weiss JM (2011) Declining coral skeletal extension for forereef colonies of *Siderastrea siderea* on the Mesoamerican barrier reef system, southern Belize. PLoS One 6(2):e14615

Castillo KD, Ries JB, Weiss JM, Lima FP (2012) Decline of forereef corals in response to recent warming linked to history of thermal exposure. Nat Clim Chang 2(10):756–760

Chalker BE, Barnes DJ (1990) Gamma-densitometry for the measurement of skeletal densty. Coral Reefs 9(1):11–23

Chalker B, Barnes D, Isdale P (1985) Calibration of x-ray densitometry for the measurement of coral skeletal density. Coral Reefs 4(2):95–100

Chen J, Edwards RL, Wasserburg G (1986) $^{238}U$, $^{234}U$ and $^{232}Th$ in seawater. Earth Planet Sci Lett 80(3):241–251

Clark TR, Roff G, Zhao J-x, Feng Y-x, Done TJ, Pandolfi JM (2014) Testing the precision and accuracy of the U–Th chronometer for dating coral mortality events in the last 100 years. Quat Geochronol 23:35–45

Cohen AL, Hart SR (1997) The effect of colony topography on climate signals in coral skeleton. Geochim Cosmochim Acta 61(18):3905–3912

Cohen AL, Owens KE, Layne GD, Shimizu N (2002) The effect of algal symbionts on the accuracy of Sr/Ca paleotemperatures from coral. Science 296(5556):331–333

Cooper TF, O'Leary RA, Lough JM (2012) Growth of western Australian corals in the anthropocene. Science 335(6068):593–596

Corrège T (2006) Sea surface temperature and salinity reconstruction from coral geochemical tracers. Palaeogeogr Palaeoclimatol Palaeoecol 232(2–4):408–428

D'Croz L, O'Dea A (2007) Variability in upwelling along the Pacific shelf of Panama and implications for the distribution of nutrients and chlorophyll. Estuar Coast Shelf S 73(1):325–340

D'Olivo J, McCulloch M, Judd K (2013) Long-term records of coral calcification across the central Great Barrier Reef: assessing the impacts of river runoff and climate change. Coral Reefs 32(4):999–1012

de Villiers S, Nelson BK, Chivas AR (1995) Biological controls on coral Sr/Ca and $\delta^{18}O$ reconstructions of sea surface temperatures. Science 269(5228):1247–1249

De'ath G, Lough JM, Fabricius KE (2009) Declining coral calcification on the Great Barrier Reef. Science 323(5910):116–119

DeCarlo TM, Cohen AL, Barkley HC, Cobban Q, Young C, Shamberger KE, Brainard RE, Golbuu Y (2015) Coral macrobioerosion is accelerated by ocean acidification and nutrients. Geology 43(1):7–10

DeLong K, Quinn TM, Taylor FW, Shen C-C, Lin K (2013) Improving coral-base paleoclimate reconstructions by replicating 350 years of coral Sr/Ca variations. Palaeogeogr Palaeoclimatol Palaeoecol 373:6–24

Deser C, Alexander MA, Timlin MS (1999) Evidence for a wind-driven intensification of the Kuroshio Current Extension from the 1970s to the 1980s. J Clim 12(6):1697–1706

Devlin M, Brodie J (2005) Terrestrial discharge into the Great Barrier Reef Lagoon: nutrient behavior in coastal waters. Mar Pollut Bull 51(1):9–22

Di Lorenzo E, Schneider N, Cobb K, Franks P, Chhak K, Miller A, McWilliams J, Bograd S, Arango H, Curchitser E (2008) North Pacific Gyre Oscillation links ocean climate and ecosystem change. Geophys Res Lett 35(8):L08607

Dietzel M, Gussone N, Eisenhauer A (2004) Co-precipitation of $Sr^{2+}$ and $Ba^{2+}$ with aragonite by membrane diffusion of $CO_2$ between 10 and 50°C. Chem Geol 203(1):139–151

Ding Y, Chan JC (2005) The East Asian summer monsoon: an overview. Meteorog Atmos Phys 89(1–4):117–142

Dishon G, Fisch J, Horn I, Kaczmarek K, Bijma J, Gruber DF, Nir O, Popovich Y, Tchernov D (2015) A novel paleo-bleaching proxy using boron isotopes and high-resolution laser ablation to reconstruct coral bleaching events. Biogeosciences 12(19):5677–5687

Dodge RE, Jickells T, Knap AH, Boyd S, Bak R (1984) Reef-building coral skeletons as chemical pollution (phosphorus) indicators. Mar Pollut Bull 15(5):178–187

Douville E, Paterne M, Cabioch G, Louvat P, Gaillardet J, Juillet-Leclerc A, Ayliffe L (2010) Abrupt sea surface pH change at the end of the Younger Dryas in the central sub-equatorial Pacific inferred from boron isotope abundance in corals (*Porites*). Biogeosciences 7(8):2445–2459

Dustan P (1975) Growth and form in the reef-building coral *Montastrea annularis*. Mar Biol 33(2):101–107

Enmar R, Stein M, Bar-Matthews M, Sass E, Katz A, Lazar B (2000) Diagenesis in live corals from the Gulf of Aqaba. I The effect on paleo-oceanography tracers. Geochim Cosmochim Acta 64(18):3123–3132

Erler DV, Wang XT, Sigman DM, Scheffers SR, Shepherd BO (2015) Controls on the nitrogen isotopic composition of shallow water corals across a tropical reef flat transect. Coral Reefs 34(1):329–338

Erler DV, Wang XT, Sigman DM, Scheffers SR, Martínez-García A, Haug GH (2016) Nitrogen isotopic composition of organic matter from a 168 year-old coral skeleton: implications for coastal

nutrient cycling in the Great Barrier Reef Lagoon. Earth Planet Sci Lett 434:161–170
Fabricius KE (2005) Effects of terrestrial runoff on the ecology of corals and coral reefs: review and synthesis. Mar Pollut Bull 50(2):125–146
Fabricius KE (2011) Factors determining the resilience of coral reefs to eutrophication: a review and conceptual model. In: Coral reefs: an ecosystem in transition. Springer, pp 493–505
Fabricius KE, Cséke S, Humphrey C, De'ath G (2013) Does trophic status enhance or reduce the thermal tolerance of Scleractinian corals? A review, experiment and conceptual framework. PLoS One 8(1):e54399
Fairchild I, Killawee J (1995) Selective leaching in glacierized terrains and implications for retention of primary chemical signals in carbonate rocks. In: Proc 8th Int Sympo Water-Rock Interaction-WRI-8 Vladivostok, Russia, 15–19 August 1995. Water-Rock Interaction. A.A. Balkema, p 79–82
Fallon S (2011) Radiocarbon ($^{14}C$): dating and corals. In: Hopley D (ed) Encyclopedia of modern coral reefs: structure, form and process. Springer, Netherlands, pp 829–834
Fallon SJ, McCulloch MT, van Woesik R, Sinclair DJ (1999) Corals at their latitudinal limits: laser ablation trace element systematics in *Porites* from Shirigai Bay, Japan. Earth Planet Sci Lett 172(3):221–238
Felis T, Patzold J, Loya Y (2003) Mean oxygen-isotope signatures in *Porites* spp. corals: inter-colony variability and correction for extension-rate effects. Coral Reefs 22(4):328–336
Felis T, Suzuki A, Kuhnert H, Dima M, Lohmann G, Kawahata H (2009) Subtropical coral reveals abrupt early-twentieth-century freshening in the western North Pacific Ocean. Geology 37(6):527–530
Felis T, Suzuki A, Kuhnert H, Rimbu N, Kawahata H (2010) Pacific decadal oscillation documented in a coral record of North Pacific winter temperature since 1873. Geophys Res Lett 37(14):L14605
Fujioka Y (1999) Mass destruction of the hermatypic corals during a bleaching event in Ishigaki Island, southwestern Japan. Galaxea, JCRS 1:41–50
Fujita K, Hongo C, Koutoku M, Ito S, Asami R, Reimer J (2015) Buried fossil microatolls off the Kaichu-Doro Causeway, between Okinawa and Henza Islands. Fauna Ryukyuana 24:13–15
Gaetani GA, Cohen AL (2006) Element partitioning during precipitation of aragonite from seawater: a framework for understanding paleoproxies. Geochim Cosmochim Acta 70(18):4617–4634
Gagan MK (1998) Temperature and surface-ocean water balance of the mid-Holocene tropical Western Pacific. Science 279(5353):1014–1018
Gagan MK, Ayliffe LK, Opdyke BN, Hopley D, Scott-Gagan H, Cowley J (2002) Coral oxygen isotope evidence for recent groundwater fluxes to the Australian Great Barrier Reef. Geophys Res Lett 29(20):1982
Gagan MK, Dunbar GB, Suzuki A (2012) The effect of skeletal mass accumulation in *Porites* on coral Sr/Ca and $\delta^{18}O$ paleothermometry. Paleoceanography 27(1):PA1203
Ghosh P, Adkins J, Affek H, Balta B, Guo W, Schauble EA, Schrag D, Eiler JM (2006) $^{13}C-^{18}O$ bonds in carbonate minerals: a new kind of paleothermometer. Geochim Cosmochim Acta 70(6):1439–1456
Goto K, Okada K, Imamura F (2009) Characteristics and hydrodynamics of boulders transported by storm waves at Kudaka Island, Japan. Mar Geol 262(1):14–24
Goto K, Kawana T, Imamura F (2010a) Historical and geological evidence of boulders deposited by tsunamis, southern Ryukyu Islands, Japan. Earth-Sci Rev 102(1):77–99
Goto K, Miyagi K, Kawamata H, Imamura F (2010b) Discrimination of boulders deposited by tsunamis and storm waves at Ishigaki Island, Japan. Mar Geol 269(1):34–45
Goto K, Shinozaki T, Minoura K, Okada K, Sugawara D, Imamura F (2010c) Distribution of boulders at Miyara Bay of Ishigaki Island, Japan: a flow characteristic indicator of tsunami and storm waves. Island Arc 19(3):412–426
Goto K, Miyagi K, Kawana T, Takahashi J, Imamura F (2011) Emplacement and movement of boulders by known storm waves-field evidence from the Okinawa Islands, Japan. Mar Geol 283(1):66–78
Grothe PR, Cobb KM, Bush SL, Cheng H, Santos GM, Southon JR, Lawrence Edwards R, Deocampo DM, Sayani HR (2016) A comparison of U/Th and rapid-screen $^{14}C$ dates from Line Island fossil corals. Geochem Geophys Geosyst 17(3):833–845
Grove CA, Nagtegaal R, Zinke J, Scheufen T, Koster B, Kasper S, McCulloch MT, Bergh G, Brummer GJA (2010) River runoff reconstructions from novel spectral luminescence scanning of massive coral skeletons. Coral Reefs 29(3):579–591
Grove CA, Zinke J, Scheufen T, Maina J, Epping E, Boer W, Randriamanantsoa B, Brummer G-J (2012) Spatial linkages between coral proxies of terrestrial runoff across a large embayment in Madagascar. Biogeosciences 9(8):3063–3081
Grove CA, Zinke J, Peeters F, Park W, Scheufen T, Kasper S, Randriamanantsoa B, McCulloch MT, Brummer G-JA (2013) Madagascar corals reveal Pacific multidecadal modulation of rainfall since 1708. Clim Past 9:641–656
Guzmán HM, Jiménez CE (1992) Contamination of coral reefs by heavy metals along the Caribbean coast of Central America (Costa Rica and Panama). Mar Pollut Bull 24(11):554–561
Guzman HM, Cipriani R, Jackson JBC (2008) Historical decline in coral reef growth after the Panama Canal. Ambio 37(5):342–346
Halfar J, Godinez-Orta L, Riegl B, Valdez-Holguin J, Borges J (2005) Living on the edge: high-latitude *Porites* carbonate production under temperate eutrophic conditions. Coral Reefs 24(4):582–592
Hanawa K, Watanabe T, Iwasaka N, Suga T, Toba Y (1988) Surface thermal conditions in the Western North Pacific during the ENSO Events. J Meteor Soc Jpn Ser II 66(3):445–456
Harii S, Hongo C, Ishihara M, Ide Y, Kayanne H (2014) Impacts of multiple disturbances on coral communities at Ishigaki Island, Okinawa, Japan, during a 15 year survey. Mar Ecol Prog Ser 509:171–180
Hart SR, Cohen AL (1996) An ion probe study of annual cycles of Sr/Ca and other trace elements in corals. Geochim Cosmochim Acta 60(16):3075–3084
Hasegawa H, Ichikawa K, Kobayashi M, Kobayashi T, Hoshino M, Mezaki S (1999) The mass-bleaching of coral reefs in the Ishigaki Lagoon, 1998. Galaxea, JCRS 1:31–39
Hayashi E, Suzuki A, Nakamura T, Iwase A, Ishimura T, Iguchi A, Sakai K, Okai T, Inoue M, Araoka D (2013) Growth-rate influences on coral climate proxies tested by a multiple colony culture experiment. Earth Planet Sci Lett 362:198–206
Helmle KP, Dodge RE, Swart PK, Gledhill DK, Eakin CM (2011) Growth rates of Florida corals from 1937 to 1996 and their response to climate change. Nat Commun 2:215
Hemming NG, Hanson GN (1992) Boron isotopic composition and concentration in modern marine carbonates. Geochim Cosmochim Acta 56(1):537–543
Hendy EJ, Gagan MK, Lough JM, McCulloch M, deMenocal PB (2007) Impact of skeletal dissolution and secondary aragonite on trace element and isotopic climate proxies in *Porites* corals. Paleoceanography 22(4):PA4101
Hirabayashi S, Yokoyama Y, Suzuki A, Kawakubo Y, Okai T, Nojima S (2013) Coral growth-rate insensitive Sr/Ca as a robust temperature recorder at the extreme latitudinal limits of *Porites*. Geochem J Express lett 47:e5
Hoegh-Guldberg O, Mumby PJ, Hooten AJ, Steneck RS, Greenfield P, Gomez E, Harvell CD, Sale PF, Edwards AJ, Caldeira K, Knowlton N, Eakin CM, Iglesias-Prieto R, Muthiga N, Bradbury RH, Dubi A, Hatziolos ME (2007) Coral reefs under rapid climate change and ocean acidification. Science 318(5857):1737–1742

Holcomb M, DeCarlo T, Gaetani G, McCulloch M (2016) Factors affecting B/Ca ratios in synthetic aragonite. Chem Geol 437:67–76

Hongo C (2010) Sea-level standstill and dominant hermatypic coral from the Holocene raised reef terraces at the Kikai Island, Ryukyu Islands. J Geogr (Chigaku Zasshi) 119(5):860–871

Hongo C (2012) Holocene key coral species in the Northwest Pacific: indicators of reef formation and reef ecosystem responses to global climate change and anthropogenic stresses in the near future. Quat Sci Rev 35:82–99

Hongo C, Yamano H (2013) Species-specific responses of corals to bleaching events on anthropogenically turbid reefs on Okinawa Island, Japan, over a 15-year period (1995–2009). PLoS One 8(4):e60952

Hönisch B, Hemming N, Grottoli A, Amat A, Hanson G, Bijma J (2004) Assessing scleractinian corals as recorders for paleo-pH: empirical calibration and vital effects. Geochim Cosmochim Acta 68(18):3675–3685

Horta-Puga G, Carriquiry JD (2012) Coral Ba/Ca molar ratios as a proxy of precipitation in the northern Yucatan Peninsula, Mexico. Appl Geochem 27(8):1579–1586

Inoue M, Tanimizu M (2008) Anthropogenic lead inputs to the western Pacific during the 20th century. Sci Total Environ 406(1):123–130

Inoue M, Hata A, Suzuki A, Nohara M, Shikazono N, Yim WW-S, Hantoro WS, Donghuai S, Kawahata H (2006) Distribution and temporal changes of lead in the surface seawater in the western Pacific and adjacent seas derived from coral skeletons. Environ Pollut 144(3):1045–1052

Inoue M, Suwa R, Suzuki A, Sakai K, Kawahata H (2011) Effects of seawater pH on growth and skeletal U/Ca ratios of *Acropora digitifera* coral polyps. Geophys Res Lett 38(12):L12809

Inoue M, Ishikawa D, Miyaji T, Yamazaki A, Suzuki A, Yamano H, Kawahata H, Watanabe T (2014) Evaluation of Mn and Fe in coral skeletons (*Porites* spp.) as proxies for sediment loading and reconstruction of 50 yrs of land use on Ishigaki Island, Japan. Coral Reefs 33(2):363–373

Isdale PP, Daniel EE (1989) The design and deployment of lightweight submarine fixed drilling system for the acquisition of coral cores. Mar Technol Soc J 23:3–8

Jacques TG, Pilson MEQ, Cummings C, Marshall N (1977) Laboratory observations on respiration, photosynthesis and factors affecting coral calcification in the temperate coral *Astrangia danae*. In: Proc 3th Int Coral Reef Symp 2, p 455–461

Jian Z, Wang P, Saito Y, Wang J, Pflaumann U, Oba T (2000) Holocene variability of the Kuroshio current in the Okinawa Trough, northwestern Pacific Ocean. Earth Planet Sci Lett 184:305–319

Juliet-Leclerc A, Schmidt G (2001) A calibration of the oxygen isotope paleothermometer of coral aragonite from *Porites*. Geophys Res Lett 28(21):4135–4138

Kawakubo Y, Yokoyama Y, Suzuki A, Okai T, Alibert C, Kinsley L, Eggins S (2014) Precise determination of Sr/Ca by laser ablation ICP-MS compared to ICP-AES and application to multi-century temperate corals. Geochem J 48(2):145–152

Kawana T, Kan H (2002) Description of the new drilling cores throuth the Holocene coral reefs at the Gushichan coast in the southern Okinawa Island, the Ryukyus, Japan. Bull CE, University of the Ryukyus 60 (60), pp 235–244 (in Japanese)

Kawana T, Nakata T (1994) Timing of late Holocene tsunamis originated around the southern Ryukyu Islands, Japan, deduced from coralline tsunami deposits. J Geogr Jpn 103(4):352–376

Kayanne H, Harii S, Yamano H, Tamura M, Ide Y, Akimoto F (1999) Changes in living coral coverage before and after the 1998 bleaching event on coral reef flats of Ishigaki Island, Ryukyu Islands. Galaxea, JCRS 1999(1):73–82

Kim S-T, O'Neil JR, Hillaire-Marcel C, Mucci A (2007) Oxygen isotope fractionation between synthetic aragonite and water: influence of temperature and $Mg^{2+}$ concentration. Geochim Cosmochim Acta 71(19):4704–4715

Kinsman DJ, Holland H (1969) The co-precipitation of cations with $CaCO_3$—IV. The co-precipitation of $Sr^{2+}$ with aragonite between 16° and 96°C. Geochim Cosmochim Acta 33(1):1–17

Kiyama O, Yamada T, Nakamori T, Iryu T (2000) Early Holocene coral $\delta^{18}O$-based sea surface temperature. Quat Res 39(1):69–80

Kleypas JA (1999) Geochemical consequences of increased atmospheric carbon dioxide on coral reefs. Science 284(5411):118–120

Kleypas JA, McManus JW, Menez LAB (1999) Environmental limits to coral reef development: where do we draw the line? Am Zool 39(1):146–159

Knutson DW, Smith SV, Buddemei RW (1972) Coral chronometers: seasonal growth bands in reef corals. Science 177(4045):270–272

Kosaka Y, Nakamura H (2010) Mechanisms of meridional teleconnection observed between a summer monsoon system and a subtropical anticyclone. Part I: the Pacific-Japan pattern. J Clim 23(19):5085–5108

Krief S, Hendy EJ, Fine M, Yam R, Meibom A, Foster GL, Shemesh A (2010) Physiological and isotopic responses of scleractinian corals to ocean acidification. Geochim Cosmochim Acta 74(17):4988–5001

Kubota K, Yokoyama Y, Ishikawa T, Obrochta S, Suzuki A (2014) Larger $CO_2$ source at the equatorial Pacific during the last deglaciation. Sci Rep 4:5261

Kubota K, Yokoyama Y, Ishikawa T, Suzuki A (2015) A new method for calibrating a boron isotope paleo-pH proxy using massive *Porites* corals. Geochem Geophys Geosyst 16(9):3333–3342

Kumarsingh K, Laydoo R, Chen JK, Siung-Chang AM (1998) Historic records of phosphorus levels in the reef-building coral Montastrea annularis from Tobago, West Indies. Mar Pollut Bull 36(12):1012–1018

Kurita N, Fujiyoshi Y, Nakayama T, Matsumi Y, Kitagawa H (2015) East Asian Monsoon controls on the inter-annual variability in precipitation isotope ratio in Japan. Clim Past 11(2):339–353

Kwiatkowski L, Cox PM, Economou T, Halloran PR, Mumby PJ, Booth BB, Carilli J, Guzman HM (2013) Caribbean coral growth influenced by anthropogenic aerosol emissions. Nat Geosci 6(6):362–366

LaVigne M, Field MP, Anagnostou E, Grottoli AG, Wellington GM, Sherrell RM (2008) Skeletal P/Ca tracks upwelling in gulf of Panama coral: evidence for a new seawater phosphate proxy. Geophys Res Lett 35(5):L05604

LaVigne M, Matthews KA, Grottoli AG, Cobb KM, Anagnostou E, Cabioch G, Sherrell RM (2010) Coral skeleton P/Ca proxy for seawater phosphate: multi-colony calibration with a contemporaneous seawater phosphate record. Geochim Cosmochim Acta 74(4):1282–1293

LaVigne M, Grottoli AG, Palardy JE, Sherrell RM (2016) Multi-colony calibrations of coral Ba/Ca with a contemporaneous in situ seawater barium record. Geochim Cosmochim Acta 179:203–216

Lea D, Boyle E (1989) Barium content of benthic foraminifera controlled by bottom-water composition. Nature 338:751–753

Lee K, Kim T-W, Byrne RH, Millero FJ, Feely RA, Liu Y-M (2010) The universal ratio of boron to chlorinity for the North Pacific and North Atlantic oceans. Geochim Cosmochim Acta 74(6):1801–1811

Li Y, Wang F (2012) Spreading and salinity change of North Pacific tropical water in the Philippine Sea. J Oceanogr 68(3):439–452

Lough JM (2008) Coral calcification from skeletal records revisited. Mar Ecol Prog Ser 373:257–264

Lough JM (2010) Climate records from corals. Wiley Interdiscip Rev Clim Chang 1(3):318–331

Lough JM, Barnes DJ (1997) Several centuries of variation in skeletal extension, density and calcification in massive *Porites* colonies from the Great Barrier Reef: a proxy for seawater temperature and a background of variability against which to identify unnatural change. J Exp Mar Biol Ecol 211(1):29–67

Lough JM, Barnes DJ (2000) Environmental controls on growth of the massive coral *Porites*. J Exp Mar Biol Ecol 245(2):225–243

Lough JM, Cantin NE (2014) Perspectives on massive coral growth rates in a changing ocean. Biol Bull 226(3):187–202

Lough JM, Cooper TF (2011) New insights from coral growth band studies in an era of rapid environmental change. Earth-Sci Rev 108(3–4):170–184

Loya Y, Sakai K, Yamazato K, Nakano Y, Sambali H, van Woesik R (2001) Coral bleaching: the winners and the losers. Ecol Lett 4(2)

Luo X, Rehkämper M, Lee D-C, Halliday AN (1997) High precision $^{230}$Th/$^{232}$Th and $^{234}$U/$^{238}$U measurements using energyfiltered ICP magnetic sector multiple collector mass spectrometry. Int J Mass Spectrom Ion Process 171(1):105–117

Macintyre I (1978) A hand-operated submersible drill for coring reef substrata. In: Coral reefs: research methods UNESCO Monographs on Oceanographic Methodology 5:75–80

Mallela J, Lewis SE, Croke B (2013) Coral skeletons provide historical evidence of phosphorus runoff on the Great Barrier Reef. PLoS One 8(9):e75663

Mantua NJ, Hare SR, Zhang Y, Wallace JM, Francis RC (1997) A Pacific interdecadal climate oscillation with impacts on salmon production. Bull Am Meteor 78(6):1069–1079

Manzello DP, Enochs IC, Kolodziej G, Carlton R (2015a) Coral growth patterns of *Montastraea cavernosa* and *Porites astreoides* in the Florida Keys: the importance of thermal stress and inimical waters. J Exp Mar Biol Ecol 471:198–207

Manzello DP, Enochs IC, Kolodziej G, Carlton R (2015b) Recent decade of growth and calcification of *Orbicella faveolata* in the Florida Keys: an inshore-offshore comparison. Mar Ecol Prog Ser 521:81–89

Marcott SA, Shakun JD, Clark PU, Mix AC (2013) A reconstruction of regional and global temperature for the past 11,300 years. Science 339(6124):1198–1201

Marion GS, Dunbar RB, Mucciarone DA, Kremer JN, Lansing JS, Arthawiguna A (2005) Coral skeletal $\delta^{15}$ N reveals isotopic traces of an agricultural revolution. Mar Pollut Bull 50(9):931–944

Marshall AT, Clode P (2004) Calcification rate and the effect of temperature in a zooxanthellate and an azooxanthellate scleractinian reef coral. Coral Reefs 23(2):218–224

Matson EG (2011) Core plugs. In: Hopley D (ed) Encyclopedia of modern coral reefs: structure, form and process. Springer, Netherlands, pp 294–296

Matthews B, Jones A, Theodorou N, Tudhope A (1996) Excitation-emission-matrix fluorescence spectroscopy applied to humic acid bands in coral reefs. Mar Chem 55(3):317–332

McConnaughey T (1989) $^{13}$C and $^{18}$O isotopic disequilibrium in biological carbonates: I. Patterns. Geochim Cosmochim Acta 53(1):151–162

McCrea JM (1950) On the isotopic chemistry of carbonates and a paleotemperature scale. J Chem Phys 18(6):849–857

McCulloch MT, Gagan MK, Mortimer GE, Chivas AR, Isdale PJ (1994) A high-resolution Sr/Ca and $\delta^{18}$O coral record from the Great Barrier Reef, Australia, and the 1982–1983 El Niño. Geochim Cosmochim Acta 58(12):2747–2754

McCulloch MT, Tudhope AW, Esat TM, Mortimer GE, Chappell J, Pillans B, Chivas AR, Omura A (1999) Coral record of equatorial sea-surface temperatures during the penultimate deglaciation at Huon Peninsula. Science 283(5399):202–204

McCulloch M, Fallon S, Wyndham T, Hendy E, Lough J, Barnes D (2003) Coral record of increased sediment flux to the inner Great Barrier Reef since European settlement. Nature 421(6924):727–730

McCulloch M, Falter J, Trotter J, Montagna P (2012a) Coral resilience to ocean acidification and global warming through pH up-regulation. Nat Clim Chang 2(8):623–627

McCulloch M, Trotter J, Montagna P, Falter J, Dunbar RB, Freiwald A, Försterra G, López Correa M, Maier C, Rüggeberg A, Taviani M (2012b) Resilience of cold-water scleractinian corals to ocean acidification: boron isotopic systematics of pH and saturation state up-regulation. Geochim Cosmochim Acta 87(15):21–34

McGregor HV, Gagan MK (2003) Diagenesis and geochemistry of *Porites* corals from Papua New Guinea. Geochim Cosmochim Acta 67(12):2147–2156

McLaren SJ (2011) Aragonite. In: Hopley D (ed) Encyclopedia of modern coral reefs: structure, form and process. Springer, Netherlands, pp 47–47

McNeil BI, Matear RJ, Barnes DJ (2004) Coral reef calcification and climate change: the effect of ocean warming. Geophys Res Lett 31(22):L22309

Midorikawa T, Ishii M, Saito S, Sasano D, Kosugi N, Motoi T, Kamiya H, Nakadate A, Nemoto K, Inoue HY (2010) Decreasing pH trend estimated from 25-yr time series of carbonate parameters in the western North Pacific. Tellus B 62(5):649–659

Miller AJ, Cayan DR, White WB (1998) A westward-intensified decadal change in the North Pacific thermocline and gyre-scale circulation. J Clim 11(12):3112–3127

Min GR, Edwards RL, Taylor FW, Recy J, Gallup CD, Beck JW (1995) Annual cycles of U/Ca in coral skeletons and U/Ca thermometry. Geochim Cosmochim Acta 59(10):2025–2042

Minobe S (1997) A 50-70 year climatic oscillation over the North Pacific and North America. Geophys Res Lett 24(6):683–686

Mishima M, Kawahata H, Suzuki A, Inoue M, Okai T, Omura A (2009) Reconstruction of the East China Sea palaeoenvironment at 16 ka by comparison of fossil and modern Faviidae corals from the Ryukyus, southwestern Japan. J Quat Sci 24(8):928–936

Mishima M, Suzuki A, Nagao M, Ishimura T, Inoue M, Kawahata H (2010) Abrupt shift toward cooler condition in the earliest 20th century detected in a 165 year coral record from Ishigaki Island, southwestern Japan. Geophys Res Lett 37(15):L15609

Mitsuguchi T, Matsumoto E, Abe O, Uchida T, Isdale PJ (1996) Mg/Ca thermometry in coral skeletons. Science 274(5289):961–963

Mizuno K, White WB (1983) Annual and interannual variability in the Kuroshio current system. J Phys Oceanogr 13(10):1847–1867

Montaggioni LF, Braithwaite CJ (2009) Quaternary coral reef systems: history, development processes and controlling factors, Developments in Marine Geology, vol 5. Elsevier, Amsterdam

Montagna P, McCulloch M, Douville E, Correa ML, Trotter J, Rodolfo-Metalpa R, Dissard D, Ferrier-Pagès C, Frank N, Freiwald A (2014) Li/Mg systematics in scleractinian corals: calibration of the thermometer. Geochim Cosmochim Acta 132(1):288–310

Morimoto M, Kitagawa H, Shibata Y, Kayanne H (2004) Seasonal radiocarbon variation of surface seawater recorded in a coral from Kikai Island, subtropical northwestern Pacific. Radiocarbon 46(2):643–648

Morimoto M, Kayanne H, Abe O, McCulloch MT (2007) Intensified mid-Holocene Asian monsoon recorded in corals from Kikai Island, subtropical northwestern Pacific. Quat Res 67(2):204–214

Morrongiello JR, Thresher RE, Smith DC (2012) Aquatic biochronologies and climate change. Nat Clim Chang 2(12):849–857

Müller A, Gagan MK, McCulloch MT (2001) Early marine diagenesis in corals and geochemical consequences for paleoceanographic reconstructions. Geophys Res Lett 28(23):4471–4474

Müller A, Gagan MK, Lough JM (2004) Effect of early marine diagenesis on coral reconstructions of surface-ocean $^{13}$C/$^{12}$C and carbonate saturation state. Glob Biogeochem Cycles 18(1):GB1033

Nagtegaal R, Grove CA, Kasper S, Zinke J, Boer W, Brummer G-JA (2012) Spectral luminescence and geochemistry of coral aragonite: effects of whole-core treatment. Chem Geol 318-319:6–15

Nakamura T, Masuda K, Miyake F, Hakozaki M, Kimura K, Nishimoto H, Hitoki E (2016) High-precision age determination of Holocene samples by radiocarbon dating with accelerator mass spectrometry at Nagoya University. Quat Int 397:250–257

Nishida K, Iguchi A, Ishimura T, Sakai K, Suzuki A (2014a) Skeletal isotopic responses of the scleractinian coral *Isopora palifera* to experimentally controlled water temperatures. Geochem J 48(4):e9–e14

Nishida K, Ishikawa K, Iguchi A, Tanaka Y, Sato M, Ishimura T, Inoue M, Nakamura T, Sakai K, Suzuki A (2014b) Skeletal oxygen and carbon isotope compositions of *Acropora* coral primary polyps experimentally cultured at different temperatures. Geochem Geophys Geosyst 15(7):2840–2849

Obert JC, Scholz D, Felis T, Brocas WM, Jochum KP, Andreae MO (2016) $^{230}$Th/U dating of last interglacial brain corals from Bonaire (southern Caribbean) using bulk and theca wall material. Geochim Cosmochim Acta 178:20–40

Okai T, Suzuki A, Kawahata H, Terashima S, Imai N (2002) Preparation of a new geological survey of Japan geochemical reference material: coral JCp-1. Geostand Newslett 26(1):95–99

Omata T, Okamoto M, Furushima Y (2002) Skeletal growth history of modern corals in Sekisei lagoon as a bio-indicator. JAMSTEC-R 45:17–30

Omata T, Suzuki A, Kawahata H, Nojima S, Minoshima K, Hata A (2006) Oxygen and carbon stable isotope systematics in *Porites* coral near its latitudinal limit: the coral response to low-thermal temperature stress. Glob Planet Chang 53(1–2):137–146

Orr JC, Fabry VJ, Aumont O, Bopp L, Doney SC, Feely RA, Gnanadesikan A, Gruber N, Ishida A, Joos F (2005) Anthropogenic ocean acidification over the twenty-first century and its impact on calcifying organisms. Nature 437(7059):681–686

Pelejero C, Calvo E, McCulloch MT, Marshall JF, Gagan MK, Lough JM, Opdyke BN (2005) Preindustrial to modern interdecadal variability in coral reef pH. Science 309(5744):2204–2207

Perrin C (2011) Diagenesis. In: Hopley D (ed) Encyclopedia of modern coral reefs: structure, form and process. Springer, Netherlands, pp 309–321

Pratchett MS, Anderson KD, Hoogenboom M, Windam E, Baird AH, Pandolfi J, Edmunds P, Lough J (2015) Spatial, temporal and taxonomic variation in coral growth—implications for the structure and function of coral reef ecosystems. Oceanogr Mar Biol Annu Rev 53:215–295

Prouty NG, Field ME, Stock JD, Jupiter SD, McCulloch M (2010) Coral Ba/Ca records of sediment input to the fringing reef of the southshore of Moloka'i, Hawai'i over the last several decades. Mar Pollut Bull 60(10):1822–1835

Prouty NG, Storlazzi CD, McCutcheon AL, Jenson JW (2014) Historic impact of watershed change and sedimentation to reefs along west-central Guam. Coral Reefs 33(3):733–749

Quinn TM, Sampson DE (2002) A multiproxy approach to reconstructing sea surface conditions using coral skeleton geochemistry. Paleoceanography 17(4):14.11–14.11

Quinn TM, Taylor FW (2006) SST artifacts in coral proxy records produced by early marine diagenesis in a modern coral from Rabaul, Papua New Guinea. Geophys Res Lett 33(4):L04601

Reimer PJ, Bard E, Bayliss A, Beck JW, Blackwell PG, Bronk Ramsey C, Buck CE, Cheng H, Edwards RL, Friedrich M, Grootes PM, Guilderson TP, Haflidason H, Hajdas I, Hatté C, Heaton TJ, Hoffmann DL, Hogg AG, Hughen KA, Kaiser KF, Kromer B, Manning SW, Niu M, Reimer RW, Richards DA, Scott EM, Southon JR, Staff RA, Turney CSM, van der Plicht J (2013) IntCal13 and Marine13 radiocarbon age calibration curves 0–50,000 years cal BP. Radiocarbon 55(4):1869–1887

Reynaud S, Hemming NG, Juillet-Leclerc A, Gattuso J-P (2004) Effect of $pCO_2$ and temperature on the boron isotopic composition of the zooxanthellate coral *Acropora* sp. Coral Reefs 23(4):539–546

Rosenfeld M, Shemesh A, Yam R, Sakai K, Loya Y (2006) Impact of the 1998 bleaching event on $\delta^{18}O$ records of Okinawa corals. Mar Ecol Prog Ser 314:127–133

Sadler J, Webb GE, Nothdurft LD, Dechnik B (2014) Geochemistry-based coral palaeoclimate studies and the potential of 'non-traditional'(non-massive *Porites*) corals: recent developments and future progression. Earth-Sci Rev 139:291–316

Sagawa T, Kuwae M, Tsuruoka K, Nakamura Y, Ikehara M, Murayama M (2014) Solar forcing of centennial-scale East Asian winter monsoon variability in the mid-to late Holocene. Earth Planet Sci Lett 395:124–135

Schoepf V, McCulloch MT, Warner ME, Levas SJ, Matsui Y, Aschaffenburg MD, Grottoli AG (2014) Short-term coral bleaching is not recorded by skeletal boron isotopes. PLoS One 9(11) e112011

Schöne BR, Oschmann W, Tanabe K, Dettman D, Fiebig J, Houk SD, Kanie Y (2004) Holocene seasonal environmental trends at Tokyo Bay, Japan, reconstructed from bivalve mollusk shells—implications for changes in the East Asian monsoon and latitudinal shifts of the Polar Front. Quat Sci Rev 23(9–10):1137–1150

Schroeder JH (1969) Experimental dissolution of calcium, magnesium, and strontium from recent biogenic carbonates: a model of diagenesis. J Sediment Res 39(3):1057–1073

Scoffin TP, Tudhope AW, Brown BE, Chansang H, Cheeney RF (1992) Patterns and possible environmental controls of skeletogenesis of *Porites lutea*, South Thailand. Coral Reefs 11(1):1–11

Seki A, Yokoyama Y, Suzuki A, Kawakubo Y, Okai T, Miyairi Y, Matsuzaki H, Namizaki N, Kan H (2012) Mid-Holocene sea-surface temperature reconstruction using fossil corals from Kume Island, Ryukyu, Japan. Geochem J Express Lett 46(6):E27–E32

Seo I, Lee YI, Watanabe T, Yamano H, Shimamura M, Yoo CM, Hyeong K (2013) A skeletal Sr/Ca record preserved in *Dipsastraea (Favia) speciosa* and implications for coral Sr/Ca thermometry in mid-latitude regions. Geochem Geophys Geosyst 14(8):2873–2885

Shaw TJ, Francois R (1991) A fast and sensitive ICP-MS assay for the determination of $^{230}$Th in marine sediments. Geochim Cosmochim Acta 55(7):2075–2078

Shen GT, Boyle EA (1988) Determination of lead, cadmium and other trace metals in annually-banded corals. Chem Geol 67(1):47–62

Shen C-C, Lee T, Chen C-Y, Wang C-H, Dai C-F, Li L-A (1996) The calibration of D [Sr/Ca] versus sea surface temperature relationship for *Porites* corals. Geochim Cosmochim Acta 60(20):3849–3858

Shen C-C, Wu C-C, Cheng H, Edwards RL, Hsieh Y-T, Gallet S, Chang C-C, Li T-Y, Lam DD, Kano A (2012) High-precision and high-resolution carbonate $^{230}$Th dating by MC-ICP-MS with SEM protocols. Geochim Cosmochim Acta 99:71–86

Shimamura M, Hyeong K, Yoo CM, Watanabe T, Irino T, Jung H-S (2008) High resolution stable isotope records of scleractinian corals near Ishigaki Island: their implication as a potential paleoclimatic recorder in middle latitude regions. Geosci J 12(1):25–31

Shinjo R, Asami R, Huang K-F, You C-F, Iryu Y (2013) Ocean acidification trend in the tropical North Pacific since the mid-20th century reconstructed from a coral archive. Mar Geol 342:58–64

Shotyk W, Immenhauser-Potthast I, Vogel HA (1995) Determination of nitrate, phosphate and organically bound phosphorus in coral skeletons by ion chromatography. J Chromatogr A 706(1):209–213

Shotyk W, Weiss D, Appleby P, Cheburkin A, Frei R, Gloor M, Kramers J, Reese S, Van Der Knaap W (1998) History of atmospheric lead deposition since 12,370 $^{14}$C yr BP from a peat bog, Jura Mountains, Switzerland. Science 281(5383):1635–1640

Shuto K (1997) Interannual variations of water temperature and salinity along the 137° E meridian. Oceanogr Lit Rev 44(5):425–425

Sinclair DJ, Kinsley LP, McCulloch MT (1998) High resolution analysis of trace elements in corals by laser ablation ICP-MS. Geochim Cosmochim Acta 62(11):1889–1901

Smith S, Buddemeier R, Redalje R, Houck J (1979) Strontium-calcium thermometry in coral skeletons. Science 204(4391):404–407

Smodej J, Reuning L, Wollenberg U, Zinke J, Pfeiffer M, Kukla PA (2015) Two-dimensional X-ray diffraction as a tool for the rapid, nondestructive detection of low calcite quantities in aragonitic corals. Geochem Geophys Geosyst 16(10):3778–3788

Sowa K, Watanabe T, Motai S, Seto Y, Nagai T (2008) Mineral phase of COCs and fibers in coral skeletons. In: Proc 11th Int Coral Reef Symp, pp 74–77

Sowa K, Watanabe T, Kan H, Yamano H (2014) Influence of land development on Holocene *Porites* coral calcification at Nagura Bay, Ishigaki Island, Japan. PLoS One 9(2):e88790

Speer JH (2010) Fundamentals of tree-ring research. University of Arizona Press, Tucson

Spencer T, Teleki KA, Bradshaw C, Spalding MD (2000) Coral bleaching in the southern Seychelles during the 1997–1998 Indian Ocean warm event. Mar Pollut Bull 40(7):569–586

Stocker T, Qin D, Plattner G-K, Tignor M, Allen SK, Boschung J, Nauels A, Xia Y, Bex V, Midgley PM (2013) Climate change 2013: the physical science basis. Contribution of Working Group I to the Fifth Assessment Report of the Intergovernmental Panel on Climate Change. Cambridge University Press New York

Storz D, Gischler E (2011) Coral extension rates in the NW Indian Ocean I: reconstruction of 20th century SST variability and monsoon current strength. Geo-Mar Lett 31(3):141–154

Strom A, Francis RC, Mantua NJ, Miles EL, Peterson DL (2004) North Pacific climate recorded in growth rings of geoduck clams: a new tool for paleoenvironmental reconstruction. Geophys Res Lett 31(6):L06206

Suga T, Kato A, Hanawa K (2000) North pacific tropical water: its climatology and temporal changes associated with the climate regime shift in the 1970s. Prog Oceanogr 47(2):223–256

Sugihara K, H. Yamano (2004) Tsushima archipelago. In: Coral reefs of Japan. MOE (Japan), Tokyo, pp 245–247

Sugihara K, Nakamori T, Iryu Y, Sasaki K, Blanchon P (2003) Holocene sea-level change and tectonic uplift deduced from raised reef terraces, Kikai-jima, Ryukyu Islands, Japan. Sediment Geol 159(1–2):5–25

Susic M, Boto K, Isdale P (1991) Fluorescent humic acid bands in coral skeletons originate from terrestrial runoff. Mar Chem 33(1):91–104

Suzuki A, Yukino I, Kawahata H (1999) Temperature-skeletal $\delta^{18}O$ relationship of *Porites australiensis* from Ishigaki Island, the Ryukyus, Japan. Geochem J 33(6):419–428

Suzuki A, Gagan MK, De Deckker P, Omura A, Yukino I, Kawahata H (2001) Last interglacial coral record of enhanced insolation seasonality and seawater $^{18}O$ enrichment in the Ryukyu Islands, northwest Pacific. Geophys Res Lett 28(19):3685–3688

Suzuki A, Gagan MK, Fabricius K, Isdale PJ, Yukino I, Kawahata H (2003) Skeletal isotope microprofiles of growth perturbations in *Porites* corals during the 1997–1998 mass bleaching event. Coral Reefs 22(4) 357–369

Suzuki A, Yokoyama Y, Kan H, Minoshima K, Matsuzaki H, Hamanaka N, Kawahata H (2008) Identification of 1771 Meiwa tsunami deposits using a combination of radiocarbon dating and oxygen isotope microprofiling of emerged massive *Porites* boulders. Quat Geochronol 3(3):226–234

Tachibana Y, Honda M, Takeuchi K (1996) The abrupt decrease of the sea ice over the southern part of the Sea of Okhotsk in 1989 and its relation to the recent weakening of the Aleutian Low. J Meteor Soc Jpn 74:579–584

Tanaka K, Holcomb M, Takahashi A, Kurihara H, Asami R, Shinjo R, Sowa K, Rankenburg K, Watanabe T, McCulloch M (2015) Response of Acropora digitifera to ocean acidification: constraints from $\delta^{11}B$, Sr, Mg, and Ba compositions of aragonitic skeletons cultured under variable seawater pH. Coral Reefs 34(4):1139–1149

Taniguchi H, Iwao K, Omori M (1999) Coral bleaching around Akajima, Okinawa:I. A report of the September 1998 survey. Galaxea, JCRS 1999(1):59–64

Tanzil JTI, Brown BE, Tudhope AW, Dunne RP (2009) Decline in skeletal growth of the coral *Porites lutea* from the Andaman Sea, South Thailand between 1984 and 2005. Coral Reefs 28(2):519–528

Tanzil JTI, Brown BE, Dunne RP, Lee JN, Kaandorp JA, Todd PA (2013) Regional decline in growth rates of massive Porites corals in Southeast Asia. Glob Chang Biol 19(10):3011–3023

Tatsumoto M, Patterson CC (1963) The concentration of common lead in seawater. In: Geiss J, Goldberg E (eds) Earth science and meteoritics. Noth-Holland, Amsterdam, pp 74–89

Trotter J, Montagna P, McCulloch M, Silenzi S, Reynaud S, Mortimer G, Martin S, Ferrier-Pagès C, Gattuso J-P, Rodolfo-Metalpa R (2011) Quantifying the pH 'vital effect' in the temperate zooxanthellate coral *Cladocora caespitosa*: validation of the boron seawater pH proxy. Earth Planet Sci Lett 303(3–4):163–173

Tsuchiya M (1968) Upper waters of the intertropical Pacific Ocean. In: The Johns Hopkins Oceanographic Studies, No. 4. The Johns Hopkins Press, Baltimore, p 50

Tsunoda T, Kawahata H, Suzuki A, Minoshima K, Shikazono N (2006) Winter sea surface temperature variations based on coral oxygen isotope record from Ishigaki Island, the Ryukyus, Japan and its relationship to 1988/1989 climate regime shift. Chikyukagaku (Geochemistry) 40(4):301–311

Tsunoda T, Kawahata H, Suzuki A, Minoshima K, Shikazono N (2008) East Asian monsoon to El Niño/Southern Oscillation: a shift in the winter climate of Ishigaki Island accompanying the 1988/1989 regime shift, based on instrumental and coral records. Geophys Res Lett 35(13):L13708

Uchikawa J, Penman DE, Zachos JC, Zeebe RE (2015) Experimental evidence for kinetic effects on B/Ca in synthetic calcite: implications for potential B (OH) $_4^-$ and B (OH) $_3$ incorporation. Geochim Cosmochim Acta 150:171–191

Uppström LR (1974) The boron/chlorinity ratio of deep-sea water from the Pacific Ocean. Deep-Sea Res Oceanogr Abstr 21(2):161–162

Urabe Y, Maeda S (2014) The relationship between Japan's recent temperature and decadal variability. SOLA 10:176–179

Urey H (1947) The thermodynamic properties of isotopic substances. J Chem Soc (Resumed):562

Venn A, Tambutté E, Holcomb M, Allemand D, Tambutté S (2011) Live tissue imaging shows reef corals elevate pH under their calcifying tissue relative to seawater. PLoS One 6(5):e20013

Venn AA, Tambutté E, Holcomb M, Laurent J, Allemand D, Tambutté S (2013) Impact of seawater acidification on pH at the tissue–skeleton interface and calcification in reef corals. PNAS 110(5):1634–1639

Veron J (1992) Environmental control of Holocene changes to the world's most northern hermatypic coral outcrop. Pac Sci 46(4):405–425

Wang Y, Cheng H, Edwards RL, He Y, Kong X, An Z, Wu J, Kelly MJ, Dykoski CA, Li X (2005) The Holocene Asian monsoon: links to solar changes and North Atlantic climate. Science 308(5723):854–857

Wang X, Sigman D, Cohen A, Sinclair D, Sherrell R, Weigand M, Erler DV, Ren H (2015) Isotopic composition of skeleton-bound organic nitrogen in reef-building symbiotic corals: a new method and proxy evaluation at Bermuda. Geochim Cosmochim Acta 148:179–190

Watanabe Y, Nakai S (2006) U-Th radioactive disequilibrium analyses for JCp-1, coral reference distributed by the Geological Survey of Japan. Geochem J 40(5):537–541

Watanabe T, Suzuki A, Minobe S, Kawashima T, Kameo K, Minoshima K, Aguilar YM, Wani R, Kawahata H, Sowa K, Nagai T, Kase T (2011) Permanent El Nino during the Pliocene warm period not supported by coral evidence. Nature 471(7337):209–211

Watanabe T, Kawamura T, Yamazaki A, Murayama M, Yamano H (2014) A 106 year monthly coral record reveals that the East Asian summer monsoon modulates winter PDO variability. Geophys Res Lett 41(10):3609–3614

Wei G, McCulloch MT, Mortimer G, Deng W, Xie L (2009) Evidence for ocean acidification in the Great Barrier Reef of Australia. Geochim Cosmochim Acta 73(8):2332–2346

Wellington G, Dunbar R, Merlen G (1996) Calibration of stable oxygen isotope signatures in Galapagos corals. Paleoceanography 11(4):467–480

Wen C, Graf H-F, Ronghui H (2000) The interannual variability of East Asian winter monsoon and its relation to the summer monsoon. Adv Atmos Sci 17(1):48–60

Wu B, Wang J (2002) Winter Arctic oscillation, Siberian high and East Asian winter monsoon. Geophys Res Lett 29(19):1897

Wu HC, Moreau M, Linsley BK, Schrag DP, Corrège T (2014) Investigation of sea surface temperature changes from replicated coral Sr/Ca variations in the eastern equatorial Pacific (Clipperton Atoll) since 1874. Palaeogeogr Palaeoclimatol Palaeoecol 412:208–222

Yamano H, Hori K, Yamauchi M, Yamagawa O, Ohmura A (2001) Highest-latitude coral reef at Iki Island, Japan. Coral Reefs 20(1):9–12

Yamano H, Sugihara K, Nakai T, Yamagawa O (2004) Iki Islands. In: Coral reefs of Japan. MOE (Japan), Tokyo, pp 242–244

Yamano H, Sugihara K, Nomura K (2011) Rapid poleward range expansion of tropical reef corals in response to rising sea surface temperatures. Geophys Res Lett 38(4):L04601

Yamazaki A, Watanabe T, Sowa K, Nakachi S, Yamano H, Iwase F (2009) Reconstructing palaeoenvironments of temperate regions based on high latitude corals at Tatsukushi Bay in Japan. J Jpn Coral Reef Soc 11:91–107

Yamazaki A, Watanabe T, Ogawa NO, Ohkouchi N, Shirai K, Toratani M, Uematsu M (2011a) Seasonal variations in the nitrogen isotope composition of Okinotori coral in the tropical western Pacific: a new proxy for marine nitrate dynamics. J Geophys Res 116:G04005

Yamazaki A, Watanabe T, Tsunogai U (2011b) Nitrogen isotopes of organic nitrogen in reef coral skeletons as a proxy of tropical nutrient dynamics. Geophys Res Lett 38:L19605

Yamazaki A, Watanabe T, Tsunogai U, Hasegawa H, Yamano H (2015) The coral $\delta^{15}N$ record of terrestrial nitrate loading varies with river catchment land use. Coral Reefs 34(1):353–362

Yamazato K (1999) Coral bleaching in Okinawa, 1980 vs 1998. Galaxea, JCRS 1999(1):83–87

Yasunaka S, Hanawa K (2002) Regime shifts found in the Northern Hemisphere SST field. J Meteor Soc 80(1):119–135

Yasunaka S, Hanawa K (2005) Regime shift in the global sea-surface temperatures: its relation to El Niño–southern oscillation events and dominant variation modes. Int J Climatol 25(7):913–930

Yu J, Elderfield H (2007) Benthic foraminiferal B/Ca ratios reflect deep water carbonate saturation state. Earth Planet Sci Lett 258(1):73–86

Zachos JC, Röhl U, Schellenberg SA, Sluijs A, Hodell DA, Kelly DC, Thomas E, Nicolo M, Raffi I, Lourens LJ (2005) Rapid acidification of the ocean during the Paleocene-Eocene thermal maximum. Science 308(5728):1611–1615

Zhang Y, Wallace JM, Battisti DS (1997) ENSO-like interdecadal variability: 1900–93. J Clim 10(5):1004–1020

Zhao J-X, Collins LB (2011) Uranium series dating. In: Hopley D (ed) Encyclopedia of modern coral reefs: structure, form and process. Springer, Netherlands, pp 1128–1132

# Coral Reproduction in Japan

7

Naoko Isomura and Hironobu Fukami

**Abstract**

In total, 78 genera and 415 species of zooxanthellate scleractinian corals have been recorded in Japanese waters. The waters of Japan can be broadly divided into three latitudinal regions: the coral reef region (24–30°N, coral reefs with high coral species diversity), the non-coral reef region (30–33°N, coral assemblages with moderate coral species diversity and without coral reef structure), and the peripheral region (33–35°N, undeveloped coral assemblages with low coral species diversity). In this chapter, we review the Japanese and English literatures on coral reproduction in these three regions, focusing upon the timing of spawning and fertilization within and between species. In the cases of *Acropora* and *Montipora*, the higher the latitude of a given place, the later the spawning occurs in the year. For example, spawning occurs from May to June in the coral reef region, June to August in the non-coral reef region, and August to September in the peripheral region. The date of spawning seems to be determined in part by water temperature and light intensity that decrease with increasing latitude. Conversely, coral species in the family Merulinidae spawn on similar dates (June to August) in all regions. In addition, several fertilization studies have been performed in Japan to understand the basic mechanisms of fertilization in corals, to identify species boundaries or cryptic species, and to establish efficient seeding techniques. Furthermore, hybridization studies have been conducted to clarify the basis for the high level of *Acropora* species diversity. Summary of these studies in Japan will contribute largely to understanding coral reproduction on other region in the world.

**Keywords**

Coral reef • Non-coral reef region • Peripheral region • Spawning • Synchrony • Fertilization • Hybridization

## 7.1 Introduction to Coral Reproduction

In total, 78 genera and 415 species of zooxanthellate scleractinian corals (hereafter "coral") have been reported from Japan (Nishihira and Veron 1995; Veron 2000a, b, c). Most species inhabit southern regions of Japan, such as the Ryukyu Archipelago, and the number of species decreases with increasing latitude owing to lower temperatures. Japanese waters can be broadly divided into three latitudinal regions:

N. Isomura (✉)
Department of Bioresources Engineering, National Institute of Technology, Okinawa College,
905 Henoko, Nago City, Okinawa 905-2192, Japan
e-mail: iso@okinawa-ct.ac.jp

H. Fukami
Faculty of Agriculture, Department of Marine Biology and Environmental Science, University of Miyazaki,
Miyazaki 889-2192, Japan


A. Iguchi, C. Hongo (eds.), *Coral Reef Studies of Japan*, Coral Reefs of the World 13,
https://doi.org/10.1007/978-981-10-6473-9_7

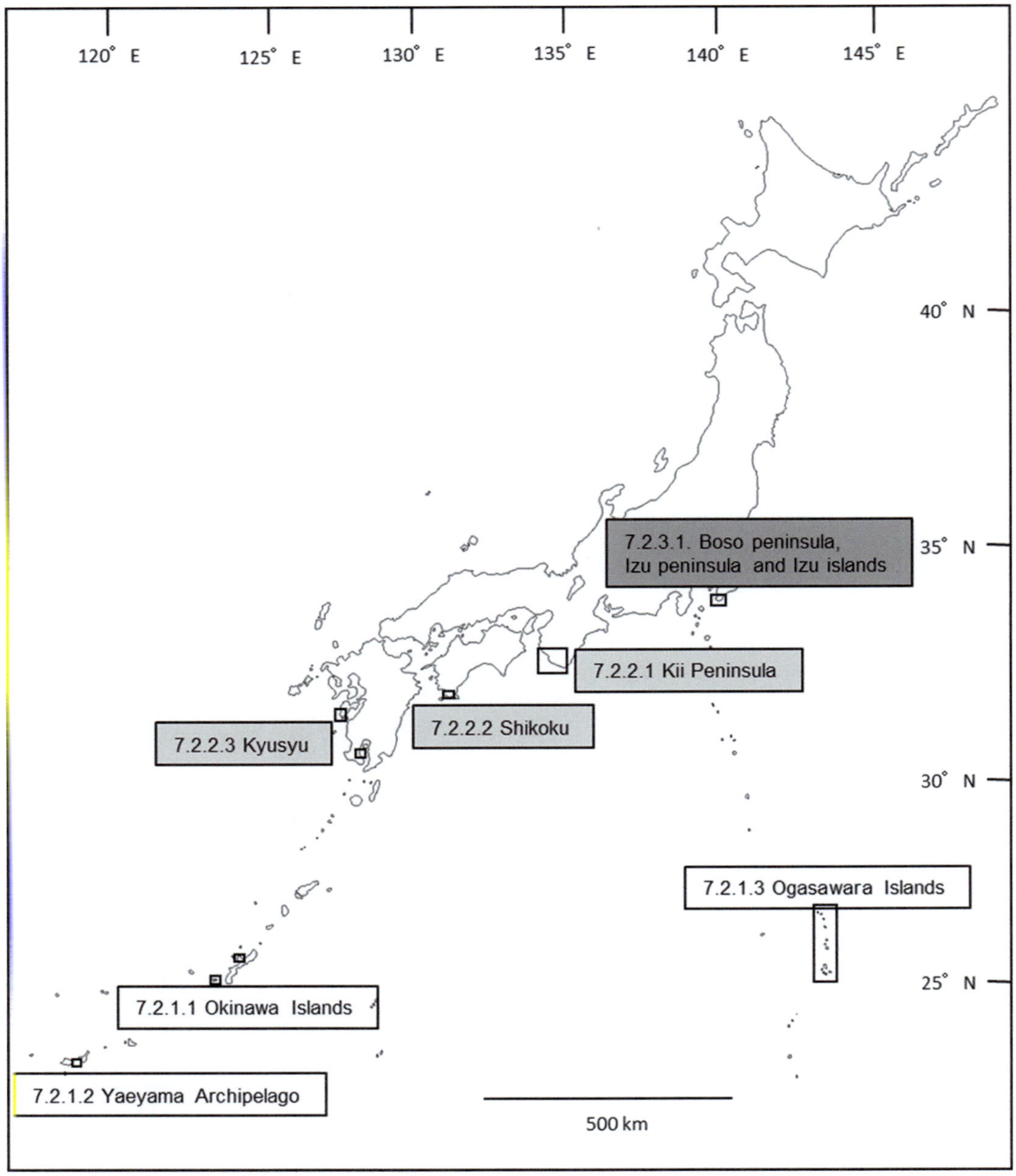


**Fig. 7.1** Map of Japan with three coral regions (coral reef region, non-coral reef region, and peripheral region). Section 7.2.1 shows coral reef region, 7.2.2 shows non-coral reef region, and 7.2.3 shows peripheral region. *Squares* surrounded by *black line* indicate fields where reproduction studies were performed, and detailed information of these fields are explained in text

The coral reef region (24–30°N; coral reefs with high coral species diversity: the Yaeyama Archipelago, Okinawa Islands, Amami and Tokara Archipelago, and Ogasawara Islands)

The non-coral reef region (30–33°N; coral assemblages with moderate diversity of coral species without coral reef structure: Kyushu, Shikoku, and Kii Peninsula)

The peripheral region (33–35°N; undeveloped coral assemblages with low coral species diversity: Izu Peninsula, Boso Peninsula) (Fig. 7.1)

The number of species in each region is estimated to be 415, 200, and 55, respectively.

Corals have two sexual patterns (hermaphroditic and gonochoric) and two modes of development (brooding and broadcast spawning). The sexual reproduction of corals worldwide has been reviewed in details (Harrison and Wallace 1990; Harrison 2011). These reviews, however, included only a few studies of coral reproduction in Japan because there are few studies published in English about the Japanese knowledge. In contrast, there are many published studies on coral reproduction written in Japanese, which have been produced by aquarium and marine stations in Japan for many years.

In this chapter, we summarize the major findings on coral reproduction from three regions (coral reef region, non-coral reef region, peripheral region), pertaining to topics such as timing and synchrony of spawning and intraspecific and interspecific fertilization. Author names and titles for reports and papers written in Japanese were translated into English by the authors of this chapter. In this chapter, we use the most recent scientific names of corals (see Fukami 2016 in detail).

## 7.2 Timing and Synchrony of Coral Reproduction in Japan

### 7.2.1 Coral Reef Region (24–30°N)

The coral reef region includes the Yaeyama Archipelago, Okinawa Islands, Amami and Tokara Archipelago, and Ogasawara Islands, but there are no data from the Amami and Tokara Archipelago on coral reproduction.

#### 7.2.1.1 Okinawa Islands

**Sesoko Islands**

Approximately 340 coral species have been reported from the Okinawa Islands (Nishihira and Veron 1995). The number of species, however, has decreased dramatically in recent years owing to mass bleaching events in 1998 and 2001 and serious damage by crown-of-thorns starfish (COTS) since the 1970s (Loya et al. 2001; van Woesik et al. 2011).

Reproductive studies of corals in the Okinawa Islands have been performed at the Sesoko station of the University of the Ryukyus, which is located on Sesoko Island, Motobu, Okinawa. The main corals studied from Sesoko Island are the families Acroporidae, Merulinidae (including the species formerly grouped within Faviidae), Pocilloporidae, Poritidae, and Fungiidae. Heyward et al. (1987) first recorded the date and time of spawning for 18 species belonging to several families, including Acroporidae and Merulinidae, in 1986 and 1987 around Sesoko Island. They retrieved samples of mature colonies and observed spawning in flow-through aquaria, showing that most species spawned in June, but some also spawned during July, August, and September from the full moon until the last quarter moon. Hirose and Hidaka (2006) also observed spawning in two branching species, *Porites cylindrica* and *Montipora digitata*, using mature colonies collected from the field. They showed that *P. cylindrica* released sperm or eggs repeatedly for 2–6 days after the full moon in June 2000 and that *M. digitata* spawned egg-sperm bundles continuously for 3–10 days after or before the full moon during May–July 1999 and June 2000. In *Acropora*, Prasetia et al. (2015) reported that *Acropora tenella* living at 40 m water depth would spawn at the same time as congeners that live in shallow water. Ohki et al. (2015) reconfirmed a difference in the timing of spawning between *A. digitifera* and the cryptic species *A.* sp.1, which was originally reported by Hayashibara and Shimoike (2002) at Akajima Island. Namely, *A. digitifera* spawned from the full moon to the new moon in June, whereas *A.* sp.1 spawned after the full moon in August. Nakajima et al. (2012) revealed that *A.* sp. 1 is genetically distinct from *A. digitifera* by population genetic approach using microsatellite markers. Within the Merulinidae, Sakai (1997) reported unusual spawning in *Coelastrea aspera* (synonymous with *Goniastrea aspera*). According to his paper, this species spawned mainly 3–6 days after the full moon in June or July. *C. aspera* first released sperm and then eggs from a single colony. The same colony also released planula larvae a few days after spawning for approximately 1 month. This reproductive manner was also reconfirmed by Nozawa and Harrison (2005). For Pocilloporidae, Yamazato et al. (2008) reported the timing of the release of planulae in three brooding species (*Pocillopora damicornis*, *Seriatopora hystrix*, *Stylophora pistillata*). *P. damicornis* released planula larvae from May to October, and the other two species released planulae from May to September from 1993 to 1995. The peak of planula release was not closely related to the phase of the moon. For example, *P. damicornis* had two peaks: the first peak was from the new moon to the full moon, and the second peak was from the last quarter moon to the new moon. *S. pistillata* also had two peaks: the first and last quarter moon in between new moons. *S. hystrix* had only one peak, during the first quarter moon to the last quarter moon. Kinzie (1993) and Hirose et al. (2000) also reported that the broadcast-spawning species *P. eydouxi* and *P. verrucosa* released sperm and nonfloating eggs in the early morning, from 6:00 to 7:00 am, during the full moon or new moon. For Fungiidae, Loya et al. (2009) reported that two fungiid species spawned in the early morning: *Lithophyllon repanda* (synonymous with *Fungia repanda*) spawned around 5:00 am while *Ctenactis echinata* spawned around 2:00 am.

A summary of the timing and synchrony of coral reproduction in Sesoko Island is shown in Table 7.1, including information by Heyward et al. (1987), Sakai (1997), and Hirose and Hidaka (2006). Most coral species tended to release their gametes during the period between the full moon and the last quarter moon, although it should be noted that these studies were conducted in different years. Some species (e.g., *Platygyra ryukyuensis*, *Favites chinensis*, *C. aspera*, and *Galaxea fascicularis*) spawned at the same timing related to lunar phase but in different months. Moreover, some species (e.g., *M. digitata*, *Platygyra pini*, and *F. chinensis*) spawned at nearly the same timing in different years.

**Table 7.1** The spawning timing of corals in 1986, 1987, 1993, 1994, 1999, and 2000 in Sesoko Island, Okinawa

| Year | Species | | References | Type of observation |
|---|---|---|---|---|
| 1986 | *Montipora* | *digitata* | Heyward et al. (1987) | A |
| | *Acropora* | sp. | | A |
| | *Platygyra* | *pini* | | A and F |
| | *Favites* | sp.1 | | A |
| | *Favites* | *chinensis* | | A and F |
| | *Favites* | sp.2 | | A |
| | *Coelastrea* | *aspera* | | A |
| | *Porites* | sp. | | F |
| | *Galaxea* | *fascicularis* | | A and F |
| 1987 | *Montipora* | *digitata* | Heyward et al. (1987) | A or F |
| | *Montipora* | *aequituberculata* | | A or F |
| | *Montipora* | *turgescens* | | A or F |
| | *Montipora* | *effusa* | | A or F |
| | | | | A or F |
| | *Acropora* | *grandis* | | A or F |
| | *Acropora* | *cytheria* | | A or F |
| | *Acropora* | *microphthalma* | | A or F |
| | *Acropora* | *latistella* | | A or F |
| | *Acropora* | *florida* | | A or F |
| | *Acropora* | *muricata* | | A or F |
| | *Platygyra* | *pini* | | A or F |
| | *Platygyra* | *ryukyuensis* | | A or F |
| | | | | A or F |
| | *Favites* | *chinensis* | | A or F |
| | | | | A or F |
| | *Coelastrea* | *aspera* | | A or F |
| | | | | A or F |
| | *Dipsastraea* | *pallida* | | A or F |
| | *Lobophyllia* | *corymbosa* | | A or F |
| | *Fungia* | sp. | | A or F |
| | *Galaxea* | *fascicularis* | | A or F |
| | | | | A or F |
| | | | | A or F |
| 1993 | *Coelastrea* | *aspera* | Sakai (1997) | A |
| | | | | A and F |
| 1994 | *Coelastrea* | *aspera* * | Sakai (1997) | A |
| 1999 | *Montipora* | *digitata* | Hirose and Hidaka (2006) | A |
| 2000 | *Montipora* | *digitata* | Hirose and Hidaka (2006) | A |
| 2000 | *Porites* | *cylyndrica* | Hirose and Hidaka (2006) | A |

References for data of each year were shown in right column. *Top line* shows moon phase. *F* full moon, *LQ* last quarter moon, *N* new moon, *FQ* first quarter moon. Spawning months: May, June, July, August. Types of observation: *A* aquarium observation, *F* field observation. Asterisks show releasing planulae

## Kerama Islands

The Kerama Islands include 30 islands of various sizes, 20–40 km away from Okinawa Island. A total of 14 families, 59 genera, and 248 species of corals has been recorded in the vicinity of the Kerama Islands (mainly Akajima Island) (Hayashibara 1995).

In the Kerama Islands, the center for research on coral reproduction is the Akajima Marine Science Laboratory, which is located on Akajima Island. Hayashibara et al. (1993) investigated the date of spawning for 10 families, 27 genera (as of April 2016, 8 family, 27 genera in the latest taxonomic classification), and 85 species during a 3-year period. They found that most species of *Acropora* spawned synchronously during 10-day periods before and after full moons in May–June. On the other hand, some corals spawned mainly 2–8 days after full moons in July–August. Hayashibara et al. (1993) also inferred that the trigger for synchronous spawning among multiple species must be some environmental factors, most likely annual variation in seawater temperature. The relationship between the date of spawning and lunar phase was unclear. Shimoike (1999) investigated coral spawning dates for 18 species from 1994 to 1998 and found that *A. latistella* and *A. divaricata* spawned in August around the time of the new moon. Moreover, Hayashibara and Shimoike (2002) reported that *Acropora* sp.1, a cryptic species of *A. digitifera*, spawned 2 months after *A. digitifera*. Thus, this study demonstrated the timing of coral spawning of some species differed from most other species in *Acropora*.

There is also a report of a brooding coral in Akajima Island. Okubo et al. (2007) reported that a brooding coral, *Isopora brueggemanni* (synonymous with *Acropora bureggemanni*), released planula larvae in September–December and February (from late summer to late winter) without apparent relation to lunar phase.

The timing of coral reproduction in Akajima Island is summarized in Table 7.2, including information from Hayashibara et al. (1993) and Shimoike (1999). Synchronous spawning in *Acropora* was relatively common in each year. However, *A. divaricata* and *A.* sp.1 spawned 1 or 2 months later than most congeners. Merulinedae species tended to spawn 1 or 2 months later than Acroporidae species, with some exceptions.

**Table 7.2** The spawning timing of corals in 1989–1991, 1994, 1995, 1997, and 1998 in Akajima, Okinawa

| Year | Species | | -6 | -5 | -4 | -3 | -2 | -1 | F | 1 | 2 | 3 | 4 | 5 | 6 | LQ | -6 | -5 | -4 | -3 | -2 | -1 | N | 1 | 2 | 3 | 4 | 5 | 6 | FQ | References | Type of observation |
|---|---|---|---|---|---|---|---|---|---|---|---|---|---|---|---|---|---|---|---|---|---|---|---|---|---|---|---|---|---|---|---|---|
| 1989 | *Montipora* | *digitaga* | | | | | | | | | | | | | | | | | | | | | | | | | | | | | Hayashibara et al. (1993) | A |
| | *Acropora* | *aspera* | | | | | | | | | | | | | | | | | | | | | | | | | | | | | | F and S |
| | *Acropora* | *cytheria* | | | | | | | | | | | | | | | | | | | | | | | | | | | | | | S |
| | *Acropora* | *digitifera* | | | | | | | | | | | | | | | | | | | | | | | | | | | | | | F |
| | *Acropora* | *exquista* | | | | | | | | | | | | | | | | | | | | | | | | | | | | | | S |
| | *Acropora* | *muricata* | | | | | | | | | | | | | | | | | | | | | | | | | | | | | | S |
| | | | | | | | | | | | | | | | | | | | | | | | | | | | | | | | | F |
| | *Acropora* | *grandis* | | | | | | | | | | | | | | | | | | | | | | | | | | | | | | S |
| | | | | | | | | | | | | | | | | | | | | | | | | | | | | | | | | F |
| | *Acropora* | *hyacinthus* | | | | | | | | | | | | | | | | | | | | | | | | | | | | | | S |
| | | | | | | | | | | | | | | | | | | | | | | | | | | | | | | | | F |
| | *Acropora* | *nasuta* | | | | | | | | | | | | | | | | | | | | | | | | | | | | | | S |
| | | | | | | | | | | | | | | | | | | | | | | | | | | | | | | | | F |
| | *Acropora* | *intermedia* | | | | | | | | | | | | | | | | | | | | | | | | | | | | | | S |
| | | | | | | | | | | | | | | | | | | | | | | | | | | | | | | | | F |
| | *Acropora* | *pulchra* | | | | | | | | | | | | | | | | | | | | | | | | | | | | | | F |
| | *Acropora* | *robusta* | | | | | | | | | | | | | | | | | | | | | | | | | | | | | | S |
| | *Acropora* | *tenuis* | | | | | | | | | | | | | | | | | | | | | | | | | | | | | | S |
| | *Acropora* | *valenciennesi* | | | | | | | | | | | | | | | | | | | | | | | | | | | | | | S |
| | *Acropora* | *valida* | | | | | | | | | | | | | | | | | | | | | | | | | | | | | | F |
| | *Galaxea* | *fascicularis* | | | | | | | | | | | | | | | | | | | | | | | | | | | | | | A |
| | *Oxypora* | *lacera* | | | | | | | | | | | | | | | | | | | | | | | | | | | | | | F |
| | *Lobophyllia* | *corymbosa* | | | | | | | | | | | | | | | | | | | | | | | | | | | | | | F |
| | *Depsastrea* | *pallida* | | | | | | | | | | | | | | | | | | | | | | | | | | | | | | F and S |
| | *Depsastrea* | *speciosa* | | | | | | | | | | | | | | | | | | | | | | | | | | | | | | F |
| | *Depsastrea* | *favus* | | | | | | | | | | | | | | | | | | | | | | | | | | | | | | A |
| | *Favites* | *chinensis* | | | | | | | | | | | | | | | | | | | | | | | | | | | | | | A |
| | *Coelastrea* | *aspera* | | | | | | | | | | | | | | | | | | | | | | | | | | | | | | A |
| | | | | | | | | | | | | | | | | | | | | | | | | | | | | | | | | A |
| | *Platygyra* | *pini* | | | | | | | | | | | | | | | | | | | | | | | | | | | | | | A |
| | *Platygyra* | *daedalea* | | | | | | | | | | | | | | | | | | | | | | | | | | | | | | F |
| 1990 | *Astreopora* | *myriophthalma* | | | | | | | | | | | | | | | | | | | | | | | | | | | | | Hayashibara et al. (1993) | S |
| | *Montipora* | *digitaga* | | | | | | | | | | | | | | | | | | | | | | | | | | | | | | F |
| | *Montipora* | *informis* | | | | | | | | | | | | | | | | | | | | | | | | | | | | | | F |
| | *Montipora* | *turgescens* | | | | | | | | | | | | | | | | | | | | | | | | | | | | | | F |
| | *Acropora* | *cytheria* | | | | | | | | | | | | | | | | | | | | | | | | | | | | | | F |
| | *Acropora* | *danai* | | | | | | | | | | | | | | | | | | | | | | | | | | | | | | F |
| | *Acropora* | *digitifera* | | | | | | | | | | | | | | | | | | | | | | | | | | | | | | F |
| | *Acropora* | *florida* | | | | | | | | | | | | | | | | | | | | | | | | | | | | | | F |
| | *Acropora* | *muricata* | | | | | | | | | | | | | | | | | | | | | | | | | | | | | | F |
| | *Acropora* | *gemmifera* | | | | | | | | | | | | | | | | | | | | | | | | | | | | | | F |
| | *Acropora* | *hyacinthus* | | | | | | | | | | | | | | | | | | | | | | | | | | | | | | F |
| | *Acropora* | *longicyathus* | | | | | | | | | | | | | | | | | | | | | | | | | | | | | | S |
| | *Acropora* | *microphthalma* | | | | | | | | | | | | | | | | | | | | | | | | | | | | | | F |
| | *Acropora* | *millepora* | | | | | | | | | | | | | | | | | | | | | | | | | | | | | | F |
| | *Acropora* | *nasuta* | | | | | | | | | | | | | | | | | | | | | | | | | | | | | | F |
| | *Acropora* | *intermedia* | | | | | | | | | | | | | | | | | | | | | | | | | | | | | | F |
| | | | | | | | | | | | | | | | | | | | | | | | | | | | | | | | | F |
| | *Acropora* | *pulchra* | | | | | | | | | | | | | | | | | | | | | | | | | | | | | | S |
| | *Acropora* | *valida* | | | | | | | | | | | | | | | | | | | | | | | | | | | | | | F |
| | *Porites* | *lutea* | | | | | | | | | | | | | | | | | | | | | | | | | | | | | | F |
| | *Pachyseris* | *speciosa* | | | | | | | | | | | | | | | | | | | | | | | | | | | | | | A and S |
| | *Fungia* | *repanda* | | | | | | | | | | | | | | | | | | | | | | | | | | | | | | A |
| | *Sandalolitha* | *robusta* | | | | | | | | | | | | | | | | | | | | | | | | | | | | | | A |
| | *Galaxea* | *fascicularis* | | | | | | | | | | | | | | | | | | | | | | | | | | | | | | F and S |
| | *Echinophyllia* | *aspera* | | | | | | | | | | | | | | | | | | | | | | | | | | | | | | A |
| | *Pectinia* | *lactuca* | | | | | | | | | | | | | | | | | | | | | | | | | | | | | | S |
| | *Merulina* | *ampliata* | | | | | | | | | | | | | | | | | | | | | | | | | | | | | | A, F and S |
| | | | | | | | | | | | | | | | | | | | | | | | | | | | | | | | | A, F and S |
| | *Merulina* | *scabricula* | | | | | | | | | | | | | | | | | | | | | | | | | | | | | | A |
| | *Depsastrea* | *pallida* | | | | | | | | | | | | | | | | | | | | | | | | | | | | | | A and S |
| | *Depsastrea* | *pallida* | | | | | | | | | | | | | | | | | | | | | | | | | | | | | | A and S |
| | *Depsastrea* | *pallida* | | | | | | | | | | | | | | | | | | | | | | | | | | | | | | A and S |
| | *Depsastrea* | *speciosa* | | | | | | | | | | | | | | | | | | | | | | | | | | | | | | S |
| | *Depsastrea* | *lizardensis* | | | | | | | | | | | | | | | | | | | | | | | | | | | | | | S |
| | *Depsastrea* | *matthaii* | | | | | | | | | | | | | | | | | | | | | | | | | | | | | | S |
| | *Depsastrea* | *favus* | | | | | | | | | | | | | | | | | | | | | | | | | | | | | | A and F |
| | *Depsastrea* | *stelligera* | | | | | | | | | | | | | | | | | | | | | | | | | | | | | | A and S |
| | *Depsastrea* | *veroni* | | | | | | | | | | | | | | | | | | | | | | | | | | | | | | A |
| | *Favites* | *halicora* | | | | | | | | | | | | | | | | | | | | | | | | | | | | | | A |
| | *Favites* | *abdita* | | | | | | | | | | | | | | | | | | | | | | | | | | | | | | F |
| | *Favites* | *flexuosa* | | | | | | | | | | | | | | | | | | | | | | | | | | | | | | A and F |
| | *Goniastrea* | *pectinata* | | | | | | | | | | | | | | | | | | | | | | | | | | | | | | F |
| | *Platygyra* | *pini* | | | | | | | | | | | | | | | | | | | | | | | | | | | | | | A and S |
| | *Platygyra* | *lamellina* | | | | | | | | | | | | | | | | | | | | | | | | | | | | | | F |
| | *Favites* | *curta* | | | | | | | | | | | | | | | | | | | | | | | | | | | | | | A |
| | *Favites* | *magnistellata* | | | | | | | | | | | | | | | | | | | | | | | | | | | | | | A |
| | *Favites* | *valenciennesi* | | | | | | | | | | | | | | | | | | | | | | | | | | | | | | S |
| | *Diploastrea* | *heliopora* | | | | | | | | | | | | | | | | | | | | | | | | | | | | | | S |
| | *Leptastrea* | *purpurea* | | | | | | | | | | | | | | | | | | | | | | | | | | | | | | A |
| | *Cyphastrea* | *chalcidicum* | | | | | | | | | | | | | | | | | | | | | | | | | | | | | | F |
| | *Euphyllia* | *divisa* | | | | | | | | | | | | | | | | | | | | | | | | | | | | | | A |

(continued)

**Table 7.2** (continued)

| Year | Species | | References | Type of observation |
|---|---|---|---|---|
| 1991 | *Astreopora* | *myriophthalma* | Hayashibara et al. (1993) | F |
| | *Montipora* | *digitaga* | | F |
| | *Montipora* | *aequituberculata* | | F |
| | *Montipora* | *informis* | | F |
| | *Montipora* | *turgescens* | | F |
| | *Montipora* | *flowen?* | | F |
| | *Montipora* | *venosa* | | F |
| | *Montipora* | *efflorescens* | | F |
| | *Acropora* | *anthocercis* | | F |
| | | | | S |
| | *Acropora* | *cytheria* | | F |
| | *Acropora* | *digitifera* | | F |
| | *Acropora* | *digitifera* | | F |
| | *Acropora* | *divaricata* | | S |
| | | | | S |
| | *Acropora* | *donei* | | F |
| | *Acropora* | *florida* | | F |
| | | | | F |
| | *Acropora* | *muricata* | | F |
| | *Acropora* | *gemmifera* | | F |
| | *Acropora* | *gemmifera* | | F |
| | *Acropora* | *humilis* | | F |
| | *Acropora* | *hyacinthus* | | F |
| | | | | F |
| | *Acropora* | *latistella* | | S |
| | *Acropora* | *longicyathus* | | F |
| | *Acropora* | *loripes* | | F |
| | *Acropora* | *microclados* | | F |
| | *Acropora* | *microphthalma* | | F |
| | *Acropora* | *monticulosa* | | F |
| | *Acropora* | *nasuta* | | F |
| | *Acropora* | *intermedia* | | F |
| | | | | F |
| | *Acropora* | *robusta* | | F |
| | *Acropora* | *samoensis* | | F |
| | | | | S |
| | *Acropora* | *tenuis* | | F |
| | *Acropora* | *valenciennesi* | | F |
| | *Acropora* | *valida* | | S |
| | | | | S |
| | *Acropora* | *verweyi* | | S |
| | *Acropora* | sp.1 | | F |
| | *Acropora* | sp.2 | | F |
| | *Porites* | *lutea* | | F |
| | *Galaxea* | *fascicularis* | | F |
| | *Hydnophora* | *rigida* | | F |
| | *Hydnophora* | *exesa* | | F |
| | *Merulina* | *ampliata* | | F |
| | *Caulastraea* | *furcata* | | A and F |
| | *Dipsastraea* | *pallida* | | F |
| | *Dipsastraea* | *matthaii* | | F |
| | *Dipsastraea* | *favus* | | A and F |
| | *Dipsastraea* | *amicorum* | | F |
| | *Favites* | *halicora* | | A and F |
| | *Favites* | *abdita* | | F |
| | *Favites* | *flexuosa* | | F |
| | *Goniastrea* | *retiformis* | | F |
| | | | | F |
| | *Goniastrea* | *pectinata* | | A and F |
| | | | | A and F |
| | *Platygyra* | *pini* | | A and F |
| | *Platygyra* | *ryukyuensis* | | F |
| | *Platygyra* | *sinensis* | | F |
| | *Platygyra* | *daedalea* | | F |
| | *Platygyra* | *contorta* | | F |
| | *Astrea* | *curta* | | A |
| | *Favites* | *magnistellata* | | F |
| | *Favites* | *valenciennesi* | | F |
| | *Cyphastrea* | *chalcidicum* | | F |
| | *Cyphastrea* | *seraillia* | | F |
| | *Echinopora* | *gemmacea* | | F |
| | *Echinopora* | *pacificus* | | F |
| 1994 | *Dipsastraea* | *laxa* | Shimoike (1999) | F |
| 1995 | *Turbinaria* | *retiformis* | Shimoike (1999) | F |
| 1997 | *Acropora* | *yongei* | Shimoike (1999) | F |
| 1998 | *Montipora* | *foveolata* | Shimoike (1999) | F |
| | *Montipora* | *grisea* | | F |
| | *Acropora* | *monticulosa* | | F |
| | *Acropora* | *palmera* | | F |
| | *Acropora* | *cerealis* | | H |
| | *Acropora* | *granulosa* | | H |
| | *Acropora* | *loripes* | | H |
| | *Acropora* | *paniculata* | | H |
| | *Acropora* | *subulata* | | H |
| | *Acropora* | *willisae* | | H |
| | *Porites* | *cylindrica* | | F |
| | *Porites* | *mayeri* | | F |
| | *Echinophyllia* | *echinata* | | F |
| | *Goniastrea* | *edwardsi* | | F |
| | *Scapophyllia* | *cylindrica* | | F |

*Top line* shows moon phase. *F* full moon, *LQ* last quarter moon, *N* new moon, *FQ* first quarter moon. Spawning months: May, June, July, August. Types of observation: *A* aquarium observation, *F* field observation, *H* histological observation, *S* sequential field sampling more than 2 days apart

#### 7.2.1.2 Yaeyama Archipelago

The Yaeyama Archipelago, which contains Ishigaki Island, Iriomote Island, Kohama Island, Taketomi Island, Yonaguni Island, Hateruma Island, Hatoma Island, and Kuroshima Island, is located 430 km southwest of the Okinawa Islands (Fig. 7.1).

Sekisei lagoon is located between Ishigaki Island and Iriomote Island and is 15 km in length (north–south) and 20 km in width (east–west). Here, a total of 363 species have been recorded, the highest level of coral species diversity in Japan (Nishihira and Veron 1995). Nevertheless, there are only two published reports about coral reproduction. Misaki (1994) investigated spawning of a total of 28 species found 700 m offshore of Kuroshima Island for 3 years (16 species in 1991, 10 species in 1992, 17 species in 1993). Misaki (1994) showed that coral spawning occurred most commonly 2–3 days after a full moon and within 1 h of the highest tide. Notably, this relationship between environmental factors and spawning time mirrored that seen in Kushimoto and Wakayama, the non-coral reef region.

In Amitori Bay, Iriomote Island, Murakami et al. (2015) reported that many *Acropora* spp., *Dipsastraea favus* (synonymous with *Favia favus*), *Goniastrea pectinata*, *G. retiformis*, *Montipora stellata*, and *Porites lutea* spawned in May and June between high and low tide, excluding *A. secale* and *A. divaricata* that spawned in July and August, respectively, at low tide.

The timing of coral reproduction in Kuroshima and Iriomote is summarized in Table 7.3, including information from Misaki (1994) and Murakami et al. (2015). The level of synchrony in spawning for *Acropora* was relatively high in each year, and most species spawned in May.

#### 7.2.1.3 Ogasawara Islands

The Ogasawara Islands are located 1000 km south of Tokyo and consist of 30 islands, including the Chichijima and Hahajima Archipelagos. In total, approximately 200 species of corals have been reported from the vicinity of the Ogasawara Islands (Tachikawa et al. 1991).

**Table 7.3** The spawning timing of corals in 1991–1993 in Kuroshima and in 2015 in Amitori Bay, Iriomote Island, Okinawa

| Year | Species | | -6 -5 -4 -3 -2 -1 F 1 2 3 4 5 6 LQ -6 -5 -4 -3 -2 -1 N 1 2 3 4 5 6 FQ | References | Type of observation |
|---|---|---|---|---|---|
| 1991 | *Montipora* | *hispida* | | Misaki (1994) | F |
| | *Montipora* | *foliosa* | | | F |
| | *Cyphastrea* | *chalcidicum* | | | F |
| | *Dipsastraea* | *pallida* | | | F |
| | *Goniastrea* | *stelligera* | | | F |
| | *Goniastrea* | *pectinata* | | | F |
| | *Platygyra* | *sinensis* | | | F |
| 1992 | *Acropora* | *caruduus* | | Misaki (1994) | F |
| | *Cyphastrea* | *serailia* | | | F |
| | *Dipsastraea* | *speciosa* | | | F |
| 1993 | *Acropora* | sp. | | Misaki (1994) | F |
| | *Acropora* | *humilis* | | | F |
| | *Acropo* | *selago* | | | F |
| | *Acropora* | *polystoma* | | | F |
| | *Acropora* | *scuminata* | | | F |
| | *Acropora* | *robusta* | | | F |
| | *Acropora* | *monticulosa* | | | F |
| 1991-1992 | *Montipora* | *digitata* | | Misaki (1994) | F |
| 1991, 1993 | *Montipora* | *informis* | | Misaki (1994) | F |
| | *Acriopora* | *gemmifera* | | | F |
| | *Acropora* | *digitifera* | | | F |
| | *Cyphastrea* | *microphthalma* | | | F |
| 1992-1993 | *Acropora* | *muricata* | | Misaki (1994) | F |
| | *Acropora* | *intermedia* | | | F |
| 1991-1993 | *Acropora* | *cytherea* | | Misaki (1994) | F |
| | *Acropora* | *hyacinthus* | | | F |
| | *Acropora* | *nasuta* | | | F |
| | *Acropora* | *cerealis* | | | F |
| 2014 | *Acropora* | *grandis* | | Murakami et al. (2015) | F |
| | *Acropora* | *millepora* | | | F |
| | *Acropora* | *muricata* | | | F |
| | *Acropora* | *microphthalma* | | | F |
| | *Acropora* | *florida* | | | F |
| | *Acropora* | *digitifera* | | | F |
| | *Acropora* | *gemmifera* | | | F |
| | | | | | F |
| | *Goniastrea* | *retiformis* | | | F |
| | *Montipora* | *stellata* | | | F |
| | *Porites* | *lutea* | | | F |
| | *Goniastrea* | *pectinata* | | | F |
| | *Dipsastraea* | *favus* | | | F |
| | *Acropora* | *secale* | | | F |
| | *Acropora* | *divaricata* | | | F |

*Top line* shows moon phase. *F* full moon, *LQ* last quarter moon, *N* new moon, *FQ* first quarter moon. Spawning month: ☐ May, June, July, August. Types of observation: *F* field observation

Coral reproduction in the Ogasawara Islands has been studied only for Chichijima. Inaba (2003) reported that *Acropora muricata* (synonymous with *A. formosa*) and *A. intermedia* (written as *A. nobilis*) spawned several times during low tide in May–June, while other species spawned synchronously within each species during low tide in August. Inaba (2003) also reported that the timing of spawning differed among species. For instance, spawning occurred near the time of sunset for *Lobophyllia hemprichii*, *Symphyllia agaricia*, *Platygyra deadalea*, and *Euphyllia ancora*; around 19:00 (local time) for *A. donei* and *Galaxea fascicularis*; and 20:00–22:00 (local time) for *A. hyacinthus*, *A. florida*, *Echinophyllia aspera*, *Goniastrea pectinata*, and *Cyphastrea serailia*. Based on observations of spawning slick, the dates of spawning for two other islands (Hahajima and Mukoujima) were similar to those of Chichijima, with occasional time lags of 1 day (Inaba 2003).

## 7.2.2 Non-coral Reef Region (30–34°N)

### 7.2.2.1 Kii Peninsula

The Kii Peninsula is located at the southernmost part (34°N, 136°E) of the Japanese mainland. Off the coast west of the Kii Peninsula, the marine environment is strongly affected by the branch flow of the Kuroshio, which transports the larvae of tropical animals from the southern parts of Japan, such as the Okinawa Islands. Coral reproduction in the Kii Peninsula has been investigated mainly in the Kushimoto Marine Park in Kushimoto, as well as the Seto Marine Biological Laboratory (Kyoto University) in Shirahama. Near Kushimoto, which is located at the southern end of the Kii Peninsula, coral reef animals are abundant: 109 species of corals have been recorded (Nomura 2004). The water temperature is also relatively high; for example, at Sabiura Bay in Kushimoto, the average annual water temperature is 21.1 °C. In the coldest month (February), the average is 16.3 °C, and in the hottest month (August), the average is 27.2 °C (Nomura et al. 2003). Shirahama is 100 km north of Kushimoto and is home to 77 coral species (Nomrua 2004).

At Sabiura Bay in Kushimoto, coral spawning has been investigated in the field from 1989 to 2008 (Misaki 1989, 1995a, b, 1996, 1997, 1998b, 1999, 2000, 2001, 2003, 2004, 2005, 2006a, b, 2007, 2008). These studies have focused mainly on *Acropora* species. All spawning data for *Acropora* species from 1989 to 2008 are summarized in Table 7.4. Notably, the dates of spawning for three species (*A. hyacinthus*, *A. solitaryensis*, and *A. japonica*) have been monitored for approximately 20 years. During the study period, *A. hyacinthus* spawned mainly from the full moon to the last quarter moon in July and August; however, this species also spawned between the new moon and the first quarter moon in 4 years (1999, 2001, 2007, 2008). *A. japonica* spawned on dates similar to *A. hyacinthus* in July and August, although the former species also spawned around the time of the first quarter moon in 1995 and the time of the new moon in 1999. *A. solitaryensis* spawned mainly between the new moon and the first quarter moon in July and August. However, in 2001 and 2003, *A. solitaryensis* spawned around the time of the full moon, like the species above. Observations of *A. muricata* spawning began in 2003. This species spawned mainly from the full moon to the last quarter moon in June and July, but the dates of spawning were inconsistent compared to other species. The timing of spawning (hours after sunset) was similar among all *Acropora* species with the exception of *A. japonica*. From 1989 to 1996, *A. japonica* spawned at the same time as *A. hyacinthus* and other species. However, since 1997, *A. japonica* has spawned 1–2 h later than other species.

Spawning has also been monitored for other coral families in Sabiura Bay (Table 7.4). For example, spawning has been studied for one *Goniopora* and two *Porites* species within the family Poritidae. *G. lobata* released sperm 1 day after the full moon (August, 1989). *P. australiensis* spawned 1 and 2 days after the full moon (August 1989), but *Porites heronensis* spawned during the new moon (month not recorded, 1995). In the family Merulinidae, corals tended to spawn around the time of the last quarter moon. For example, *Caulastrea tumida*, *Cyphastrea microphthalma*, and *Dipsastraea speciosa* (synonymous with *Favia speciosa*) spawned 2–5 days after the full moon (August 1989), and 1–10 colonies each of *Favites valenciennesi* (synonymous with *Montastraea valenciennesi*), *Paragoniastrea deformis* (synonymous with *Goniastrea deformis*), and *P. australiensis* (synonymous with *Goniastrea australiensis*) spawned every day 6–16 days after the full moon (July 1999).

At Kushimoto Marine Park, coral spawning has been studied in *Acropora* and other genera using corals kept in an aquarium (Misaki 1995b, 1996, 1998a, b, 1999, 2000, 2001, 2003, 2004, 2005, 2006a, b, 2007, 2008). The dates and timing of spawning for corals in the aquarium were nearly the same as those for corals in the field. Here, we include only a few examples. In 1995, *Pocillopora damicornis* released planulae 5–9 days after the full moon in June. In 1998, colonies of *P. damicornis*, which grew from the planulae released in 1995, also released planulae in June. We can therefore conclude that it takes just 3 years for *P. damicornis* to reach maturity. In 2002, *Stylophora pistillata* released planulae from the night of July 26 until the morning of July 27 (2–3 days after the full moon). In 2007, *Euphyllia ancora* released sperm at 18:00–18:30 around the time of the new moon in August.

At Shirahama, Zayasu et al. (2015) investigated the spawning of *Oulastrea crispata*. They found that *O. crispata* released sperm at 23:20, 8 days after the full moon, and released eggs 20 min later. Furthermore, once one colony

**Table 7.4** The spawning timing of corals in 1989 and 1994–2008 in Kushimoto, Wakayama

| Year | Species | | -6 -5 -4 -3 -2 -1 F 1 2 3 4 5 6 LQ -6 -5 -4 -3 -2 -1 N 1 2 3 4 5 6 FQ | References | Types of observation |
|---|---|---|---|---|---|
| 1989 | *Acropora* | *hyacinthus* | | Misaki (1989) | F |
| | *Acropora* | *japonica* | | | F |
| | *Acropora* | *willisae* | | | F |
| | *Acropora* | *solitaryensis* | | | F |
| | *Porites* | sp. | | | F |
| | *Caulastraea* | *tumida* | | | F |
| | *Goniopora* | *lobata* | | | F |
| | *Cyphastrea* | *microphthalma* | | | F |
| | *Dipsastraea* | *speciosa* | | | F |
| 1994 | *Acropora* | *hyacinthus* | | Misaki (1995a) | F |
| | *Acropora* | *japonica* | | | F |
| | *Acropora* | *solitaryensis* | | | F |
| 1995 | *Acropora* | *hyacinthus* | | Misaki (1995b) | F |
| | *Acropora* | *japonica* | | | F |
| | *Acropora* | *solitaryensis* | | | F |
| | *Porites* | *heronensis* | | | F |
| | *Echinophyllia* | *aspera* | | | F |
| | *Hydnophora* | *exesa* | | | F |
| | *Caulastraea* | *tumida* | | | F |
| | *Pocillopora* | sp. | | | A |
| 1996 | *Acropora* | *hyacinthus* | | Misaki (1996) | F |
| | *Acropora* | *japonica* | | | F |
| | *Acropora* | *solitaryensis* | | | F |
| 1997 | *Acropora* | *hyacinthus* | | Misaki (1997) | F |
| | *Acropora* | *japonica* | | | F |
| | *Acropora* | *willisae* | | | F |
| 1999 | *Acropora* | *hyacinthus* | | Misaki (1999) | A and F |
| | *Acropora* | *japonica* | | | F |
| | *Acropora* | sp. A | | | F |
| | *Acropora* | *solitaryensis* | | | F |
| | *Favites* | *valenciennesi* | | | F |
| | *Paragoniastrea* | *deformis* | | | F |
| | *Paragoniastrea* | *australiensis* | | | F |
| | *Platygyra* | *contorta* | | | F |
| 2000 | *Pocillopora* | *damicornis* | | Misaki (2000) | F |
| | *Acropora* | *hyacinthus* | | | F |
| 2001 | *Acropora* | *hyacinthus* | | Misaki (2001) | F |
| | *Acropora* | *japonica* | | | F |
| | *Acropora* | *solitaryensis* | | | F |
| 2002 | *Acropora* | *hyacinthus* | | Misaki (2003) | F |
| | *Acropora* | *japonica* | | | F |
| | *Acropora* | *valida* | | | F |
| 2003 | *Acropora* | *hyacinthus* | | Misaki (2004) | F |
| | *Acropora* | *japonica* | | | F |
| | *Acropora* | *solitaryensis* | | | F |
| | *Acropora* | *valida* | | | F |
| | *Acropora* | *muricata* | | | F |
| 2004 | *Acropora* | *hyacinthus* | | Misaki (2005) | F |
| | *Acropora* | *japonica* | | | F |
| | *Acropora* | *muricata* | | | F |
| 2005 | *Acropora* | *hyacinthus* | | Misaki (2005a) | F |
| | *Acropora* | *japonica* | | | F |
| | *Acropora* | *solitaryensis* | | | F |
| | *Acropora* | *muricata* | | | F |
| 2006 | *Acropora* | *hyacinthus* | | Misaki (2005b) | F |
| | *Acropora* | *japonica* | | | F |
| | *Acropora* | *solitaryensis* | | | F |
| | *Acropora* | *muricata* | | | F |
| 2007 | *Acropora* | *hyacinthus* | | Misaki (2007) | F |
| | *Acropora* | *japonica* | | | F |
| | *Acropora* | *solitaryensis* | | | F |
| | *Acropora* | *muricata* | | | F |
| | *Dipsastrea* | *favus* | | | F |
| | *Favites* | *abdita* | | | F |
| | *Favites* | *valenciennesi* | | | F |
| 2008 | *Acropora* | *hyacinthus* | | Misaki (2008) | F |
| | *Acropora* | *japonica* | | | F |
| | *Acropora* | *solitaryensis* | | | F |
| | *Acropora* | *muricata* | | | F |

References for data of each year were shown in right column. *Top line* shows moon phase. *F* full moon, *LQ* last quarter moon, *N* new moon, *FQ* first quarter moon. Spawning months: June, July, August, July or August. Types of observation: *A* aquarium observation, *F* field observation began to release sperm, other colonies began releasing sperm as well. Therefore, in this species, the release of sperm from the first colony may serve as a chemical cue for synchronous spawning.

#### 7.2.2.2 Shikoku

In Shikoku, corals are distributed mainly off the Pacific coast of the Tokushima, Kochi, and Ehime Prefectures. In particular, many corals are distributed from Cape Ashizuri in Kochi through Cape Sata in Ehime. In the vicinity of Cape Ashizuri, 127 coral species have been reported at Tosashimizu (Nishihira and Veron 1995) and 130 at Otsuki (Nomura and Mezaki 2005). Mezaki et al. (2007) reported that 6 families, 12 genera (as of April 2016, 4 families, 13 genera in the latest taxonomic classification), and 28 species of coral spawning have been observed at Otsuki from 2002 to 2006.

According to Mezaki et al. (2007), *Montipora* spp. spawn 1–1.5 h after sunset, merulinids spawn 1–3 h after sunset, and *Acropora* spp. spawn 2–4 h after sunset. Although the time of spawning varies by about 2 h within the genus, the level of synchrony in spawning among species was high in each year of the study period. The three dominant species (*A. muricata*, *A. hyacinthus*, and *A. solitaryensis*) spawned mainly from late June to early July, late June to early August, and late July to late August, respectively. The average water temperature during each of these spawning terms was 24.3, 25.7, and 27.6 °C, respectively. The timing of spawning for other *Acropora* species was similar to the three species discussed above. Conversely, *Montipora* spp. spawned slightly later (late August to early September) than *Acropora* spp. While a few *Acropora* and *Montipora* species spawned from the first quarter moon to the full moon, most spawned during other lunar phases. Nozawa (2012) also reported that coral spawning occurred for four species (*A. hyacinthus*, *A. japonica*, *Favites pentagona*, and *Platygyra contorta*) around the time of the last quarter moon in July or August from 2006 to 2009, irrespective of high levels of annual variation in seawater temperature (from 23.7 to 29.5 °C) and weather during the spawning period. The exact timing of spawning during the spawning period, however, varied among years and was correlated with average seawater temperature during the late period of gametogenesis (0–3 months before spawning). For merulinids, synchronous spawning was observed mainly around the time of the last quarter moon in late July to mid-August. During this period, the average daily water temperature exceeded 26 °C. Only *Hydnophora exesa* spawned mainly around the time of the new moon. A summary of the timing of coral reproduction near Otsuki based on van Woesik (1995); Mezaki et al. (2007) and Nozawa (2012) is shown in Table 7.5.

Within the Agariciidae, *Pavona decussata* released sperm and, subsequently, eggs around 4:30 am in the morning 6 days after the full moon in August (Mezaki et al. 2014). Miyamoto and Yokochi (2015) reported that *Pocillopora damicornis* released planula larvae during 15 days around the time of the new moon in June and July, and *Stylophora pistillata* released planula larvae twice during a 25-day period from late June to August, showing periodicity from the first quarter moon to the new moon. Both species released larvae throughout the entire day but particularly at night.

#### 7.2.2.3 Kyushu

Corals are distributed widely through the entire Kyushu region, with the exception of northern areas such as Fukuoka and the northern part of Oita Prefecture. Corals are abundant near Amakusa and Yatsushiro on Kumamoto Prefecture, the southern parts on Oita Prefecture, and along the coasts of Miyazaki and Kagoshima Prefectures.

In Kagoshima Bay, *Cyphastrea* sp., *Montipora venosa*, *Coelastrea aspera*, and *Acropora solitaryensis* were observed to spawn around the time of the last quarter moon, from 20:00 to 21:20 (Dewa et al. 2015). This is consistent with data from Otsuki (Mezaki et al. 2007).

In Kumamoto, about 98 coral species have been reported in the vicinity of Amakusa (Nishihira and Veron 1995), and coral reproduction has been well studied in the Amakusa Marine Laboratory at Kyushu University. Nozawa et al. (2006) investigated the spawning of seven species: *A. hyacinthus*, *Cyphastrea serailia*, *Echinophyllia aspera*, *Dipsastraea favus* (synonymous with *Favia favus*), *Favites valenciennesi* (synonymous with *Montastraea valenciennesi*), *Hydnophora exesa*, and *Platygyra contorta* (Table 7.6). All species spawned during the period of the highest annual water temperatures. The relationship between spawning and lunar phase differed from year to year, and *D. favus* tended to spawn 1 week later than all other species. Most of the above species spawned synchronously within species, but in 2003, *E. aspera* and *P. contorta* performed split spawning over a period of 2 months. *Alveopora japonica*, a brooding species, released planula larvae in the morning 3 days after the full moon in September (Thamrin et al. 2001), which is consistent with data from Tateyama, Chiba (see below).

### 7.2.3 Peripheral Region (33–35°N)

#### 7.2.3.1 Boso Peninsula, Izu Peninsula, and Izu Islands

Near the northern limit of coral distribution, some corals are found near Boso, Izu, and the Izu Islands. Thirty-two coral species have been reported in the vicinity of the Boso peninsula (Nature Conservation Bureau, Environmental Agency 1994). Forty-two species near Shimoda and Nakagi have also been recorded (Nishihira and Veron 1995), as well as 60 species in Uchiura Bay in Shizuoka (Sugiyama 1937, Minegishi and Ueno 1995) and 80 species at Miyake Island (Tribble and Randall 1986).

**Table 7.5** The spawning timing of corals in 1993 in Shirigai, Kochi, and 2002–2009 in Otuski, Kochi

(Moon phase columns: -6 -5 -4 -3 -2 -1 F 1 2 3 4 5 6 LQ -6 -5 -4 -3 -2 -1 N 1 2 3 4 5 6 FQ)

| Year | Species | | References | Types of observation |
|---|---|---|---|---|
| 1993 | *Depsastrea* | *speciosa* | van Woesik (1995) | F |
| | *Depsastrea* | *favus* | | F |
| | *Favites* | *abdita* | | F |
| | *Favites* | *flexuosa* | | F |
| | *Favites* | *halicora* | | F |
| | *Favites* | *pentagona* | | F |
| | *Goniastrea* | *pectinata* | | F |
| | *Paragoniastrea* | *australensis* | | F |
| | *Oulophyllia* | *crispa* | | F |
| | *Platygyra* | *deadalea* | | F |
| | *Montipora* | *mollis* | | F |
| | *Montipora* | *turgescens* | | F |
| | *Acropora* | *muricata* | | E |
| | *Acropora* | *japonica* | | E |
| | *Acropora* | *microphthalma* | | E |
| | *Acropora* | *hyacinthus* | | E |
| | *Acoropora* | *solitaryensis* | | E |
| | *Acropora* | *willisae* | | E |
| 2002 | *Acropora* | *muricata* | Mezaki et al. (2007) | F |
| | *Acropora* | *hyacinthus* | | F |
| 2003 | *Acropora* | *dendrum* | Mezaki et al. (2007) | F |
| | *Acropora* | *muricata* | | F |
| | *Acropora* | *hyacinthus* | | F |
| | *Acropora* | *japonica* | | F |
| | *Acropora* | aff. *samoensis* red type | | F |
| | *Acropora* | aff. *samoensis* white type | | F |
| | *Acropora* | *solitaryensis* | | F |
| 2004 | *Acropora* | *dendrum* | Mezaki et al. (2007) | F |
| | *Acropora* | *muricata* | | F |
| | *Acropora* | *hyacinthus* | | F |
| | *Acropora* | *hyacinthus* | | F |
| | *Acropora* | *japonica* | | F |
| | *Acropora* | aff. *samoensis* yellow type | | F |
| | *Acropora* | aff. *samoensis* white type | | F |
| | *Acropora* | *solitaryensis* | | F |
| | *Acropora* | *tumida* | | F |
| 2005 | *Acropora* | *muricata* | Mezaki et al. (2007) | F |
| | *Acropora* | *hyacinthus* | | F |
| | *Acropora* | *japonica* | | F |
| | *Acropora* | aff. *samoensis* yellow type | | F |
| | *Acropora* | *solitaryensis* | | F |
| | *Acropora* | *tumida* | | F |
| | *Cyphastrea* | *serailia* | | F |
| | *Favites* | *abdita* | | F |
| | *Favites* | *pentagona* | | F |
| | *Paragoniastrea* | *australiensis* | | F |
| | *Paragoniastrea* | *favulus* | | F |
| | *Favites* | *valenciennesi* | | F |
| | *Platygyra* | *contorta* | | F |
| | *Platygyra* | *sinensis* | | F |
| | *Echinophyllia* | *aspera* | | F |
| | *Hydonophora* | *exesa* | | F |
| 2006 | *Montipora* | *informis* | Mezaki et al. (2007) | F |
| | *Montipora* | *venossa* | | F |
| | *Acropora* | *hyacinthus* | | F |
| | *Acropora* | *hyacinthus* | * | F |
| | *Acropora* | *japonica* | | F |
| | *Acropora* | *japonica* | * | F |
| | *Acropora* | aff. *samoensis* yellow type | | F |
| | *Acropora* | *solitaryensis* | | F |
| | *Acropora* | sp. 1 | | F |
| | *Acropora* | *valida* | | F |
| | *Acropora* | *willisae* | | F |
| | *Cyphastrea* | *serailia* | | F |
| | *Cyphastrea* | *chalcidicum* | | F |
| | *Dipsastraea* | *favus* | | F |
| | *Favites* | *abdita* | | F |
| | *Favites* | *pentagona* | | F |
| | *Favites* | *pentagona* | | F |
| | *Favites* | *pentagona* | * | F |
| | *Favites* | *pentagona* | * | F |
| | *Coelastrea* | *aspera* | | F |
| | *Paragoniastrea* | *australiensis* | | F |
| | *Paragoniastrea* | *deformis* | | F |
| | *Paragoniastrea* | *deformis* | | F |
| | *Favites* | *valenciennesi* | | F |
| | *Platygyra* | *contorta* | | F |
| | *Platygyra* | *contorta* | | F |
| | *Platygyra* | *contorta* | * | F |
| | *Platygyra* | *sinensis* | | F |
| | *Echinophyllia* | *aspera* | | F |
| | *Micromussa* | *amakusensis* | | F |
| | *Hydonophora* | *exesa* | | F |
| | *Turbinaria* | spp | | F |
| 2007 | *Acropora* | *hyacinthus* | Nozawa (2012) | F |
| | *Acropora* | *japonica* | | F |
| | *Acropora* | *japonica* | | F |
| | *Favites* | *pentagona* | | F |
| | *Favites* | *pentagona* | | F |
| | *Platygyra* | *contorta* | | F |
| 2008 | *Acropora* | *hyacinthus* | Nozawa (2012) | F |
| | *Acropora* | *japonica* | | F |
| | *Favites* | *pentagona* | | F |
| | *Platygyra* | *contorta* | | F |
| 2009 | *Acropora* | *hyacinthus* | Nozawa (2012) | F |
| | *Acropora* | *japonica* | | F |
| | *Favites* | *pentagona* | | F |
| | *Platygyra* | *contorta* | | F |

References for data of each year were shown in right column. In 2006, asterisk means data from Nozawa (2012). *Top line* shows moon phase. *F* full moon, *LQ* last quarter moon, *N* new moon, *FQ* first quarter moon. Spawning months: June, July, August, September. Types of observation: *E* inferred spawning time based on examination of gonads, *F* field observation

**Table 7.6** The spawning timing of corals in 2001–2003 in Amakusa, Kumamoto

| Year | Species | | References | Types of observation |
|---|---|---|---|---|
| 2001 | *Acropora* | *hyacinthus* | Nozawa et al.(2006) | F |
| | *Cyphastrea* | *serailia* | | F |
| | *Echinophyllia* | *aspera* | | F |
| | *Dipsastraea* | *favus* | | F |
| | *Hydnophora* | *exesa* | | F |
| | *Favites* | *valenciennesi* | | F |
| 2002 | *Acropora* | *hyacinthus* | Nozawa et al.(2006) | F |
| | *Cyphastrea* | *serailia* | | F |
| | *Echinophyllia* | *aspera* | | F |
| | *Dipsastraea* | *favus* | | F |
| | *Favites* | *valenciennesi* | | F |
| | *Platygyra* | *contorta* | | F |
| 2003 | *Cyphastrea* | *serailia* | Nozawa et al.(2006) | F |
| | *Echinophyllia* | *aspera* | | F |
| | *Echinophyllia* | *aspera* | | F |
| | *Favites* | *valenciennesi* | | F |
| | *Platygyra* | *contorta* | | F |
| | *Platygyra* | *contorta* | | F |

All data were referred from Nozawa et al. (2006). *Top line* shows moon phase. *F* full moon, *LQ* last quarter moon, *N* new moon, *FQ* first quarter moon. Spawning months: June, July, August, september. Types of observation: *F* field observation

Coral reproduction has been studied only near Tateyama in the Boso Peninsula, which has focused on *Acropora pruinosa* (written as *A. tumida*) and *Alveopora japonica*. *A. pruinosa* was observed to spawn 130 min after sunset in August and September without any apparent relation to lunar phase, and no synchronous spawning was seen within this species (Hagiwara 2003). *Alv. japonica* released planula larvae with zooxanthellae in the morning in September and October (Harii 2001; Hagiwara 2003). At this site, average monthly water temperatures were highest in August (23.8 °C) and lowest in March (14.3 °C). From December to May, average monthly temperatures were lower than 18 °C.

### 7.2.4 Spawning Season (Month) for Japanese Corals

In general, *Acropora* spp. in the coral reef region of the Ryukyu archipelago tend to spawn 2 days before to 8 days after the full moon in May and/or June. However, some species (especially *A. divaricata*, *A. latistella*, and *A.* sp.1) spawned 1 month later than others. One exception was observed in the coral reef region of Ogasawara, where two species (*A. muricata* and *A. intermedia*) spawned at dates corresponding with the Ryukyu archipelago, while other *Acropora* species (such as *A. hyacinthus* and *A. florida*) spawned 2 months later, in August. On the other hand, in the non-coral reef region, most *Acropora* species spawned in July and/or August, although some (*A. muricata* and *A. hyacinthus*) sometimes spawned in late June. Finally, in the peripheral region, one *Acropora* species (*A. pruinosa*) spawned in August and September. Thus, the date of spawning becomes later with increasing latitude, i.e., May–June in the coral reef region, June–August in the non-coral reef region, and August–September in the peripheral region. Spawning date may be affected by water temperature and/or light intensity, which decrease at higher latitudes. Notably, one species with an arborescent colony form (*A. muricata*) spawned on similar dates in the coral reef and non-coral reef regions. It is possible that spawning in this species is genetically determined or may be influenced by environmental factors other than water temperature and light intensity. Unlike spawning date, the time of day at which spawning occurred was similar throughout all regions; therefore, the timing of spawning may be genetically controlled. There is one exception to this rule: since 1997, *A. japonica*, a species not specific to coral reefs, has spawned 1 or 2 h later in the day (Misaki 1997). At present, *A. japonica* spawns around 23:00–24:00 (Misaki 1997 in Wakayama and our observation in Miyazaki). Thus, the delay in the timing of spawning may be a recent evolution for this species.

For *Montipora*, a tendency to spawn at later dates at higher latitudes has also been observed. For example, *Montipora* spp. spawned from May to August in Okinawa but from August to September in Shikoku. In contrast, corals in the Merulinidae spawned on similar dates (July and August) in all regions, a pattern similar to that of *Acropora muricata*. The environmental cues for spawning date remain unknown, and further studies are required to elucidate the determinant mechanisms of coral spawning timing.

## 7.3 Fertilization and Hybridization of Corals in Japan

### 7.3.1 Fertilization Studies

Crossing experiments are the primary method for investigating coral fertilization, including sperm concentration and fertilization rates within and between species. Some studies have shown that to achieve intraspecific fertilization with close to 100 % success, the requisite sperm concentration is approximately $10^6$–$10^7$ sperm/ml (Willis et al. 1997; Marshall 2006; Nakamura et al. 2011; Nozawa et al. 2015). Omori et al. (2001) first measured sperm concentration for wild corals in Akajima, Okinawa. They revealed that within 30 min of mass spawning, sperm concentration was $10^6$ sperm/ml; concentration then gradually decreased over time. Iguchi et al. (2009) also reported that most gametes within a species achieve fertilization within 30 min, based on crossing of *Acropora digitifera* using a range of gamete contact times.

Using gametes of three *Acropora* species (*A. digitifera*, *A. gemmifera*, and *A. tenuis*) in Okinawa, Morita et al. (2006) found that coral sperm exhibited motility and was attracted to intraspecific eggs, but not to eggs from other species. No hybridization occurred among these three species in crossing experiments. In general, some evidences suggest that mechanisms exist to maintain reproductive isolation between species during mass spawning events.

Crossing experiments are useful to identify species boundaries and cryptic species. *Acropora* sp.1 is recognized as a cryptic species of *A. digitifera* because *A.* sp.1 spawns 1 month later than *A. digitifera*. Ohki et al. (2015) performed crossing experiments using cryopreserved sperm from each species and found high rates of hybrid fertilization, suggesting that differences in the timing of spawning are closely related to speciation. Similarly, Fukami and Nomura (2009) showed that *Favites valenciennesi* (written as *Montastraea valenciennesi*) contains a cryptic species that differs morphologically from the original species and exists in complete reproductive isolation. In contrast, Zayasu et al. (2009) showed that three morphological variants of *Acanthastrea hemprichii* belonged to the same species, since cross-fertilization among all three species was achieved with close to 100 % success in crossing experiments.

Crossing experiments are also used for efficient coral seeding. Using *Acropora tenuis* in Akajima, Iwao et al. (2014) showed that a mixture of gametes from more than six intraspecific colonies could achieve stable fertilization rates of greater than 95 %. Based on data for *A. intermedia* in Okinawa, Isomura et al. (2013a) reported a significant negative relationship between fertilization rate and genetic similarity. These studies may contribute to the effective seedling production of corals for conservation.

### 7.3.2 Hybridization Studies

Hybrid speciation, which is a part of the pattern of reticulate evolution proposed by Veron (1995), may be an underlying mechanism for the high species diversity of *Acropora* species. Synchronous spawning occurs among many *Acropora* species, and crossing experiments have shown that some species are able to hybridize in vitro (Wallace and Willis 1994; Willis et al. 1997; Hatta et al. 1999; Isomura et al. 2013b). Therefore, it is highly possible that hybridization between *Acropora* species occurs in nature. Indeed, it is reported that one out of every three *Acropora* species in the Caribbean is an $F_1$ hybrid (Vollmer and Palumbi 2002). In contrast, in the Indo-Pacific, there is no direct evidence to date for the existence of hybrids, although colonies with morphology intermediate between species have been reported (Hatta and Matshushima 2008; Isomura et al. 2013b).

In Japan, only a few studies on coral hybridization have been conducted. A total of eight *Acropora* species have been shown to hybridize in vitro (Hatta et al. 1999; Fukami et al. 2003). Three of these species (*A. tenuis*, *A. donei*, and *A. yongei*) are early-hour spawners, spawning 2–3 h earlier than most other species. Fukami et al. (2003) showed that the fertilization rate among these three species is over 90 %, although these three species might be in the morphological variation within a single species, considering the very high fertilization rates and genetic similarity among them (Fukami et al. 2003). The other five species, which spawn in synchrony, are *A. nasuta*-A, *A. nasuta*-C, *A. muricata*-A, *A. intermedia*, and *A. florida* (Hatta et al. 1999). *A. nasuta*-A and *A. nasuta*-C are treated as different species since while superficially similar in morphology, they have certain key morphological differences. Hatta et al. (1999) showed that hybridization occurs between *A. nasuta*-A, *A. nasuta*-C, and *A. muricata*-A and between *A. intermedia* and *A. florida*, with various fertilization rates. Later, hybridization between *A. intermedia* and *A. florida* was studied in detail by Isomura et al. (2013b). They revealed the existence of a colony with morphology intermediate between these two species in the field and concluded that in situ fertilization between the putative hybrid, *A. intermedia* and *A. florida*, is highly probable. Moreover, Isomura et al. (2016) revealed that the $F_1$ hybrid colonies, produced experimentally from *A. intermedia* and *A. florida*, have the high fecundity and high self-fertilization. This finding suggests that some products of interspecific hybridization may persist as the offspring of self-fertilizing $F_1$ hybrids.

## 7.4 Future Perspectives for Japanese Coral Reproduction Research

More than 400 coral species exist in Japan. However, the dates and timing of spawning have only been investigated in around 20 species. Moreover, studies on coral fertilization (sperm concentration and fertilization rates) are limited to more or less ten species (refer to Nozawa et al. 2015). Such studies are critical in order to elucidate regional differences among species and to understand the relationship between coral reproduction and environmental changes, such as climate change and ocean acidification. Morita et al. (2010) showed that sperm flagellar motility was seriously impaired in *A. digitifera* below pH 7.7, of which the oceans are expected to reach within the next 100 years. Iguchi et al. (2015) suggested that coral fertilization is more sensitive to global warming than ocean acidification in the reef-building coral *A. digitifera*. In Schutter et al. (2015), global warming had a more significant impact than ocean acidification on early embryonic development in *A. hyacinthus* in Kochi, Japan. Moreover, Paxton et al. (2015) suggested that higher temperatures might reduce fertilization rate and delay spawning date in *A. digitifera*. Thus, continual monitoring of corals across all regions of Japan is required in order to assess the effects of climate change on different species.

Aquariums in Japan represent a critical resource for researchers. As mentioned above, many field and/or aquarium surveys of coral reproduction have been carried out (Kagoshima Aquarium; Dewa et al. 2015, Kushimoto Marine Park; Misaki 1989, 1994, 1995a, b, 1996, 1997, 1998a, b, 1999, 2000, 2001, 2003, 2004, 2005, 2006a, b, 2007, 2008). In Okinawa Churaumi Aquarium, experimental coral studies have been conducted focusing on metamorphosis and acquisition of symbiotic algae in juvenile corals (Hirose et al. 2008), reproduction in cultured versus wild *Acropora* colonies (Okubo et al. 2010), and the relationship between genetic similarity and reproductive success in cultured *Acropora* colonies (Isomura et al. 2013a). Aquarium-based research on coral reproduction is likely to become even more important for conservation in the future.

Until date, a wide range of studies on coral reproduction has been carried out in Japan. However, given that more than 400 corals species inhabit Japanese waters, at present, we have relatively limited information about coral reproduction. Currently, the environment is changing rapidly owing to factors such as global warming and anthropogenic disturbances. Further studies on coral ecology and reproduction are more required in order to understand how corals are affected by such climate changes and to minimize future adverse impacts on coral biodiversity.

**Acknowledgments** We are grateful to Y. Nozawa for reviewing our MS and giving us very useful comments. And we thank T. Mezaki, M. Hirose, and S. Harii for providing us with information of coral reproduction in Japan including their papers. This study was partially supported by JSPS KAKENHI (#26440225) to N.I.

## References

Dewa N, Tsuchida H, Nishida K, Yamada M, Hirose J, Yamaki A, Chikuchisin M, Higashimine M (2015) First report of spawning by four scleractinian corals in Kagoshima Bay. Nat Kagoshima 41:311–313. (in Japanese)

Fukami H (2016) Taxonomy and phylogeny: recent taxonomic revision of the zooxanthellate scleractinian corals. Biol Sci 67:201–215. (in Japanese)

Fukami H, Nomura K (2009) Existence of a cryptic species of *Montastraea valenciennesi* (Milne Edwards and Haime, 1848) in Wakayama, southern Honshu, Japan. J Jpn Coral Reef Soc 11:25–31. (in Japanese with English abstract)

Fukami H, Omori M, Shimoike K, Hayashibara H, Hatta M (2003) Ecological and genetic aspects concerned with reproductive isolation by differential timing of spawning in *Acropora* corals. Mar Biol 142:679–684

Hagiwara R (2003) Distribution, environmental conditions and sexual reproduction of the zooxanthellae corals at the Banda area of Tateyama Bay, Chiba, Japan. Midoriishi 14:24–30. (in Japanese)

Harii S, Omori M, Yamakawa H, Koike Y (2001) Sexual reproduction and larval settlement of the zooxanthellate coral *Alveopora japonica* Eguchi at high latitudes. Coral Reefs 20:19–23

Harrison PL (2011) Sexual reproduction of scleractinian corals. In: Dubinsky Z, Stambler N (eds) Coral reefs: an ecosystem in transition. Springer, The Netherlands, pp 59–85

Harrison PL, Wallace CC (1990) Reproduction, dispersal and recruitment of scleractinian corals. In: Dubinsky Z (ed) Coral reefs. Elsevier, Amsterdam, pp 133–207

Hatta M, Matsushima K (2008) Presumed natural hybrids between *Acropora donei* and *Acropora tenuis*. Galaxea 10:91

Hatta M, Fukami H, Wang W, Omori M, Shimoike K, Hayashibara T, Ina Y, Sugiyama T (1999) Reproductive and genetic evidence for a reticulate evolutionary history of mass spawning corals. Mol Biol Evol 16:1607–1613

Hayashibara T (1995) Ecological studies on reef-building corals and their sexual reproduction around Akajima Island, Kerama Islands Group. Ph. D. thesis, Tokyo University of Fisheries, Tokyo, 123 pp (in Japanese)

Hayashibara T, Shimoike K (2002) Cryptic species of *Acropora digitifera*. Coral Reefs 21:224–225

Hayashibara T, Shimoike K, Kimura T, Hosaka S, Heyward A, Harrison P, Kudo K, Omori M (1993) Patterns of coral spawning at Akajima Island, Okinawa, Japan. Mar Ecol Prog Ser 101:253–262

Heyward A, Yamazato K, Yeemin T, Minei M (1987) Sexual reproduction in Okinawa. Galaxea 6:331–343

Hirose M, Hidaka M (2006) Early development of zooxanthella-containing eggs of the corals *Porites cylindrica* and *Montipora digitata* with special reference to the localization of zooxanthellae to the endoderm. Zool Sci 23:873–881

Hirose M, Kinzie RA, Hidaka M (2000) Early development of zooxanthella-containing eggs of the corals *Pocillopora verrucosa* and *P. eydouxi* with special reference to the distribution of zooxanthellae. Biol Bull 199:68–75

Hirose M, Yamamoto HH, Nonaka M (2008) Metamorphosis and acquisition of symbiotic algae in planula larvae and primary polyps of *Acropora* spp. Coral Reefs 27:247–254

Iguchi A, Morita M, Nakajima Y, Nishikawa A, Miller DJ (2009) In vitro fertilization efficiency in coral *Acropora digitifera*. Zygote. https://doi.org/10.1017/S096719940900519X

Iguchi A, Suzuki A, Sakai K, Nojiri Y (2015) Comparison of the effects of thermal stress and $CO_2$-driven acidified seawater on fertilization in coral *Acropora digitifera*. Zygote 23:631–634

Inaba M (2003) Ecological feature and status of reef-building corals in the Bonin Islands, Japan. Midoriishi 14:20–23. (in Japanese)

Isomura N, Baba Y, Nagata S, Nonaka M, Yamamoto HH (2013a) The relationship between genetic similarity and reproductive success in the branching coral *Acropora intermedia*. Mar Biol Res 9:181–188

Isomura N, Iwao K, Fukami H (2013b) Possible natural hybridization of the coral genus *Acropora*: fertilization and larva survival rates. PLoS One. https://doi.org/10.1371/journal.pone.0056701

Isomura N, Iwao K, Morita M, Fukami H (2016) Spawning and fertility of $F_1$ hybrids of the coral genus *Acropora* in the Indo-Pacific. Coral Reefs. https://doi.org/10.1007/s00338-016-1461-9

Iwao K, Wada N, Ohdera A, Omori M (2014) How many donor colonies should be cross-fertilized for nursery farming of sexually propagated corals? Nat Res 5:521–526

Kinzie RA (1993) Spawning in the reef corals *Pocillopora verrucosa* and *P. eydouxi* at Sesoko Island, Okinawa. Galaxea 11:93–105

Loya Y, Sakai K, Yamazato K, Nakano Y, Sambali H, van Woesik R (2001) Coral bleaching: the winners and the losers. Ecol Lett. https://doi.org/10.1046/j.1461–0248.2001.00203.x

Loya Y, Sakai K, Heyward A (2009) Reproductive patterns of fungiid corals in Okinawa, Japan. Galaxea 11:119–129

Marshall D (2006) Reliably estimating the effect of toxicants on fertilization success in marine broadcast spawners. Mar Pollut Bull 52:734–738

Mezaki T, Hayashi T, Iwase F, Nakachi S, Nozawa Y, Miyamoto M, Tominaga M (2007) Spawning patterns of high latitude scleractinian corals from 2002 to 2006 at Nishidomari, Otsuki, Kochi, Japan. Kuroshio Biosphere 3:33–47. (in Japanese with English abstract)

Mezaki T, Keshavmurthy S, Chen CA (2014) An old and massive colony of *Pavona decussata* is sexually active at high latitude (32°N) in Japan. Coral Reefs 33:97

Minegishi N, Ueno S (1995) Ecological distribution of scleractinian corals and their abiotic environment in the Uchiura Coast of the Suruga Bay. Bull Inst Oceanic Res & Develop, Tokai Univ 16:31–40. (in Japanese with English abstract)

Misaki H (1989) Spawning of scleractinian corals in Kushimoto. Mar Pavilion 18(10):2–3. (in Japanese)

Misaki H (1994) Spawning of scleractinian corals in Kuroshima. Mar Pavilion 23(7):4–5. (in Japanese)

Misaki H (1995a) Spawning and settlement of juveniles of scleractinian corals in 1994. Mar Pavilion 24(5):2–3. (in Japanese)

Misaki H (1995b) Spawning and settlement of juveniles of scleractinian corals in 1995. Mar Pavilion 24(11):2–3. (in Japanese)

Misaki H (1996) Spawning and settlement of juveniles of scleractinian corals in 1996. Mar Pavilion 25(12):14–15. (in Japanese)

Misaki H (1997) Spawning and settlement of juveniles of scleractinian corals in 1997. Mar Pavilion 26(11):2–3. (in Japanese)

Misaki H (1998a) Reproduction of corals under aquarium condition. Mar Pavilion 27(7):4. (in Japanese)

Misaki H (1998b) Spawning and settlement of juveniles of scleractinian corals in 1998. Mar Pavilion 27(12):2. (in Japanese)

Misaki H (1999) Spawning of scleractinian corals in 1999. Mar Pavilion 18(11):2–3. (in Japanese)

Misaki H (2000) Spawning of scleractinian corals in 2000. Mar Pavilion 29(12):2. (in Japanese)

Misaki H (2001) Spawning of scleractinian corals in 2001. Mar Pavilion 30(12):2–3. (in Japanese)

Misaki H (2003) Spawning of scleractinian corals in 2002. Mar Pavilion 32(4):2–3. (in Japanese)

Misaki H (2004) Spawning of scleractinian corals in 2003. Mar Pavilion 33(1):2–3. (in Japanese)

Misaki H (2005) Spawning of scleractinian corals in 2004. Mar Pavilion 34(3):2–3. (in Japanese)

Misaki H (2006a) Spawning of scleractinian corals in 2005. Mar Pavilion 35(1):2–3. (in Japanese)

Misaki H (2006b) Spawning of scleractinian corals in 2006. Mar Pavilion 35(6):2–3. (in Japanese)

Misaki H (2007) Spawning of scleractinian corals in 2007. Mar Pavilion 36(5):2–3. (in Japanese)

Misaki H (2008) Spawning of scleractinian corals in 2008. Mar Pavilion 37(6):2–3. (in Japanese)

Miyamoto M, Yokochi H (2015) Seasonal and daily planulation patterns of two pocilloporid corals at a high-latitude habitat in Southwest Kochi, Japan. J Jpn Coral Reef Soc 17:1–17. (in Japanese with English abstract)

Morita M, Nishikawa A, Nakajima A, Iguchi A, Sakai K, Takemura A, Okuno M (2006) Eggs regulate sperm flagellar motility initiation, chemotaxis and inhibition in the coral *Acropora digitifera*, *A gemmifera* and *A tenuis*. J Exp Biol. https://doi.org/10.1242/jeb.02500

Morita M, Suwa R, Iguchi A, Nakamura M, Shimada K, Sakai K, Suzuki A (2010) Ocean acidification reduces sperm flagellar motility in broadcast spawning reef invertebrates. Zygote 18:103–107

Murakami T, Kawano H, Yamamoto Y, Mizutani A, Shimokawa S (2015) Numerical simulation for initial dynamic state of bundle based on field observations of Acroporidae spawning in Amitori Bay, Iriomote Island, Japan. J JSCE 71: I_1225-I_1230 (in Japanese with English abstract)

Nakajima Y, Nishikawa A, Iguchi A, Sakai K (2012) The population genetic approach delineates the species boundary of reproductively isolated corymbose acroporid corals. Mol Phylogenet Evol 63:527–531

Nakamura R, Ando W, Yamamoto H, Kitano M, Sato A, Nakamura M, Kayanne H, Omori M (2011) Corals mass-cultured from eggs and transplanted as juveniles to their native, remote coral reef. Mar Ecol Prog Ser 436:161–168

Nature Conservation Bureau, Environmental Agency (1994) The report of the marine biotic environment survey in the 4th national survey on the natural environment. 3: Coral reefs (in Japanese)

Nishihira M, Veron JEN (1995) Hermatypic corals of Japan. Kaiyusha, Tokyo. 440 p. (in Japanese)

Nomura K (2004) Coral communities in Kushimoto (13). Marine Pavilion 32:5. (in Japanese)

Nomura K, Mezaki T (2005) Reef building corals from Otsuki, Kochi Prefecture, Japan. Kuroshio Biosph 2:29–41. (in Japanese with English abstract)

Nomura K, Kodera A, Ui S, Mori M, Uchia H (2003) Survey report on state of Kushimoto Marine Park Zone. Report on conservational activities in marine park zones, in fiscal 2002. Nature Conservation Bureau, Ministry of the Environment, pp 25–27, 43–66, 89–96 (in Japanese)

Nozawa Y (2012) Annual variation in the timing of coral spawning in a high-latitude environment: influence of temperature. Biol Bull 222:192–202

Nozawa Y, Harrison PL (2005) Temporal settlement patterns of larvae of the broadcast spawning reef coral *Favites chinensis* and the broadcast spawning and brooding reef coral *Goniastrea aspera* from Okinawa, Japan. Coral Reefs 24:274–282

Nozawa Y, Nojima S, Tokeshi M (2006) Reproduction and recruitment of scleractinian corals in a high-latitude coral community, Amakusa, southwestern Japan. Mar Biol 49:1047–1053

Nozawa Y, Isomura N, Fukami H (2015) Influence of sperm dilution and gamete contact time on fertilization rate of scleractinian corals. Coral Reefs. https://doi.org/10.1007/s00338-015-1338-3

Ohki S, Kowalsk RK, Kitanobo S, Morita M (2015) Changes in spawning time led to the speciation of the broadcast spawning corals Acropora digitifera and the cryptic species Acropora sp. 1 with similar gamete recognition systems. Coral Reefs. https://doi.org/10.1007/s00338-015-1337-4

Okubo N, Isomura N, Hidaka M, Motokawa T (2007) Possible self-fertilization in the brooding coral *Acropora* (*Isopora*) *brueggemanni*. Zool Sci 24:277–280

Okubo N, Yamamoto HH, Nakaya F, Okaji K (2010) Reproduction in cultured versus wild coral colonies: fertilization, larval oxygen consumption, and survival. Biol Bull 218:230–236

Omori M, Fukami H, Kobinata H, Hatta M (2001) Significant drop of fertilization of Acropora orals in 1999: an after-effect of heavy coral bleaching? Limnol Oceanogr 46:704–706

Paxton CW, Baria MVB, Weis VM, Harii S (2015) Effect of elevated temperature on fecundity and reproductive timing in the coral *Acropora digitifera*. Zygote. https://doi.org/10.1017/S0967199415000477

Prasetia R, Sinniger F, Harii S (2015) Gametogenesis and fecundity of *Acropora tenella* (Brook 1892) in a mesophotic coral ecosystem in Okinawa. Japan Coral Reefs. https://doi.org/10.1007/s00338-015-1348-1

Sakai K (1997) Gametogenesis, spawning, and planula brooding by the reef coral *Goniastrea aspera* (Scleractinia) in Okinawa, Japan. Mar Ecol Prog Ser 151:67–72

Schutter M, Nozawa Y, Kurihara H (2015) The effect of elevated $CO_2$ and increased temperature on in vitro fertilization success and initial embryonic development of single male: female crosses of broad-cast spawning corals at mid- and high-latitude locations. J Mar Sci Eng 3:216–239

Shimoike K (1999) Observation of coral spawning in Akajima Island with newly found spawning patterns. Midoriishi 10:29–31. (in Japanese)

Sugiyama T (1937) Reef building corals in coastal areas of Japan. Geol Paleontol Lab, Dept Sci, Tohoku Univ 26:5–15. (in Japanese)

Tachikawa H, Suganuma H, Sato F (1991) Reef-building corals in the Chichi-jima Islands and the Haha-jima Islands. Report of the second general survey on natural environment of the Ogasawara (Bonin) Islands, Tokyo Metropolitan University, pp 285–296 (in Japanese)

Thamrin S, Nojima S, Tokeshi M (2001) Experimental analysis of planula release in a scleractinian coral *Alveopora japonica*. Galaxea 3:25–30

Tribble GW, Randall RH (1986) A description of the high-latitude shallow water coral communities of Miyake-jima, Japan. Coral Reefs 4:151–160

van Woesik R (1995) Coral communities at high latitude are not pseudopopulations: evidence of spawning at 32°N, Japan. Coral Reefs 14:119–120

van Woesik R, Sakai K, Ganase A, Loya Y (2011) Revisiting the winners and the losers a decade after coral bleaching. Mar Ecol Prog Ser 434:67–76

Veron JEN (2000a) Corals of the world, vol 1. Aust Inst Mar Sci, Townsville. 463 pp

Veron JEN (2000b) Corals of the world, vol 2. Aust Inst Mar Sci, Townsville. 429 pp

Veron JEN (2000c) Corals of the world, vol 3. Aust Inst Mar Sci, Townsville. 490 pp

Volmer SV, Palumbi SR (2002) Hybridization and the evolution of reef coral diversity. Science 296:2023–2025

Wallace CC, Willis BL (1994) The systematics of *Acropora*: the effect of new biological findings on species concepts. Annu Rev Ecol Syst 25:237–262

Willis BL, Babcock RC, Harrison PL, Wallace CC (1997) Experimental hybridization and breeding incompatibilities within the mating systems of mass spawning reef corals. Coral Reefs 16:S53–S65

Yamazato K, Suwardi E, Sultana S (2008) Reproductive cycle of brooding corals at high latitude. J Jpn Coral Reef Soc 10:1–11. (in Japanese with English abstract)

Zayasu Y, Nomura K, Suzuki G, Shirayama Y, Fukami H (2009) Are the morphological variations of the living colonies of *Acanthastrea hemprichii* (Ehrenberg, 1834) in Wakayama prefecture inter- or intraspecific variations? J Jpn Coral Reef Soc 11:33–37. (in Japanese with English abstract)

Zayasu Y, Miyazaki K, Lien YT, Okubo N (2015) Direct evidence of sexual reproduction in the zebra coral, Oulastrea crispata (Anthozoa, Scleractinia), in Japan. Invert Rep Dev. https://doi.org/10.1080/07924259.2015.1006340

# Population Genetics of Corals in Japan

# 8

Yuichi Nakajima

**Abstract**

Understanding how coral populations are established and maintained is important to predict how coral reef ecosystems will respond to future conditions. Population genetic analyses using DNA markers have provided useful information on how coral populations are maintained. In this chapter, I briefly introduce the history of using DNA markers in the population genetic analyses of corals. I also explain the merit of population genetic analyses to delineate the species boundaries of corals and infer how reproductive characteristics contribute to connectivity among populations. Based on previous studies on population genetic analyses of corals, I also discuss how population genetic analyses have contributed toward understanding the patterns of connectivity among coral populations and geographic variations in genetic diversity, primarily focusing on examples along the Ryukyu Archipelago. Finally, I propose future directions for the population genetics of corals in Japan, taking several aspects into consideration, including methodological information, such as seascape genetics and the development of novel molecular markers for utilizing next-generation sequencing technologies.

**Keywords**

Coral recovery • Clonality • DNA marker • Genetic diversity • Genetic structure • Population connectivity • Species diversity

## 8.1 History of Coral Population Genetics in Japan

Recent population genetic studies based on genotypic information using molecular markers have investigated genetic diversity and connectivity, clonality, as well as species diversity such as cryptic species and adaptations in coral populations (e.g., Hedgecock et al. 2007; Lowe and Allendorf 2010). Alleles, and their frequencies, vary depending on the extent of differentiation among populations caused by historical and ecological events. Gene flow is also useful for evaluating the expression of population dynamics; for example, low gene flow among sites suggests, possible, restricted larval recruitment, indicating little or no chance of long-distance larval dispersal through multiple generations of a given species. Population genetic studies are useful for estimating reproductive features and past events based on the extent of genetic differentiation in populations from different habitats. For example, if large genetic differentiation is found between pairwise sites, the mean larval distance might be shortened, even on a small geographic scale. This phenomenon is expected in brooding corals, which release developed planula larvae during the spawning season (Fig. 8.1). In contrast, species that exhibit broadcast spawning appear to maintain genetic connectivity by stochastic long-distance dispersal (Fig. 8.1). Multilocus genotyping also provides information on the extent of clonality in coral populations. If clonal diver-

Y. Nakajima (✉)
Okinawa Institute of Science and Technology Graduate University, Tancha, Onna, Okinawa 904-0495, Japan
e-mail: yuichi.nakajima@oist.jp; yuichi.nakajima@outlook.jp


A. Iguchi, C. Hongo (eds.), *Coral Reef Studies of Japan*, Coral Reefs of the World 13,
https://doi.org/10.1007/978-981-10-6473-9_8

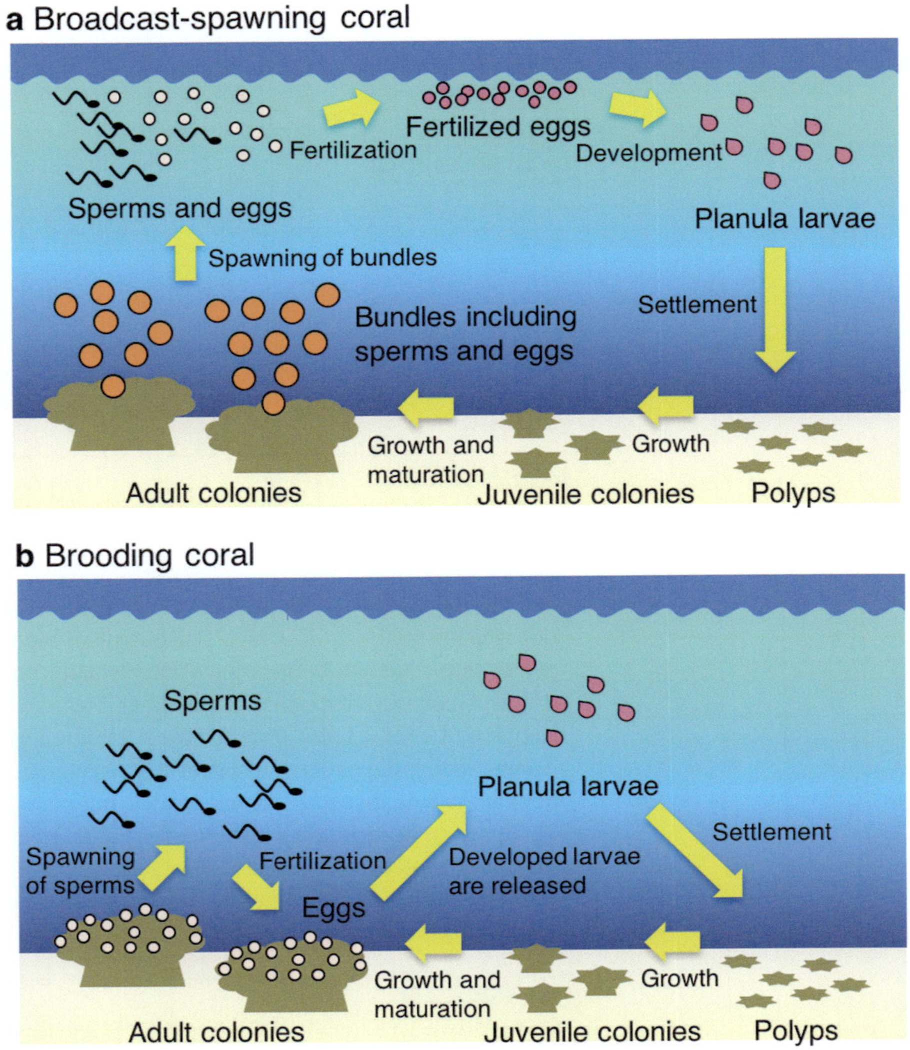


**Fig. 8.1** Schematic of the life cycle and reproductive mode for typical hermaphroditic reef-building corals. Species with larvae that have long dispersal distances appear to maintain high gene flow among reefs. (**a**) Broadcast-spawning coral. Bundles containing both sperm and eggs are released to the water column, and the sperm or eggs fertilize the eggs or sperm of other colonies. Larvae develop on the sea surface and settle on the seafloor. (**b**) Brooding coral. The released sperm fertilize the eggs of other colonies. The larvae that develop are released to the water column and settle on the seafloor

sity is low (i.e., high rate of clonemates) in a population, the effective population size is smaller than it appears, and the population is considered susceptible to environmental changes. Restricted population size promotes inbreeding and causes genetic diversity to decline.

Mitochondrial DNA usually has higher mutation rates than nucleic DNA and has been used as a genetic marker in population genetic studies focusing on how genotypes vary among individuals and their frequency among sites. However, this type of analysis is not suitable for coral species because of their low mutation rates (van Oppen et al. 1999, 2001; Fukami et al. 2000). Instead, allozymes and microsatellites have been frequently used for population genetic studies of corals. Allozymes (or isozymes) are enzymes that possess polymorphic characterization but lack functional differentiation. The allozyme locus is comparatively conserved among species; in contrast, the microsatellite (or single sequence repeat) is a highly polymorphic DNA marker, which is more convenient for genetic analyses because DNA is more stable than proteins such as enzymes. In addition, there are numerous microsatellite loci because they are widely distributed on genomic DNA. These high-resolution markers provide sufficient variability not only among species but also among individuals. Therefore, microsatellite loci are species specific and are often useful for estimating genetic differentiation among related species. Null alleles and size homoplasy might sometimes cause scoring errors; thus, suitable markers must be selected for the proper analysis of population genetic studies using microsatellite loci.

Most population genetic studies of corals have been conducted in the Great Barrier Reef (GBR), Australia (e.g., Ayre and Dufty 1994; Ayre et al. 1997; Ayre and Hughes 2000, 2004; Miller and Ayre 2008a; van Oppen et al. 2008, 2011). In addition, Baums et al. (2012) reported a population genetic study targeting *Porites lobata* across a wide geographic scale (approximately 18,000 km) in the Pacific. In Japan and the northwestern Pacific, studies on the population genetic analyses of corals remain limited; however, some studies have been reported in the Ryukyu Archipelago and the Nansei Islands (including the Osumi Islands) (see Table 8.1).

**Table 8.1** A list of principal studies in Japan conducted to determine the genetic diversity and/or differentiation of reef-building corals

| Species | Region | Marker | Scale (km) | Clonal diversity (*Ng*/*N*) | Genetic diversity | Genetic differentiation ($F_{ST}$) | Reference |
|---|---|---|---|---|---|---|---|
| *Acropora* | | | | | | | |
| *digitifera* | Ryukyu Archipelago | Allozyme | <600 | 0.81–0.96 | n.a. | 0.006–0.020 | Nishikawa and Sakai (2005a) |
| *digitifera* | Sekisei Reef | Microsatellite | <25 | 0.91–1.00 | n.a. | −0.006–0.015 | Nakajima et al. (2009b) |
| *digitifera* | Nansei Is. | Microsatellite | <1000 | 0.91–1.00 | Not decreased in peripheral population | −0.007–0.033 | Nakajima et al. (2010) |
| *digitifera* | Ryukyu Archipelago | SNPs | <530 | n.a. | Not decreased with latitude | −0.0007–0.0138 | Shinzato et al. (2015) |
| sp. 1 aff. *digitifera* | Okinawa Is.–Ogasawara Is. | Microsatellite | <1500 | 0.93–1.00 | Not decreased in isolated population | 0.001–0.043 | Nakajima et al. (2012a) |
| *papillare* | Okinawa–GBR–West Australia | Microsatellite | <5800 | 0.93–1.00 | Decreased in Okinawa | n.a. | Richards and van Oppen (2012) |
| *tenuis* | Ryukyu Archipelago | Allozyme | <600 | 0.94–1.00 | n.a. | 0.011–0.057 | Nishikawa et al. (2003) |
| *tenuis* | Ryukyu Archipelago | Microsatellite | <750 | 1.00 | Not decreased with latitude | 0.000–0.079 | Zayasu et al. (2016) |
| *Coelastrea* | | | | | | | |
| *aspera* | Ryukyu Archipelago | Allozyme | <500 | 0.82–1.00 | n.a. | 0.025–0.072 | Nishikawa and Sakai (2003) |
| *aspera* | Okinawa Is. | Allozyme | <200 | 0.92–1.00 | n.a. | 0.026 ($F_{CT}$), 0.039 ($F_{SC}$) | Nishikawa and Sakai (2005b) |
| *Galaxea* | | | | | | | |
| *fascicularis* (mt-L) | Okinawa Is. | Microsatellite | <1 (1 site) | 0.189 (*R*) | n.a. | 0.080 between mt-L and mt-S | Nakajima et al. (2015) |
| *fascicularis* (mt-S) | | | | 0.159 (*R*) | n.a. | | |
| *fascicularis* (mt-L) | Ryukyu Archipelago | Microsatellite | <750 | 0.00–1.00 (*R*) | Not decreased with latitude | 0.009–0.024 | Nakajima et al. (2016) |
| *fascicularis* (mt-S) | | | | 0.00–1.00 (*R*) | Not decreased with latitude | 0.009–0.032 | |
| *Heliopora* | | | | | | | |
| *coerulea* | Ryukyu Archipelago | Microsatellite | <500 | Clone in a site | Clone in a site | n.a. | Yasuda et al. (2012) |
| *coerulea* | Ryukyu Archipelago–Philippines | Microsatellite | <1900 | 0.03–1.00 | Not decreased with latitude | 0.002–0.499 | Yasuda et al. (2014) |
| *Plesiastrea* | | | | | | | |
| *versipora* | Ryukyu Archipelago–East Australia | ITS | <7500 | n.a. | n.a. | 1.24–3.42% (K2P distance) | Rodriguez-Lanetty and Hoegh-Guldberg (2002) |
| *Pocillopora* | | | | | | | |
| *damicornis* | Ryukyu Archipelago | Allozyme | <500 | 0.47–0.75 | n.a. | 0.027–0.092 | Adjeroud and Tsuchiya (1999) |
| *Seriatopora* | | | | | | | |
| Seriatopora-A | Ryukyu Archipelago | Microsatellite | <490 | 0.75–1.00 (*R*) | Lower than Seriatopora-B | n.a | Nakajima et al. (2017) |
| Seriatopora-B | | | | 0.91–1.00 (*R*) | Higher than Seriatopora-A and C | 0.064–0.116 | |
| Seriatopora-C | | | | 1.00 (*R*) | Lower than Seriatopora-B | n.a | |
| *Stylophora* | | | | | | | |
| *pistillata* | West Pacific Ocean | ITS | <6200 | n.a. | n.a. | 3.6 % (mean K2P distance) | Takabayashi et al. (2003) |
| *pistillata* | Ryukyu Archipelago. | Allozyme | <500 | 0.51–0.94 | n.a. | 0.070–0.260 | Nishikawa et al. (2003) |

The scale is the maximum geographic distance (km) between sampling sites and was measured by Google Earth, based on location information provided by the sources

The Nansei Islands are located in the subtropical region of Japan and are at a relatively high latitude compared to the "coral triangle" that represents the core habitat of tropical coral species. Many tropical coral species occur at the northern edge of the Nansei Islands. The Tanegashima Island area marks the boundary between the subtropical and temperate zones, with a distinct transition in species composition. The minimum seawater temperature required for coral reefs to form is around 18 °C, which occurs around the Tanegashima Island in Japan (e.g., Yamano 2008). Although corals also inhabit the temperate zone, they do not form reef there. Japan has a high-latitude coral reef formation around the southern part of the mainland, because the Kuroshio Current flows from the south to the north of the Nansei Islands. Seawater temperature in this region is comparatively higher than in other areas of similar latitude.

Several population genetic studies have been conducted on coral species in Japan. Studies on clonality, genetic diversity, and differentiation are listed in Table 8.1. The first published research was conducted by Adjeroud and Tsuchiya (1999). This study focused on *Pocillopora damicornis* in Okinawa, Kerama, and Ishigaki in the Ryukyu Archipelago, using allozyme electrophoresis. Subsequently, Nishikawa et al. (2003) provided a much-needed boost to the population genetic studies of corals in Japan, by analyzing multiple species using allozymes. This study included both broadcast-spawning (*Acropora tenuis*) and brooding species (*Stylophora pistillata*) and identified the negative relationship between larval duration and genetic differentiation among populations. Specifically, brooding species showed genetic differentiation among sites, whereas the broadcast-spawning species did not (see also Sect. 8.3). This comparison was the first attempt to determine the variability in the ecological and genetic traits of corals in Japan. Although allozyme electrophoresis was mainly used to determine the population genetics of corals in the early 2000s, the development of microsatellites caused a rapid advance in this field of research. Used as cross-species markers, the microsatellites developed from *Acropora palmata*, *Acropora millepora*, and *Acropora digitifera* (Baums et al. 2005; van Oppen et al. 2007; Nakajima et al. 2009a) proved useful for other neighboring *Acropora* species. Nakajima et al. (2010) reported significant differentiation among many sites and also detected a high level of genetic connectivity for *A. digitifera* in the Nansei Islands using microsatellite markers (see Sect. 8.4). In the case of *Acropora* sp. 1 aff. *digitifera*, which was reported as a cryptic species of *A. digitifera* (Hayashibara and Shimoike 2002), significant genetic differentiation was detected between the Okinawa and Ogasawara regions (see Sect. 8.4). Not only was there a genetic differentiation among regions but also unexpected speciation, with adequate species diversity in Japanese corals also being detected with microsatellite markers (*Acropora*, Nakajima et al. 2012b; *Galaxea*, Nakajima et al. 2015, 2016; *Heliopora*, Yasuda et al. 2014; *Seriatopora*, Nakajima et al. 2017) (see Sect. 8.2). Furthermore, population genetics is helpful for the estimation of clonality and is important for understanding reproductive strategies during population recovery. Some coral species are indicative of the high extent of clonal diversity that exists in this group (*Galaxea*, Nakajima et al. 2015, 2016; *Heliopora*, Yasuda et al. 2012) (see Sect. 8.3). In the following sections, I discuss species delimitation, reproductive characteristics, genetic differentiation, and the future of population genetics, with a focus on coral populations of Japan.

## 8.2 Species Delimitation and Genetic Lineages of Corals

The population genetics approach is useful for elucidating patterns in speciation. In general, animal species identification based on genetic evidence is extrapolated from phylogenetic studies using mitochondrial markers via DNA barcoding. However, in the case of corals, such markers are not available due to their slow mutation rates (see Sect. 8.1 for more details). Therefore, coral species identification by molecular markers is ambiguous in many cases. However, recent studies using microsatellite markers were able to differentiate the cryptic species of corals in Japan. Two sympatric *Acropora* species, *A. digitifera* and its cryptic species *Acropora* sp. 1 aff. *digitifera*, also showed significant genetic differentiation, with two genetic clusters being detected (Nakajima et al. 2012b). These two genetic clusters were probably established through reproductive isolation caused by differences in the spawning season in Japan, i.e., May–June for *A. digitifera* and August for *Acropora* sp. 1 aff. *digitifera* (Hayashibara and Shimoike 2002; Ohki et al. 2015). *Acropora hyacinthus* also had four genetic lineages based on the mitochondrial putative control region, of which one is the dominant lineage in a temperate area (Suzuki et al. 2016). *Acropora* is a highly diversified genus of corals (with 134 recognized species globally; see Richards et al. 2013 and references therein). Some *Acropora* species might have produced hybridized colonies in the field (e.g., Isomura et al. 2016; Kitanobo et al. 2016). Therefore, further studies are expected to produce more detailed delimitations on the complex patterns of the species in this genus.

The brooding octocoral species *Heliopora coerulea* was also divided into two genetic lineages, based on branch morphotypes (small-branch shape and flat shape) in Japan and the Philippines. These morphotypes are related to the genotypic composition of a nuclear gene (internal transcribed spacer region of nuclear ribosomal DNA; ITS) and microsatellite markers (Yasuda et al. 2014, 2015). These two genetic lineages were probably established through reproductive isolation, based on planula being released at different times (difference of approximately a month) within a site of the Philippines (Villanueva 2015). Seasonal differences in the release of planula between lineages are expected to be confirmed in Japan during future surveys.

The species *Galaxea fascicularis* also exhibits genetic differentiation between types, based on the intergenic region between *cytb* and *nad2* on mitochondrial DNA (Watanabe et al. 2005) and microsatellite markers (Nakajima et al. 2015, 2016) (see Fig. 8.2). These two genetic types (mt-L and mt-S) are often distinguished from the shape of the nematocysts; however, the relationship has not been completed (Watanabe et al. 2005; Abe et al. 2008). The genus *Galaxea* might also exhibit reproductive isolation caused by a shift in its spawning season (Watanabe et al. 2005); however, this suggestion is not usual, because of overlap in the spawning seasons (Abe et al. 2008). In addition, an unexpected genetic lineage was partially detected using both mitochondrial DNA and microsatellite markers (Nakajima et al. 2016). The mitochondrial haplotype of this new lineage was similar to the haplotype previously assigned as LE in Watanabe et al. (2005). However, the morphological traits among lineages were not distinguished based on the appearance of colonies. Further analysis is required to identify the species properly using detailed morphological traits (e.g., skeletal structure or shape of nematocysts) and divergent processes using more genetic loci.

In various regions, genus *Pocillopora* from the family Pocilloporidae also shows interspecific ambiguity, with high levels of morphological variability in colony shape and branch size. However, these well-known variants do not appear to be correlated with genetic variation (e.g., Pinzón and LaJeunesse 2011; Pinzón et al. 2013; Schmidt-Roach et al. 2013, 2014). Yet, genetic lineages have been distinguished using mitochondrial DNA and microsatellite regions (Pinzón and LaJeunesse 2011; Pinzón et al. 2013) using the

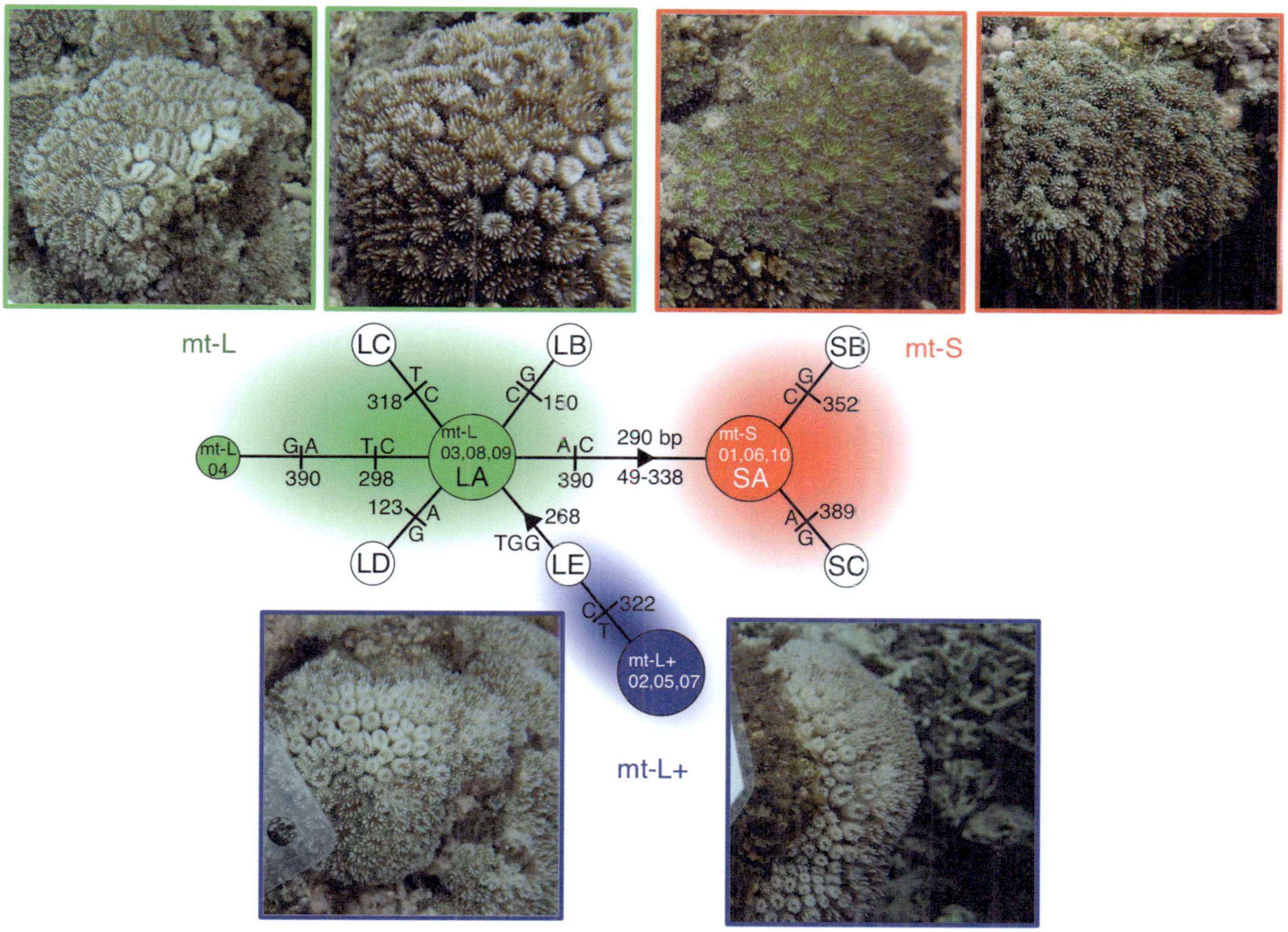


**Fig. 8.2** The relationship between morphotype and mitochondrial DNA type (or cryptic species) in *Galaxea fascicularis*. Without observing polyps under a microscope, it is difficult to distinguish type by the morphological characterization of colonies in coral species. Thus, problems have been reported distinguishing some corals (e.g., family Pocilloporidae). Genetic markers with mitochondrial and nuclear DNA are helpful for identifying cryptic type (This diagram was cited from Nakajima et al. 2016)

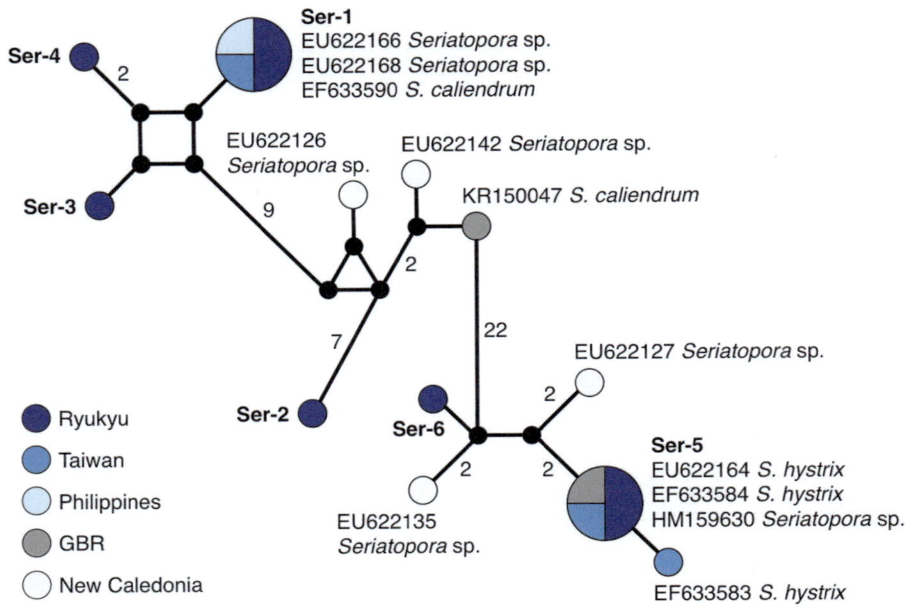


**Fig. 8.3** Six haplotypes of the mitochondrial control region of *Seriatopora* were detected by Nakajima et al. (2017) and were compared with haplotypes obtained by previous studies, including Chen et al. (2008), Flot et al. (2008), Bongaerts et al. (2010), and Warner et al. (2015). The haplotype network was drawn using a median-joining network in PopART ver. 1.7 (Leigh and Bryant 2015)

open reading frame (ORF) of unknown function between *atp6* and *nad4* and the D-loop, which is the control region between *atp8* and *cox1* (Flot and Tillier 2007). The genus *Seriatopora* morphologically similar to *Seriatopora hystrix* in the Ryukyu Archipelago also has three genetic lineages, based on mitochondrial DNA and microsatellite markers (Nakajima et al. 2017, see also Fig. 8.3). Thus, genetic identification has contributed toward classifying species and estimating the species diversity of corals. However, further studies using other genetic loci are required to elucidate speciation patterns and processes adequately.

## 8.3 Reproductive Modes and Clonality of Corals

The migration of sessile marine organisms, such as reef-building corals, usually occurs during the pelagic larval stage. Larvae are produced by sexual reproduction, and they settle on the sea bottom. Larval recruitment is important for the recovery of populations of coral species. Most coral species release gametes during the spawning season as a consequence of sexual reproduction (e.g., Harrison and Wallace 1990). This mode of reproduction promotes long-distance dispersal in broadcast-spawning coral species and maintains genetic connectivity among reefs. Modes of sexual reproduction in corals are usually differentiated into broadcast spawning and brooding. Broadcast-spawning species release sperm and eggs in the seawater, with fertilization occurring on the sea surface or in a water column. In contrast, fertilization of eggs occurs on the maternal colonies, and developed larvae are released in brooding species. Nishikawa et al. (2003) showed that the broadcast-spawning species, *A. tenuis*, exhibited lower genetic differentiation than the brooding species, *S. pistillata*. Based on these differences, the authors discussed the relationship between larval duration and genetic differentiation (Fig. 8.4). However, this pattern was not universally applicable to other coral species. For instance, the species *Goniastrea favulus* is self-compatible (Heyward and Babcock 1986), has sticky, negatively buoyant eggs and larvae, and its gametes and larvae are expected to have restricted dispersal. In comparison, *Platygyra daedalea* is self-incompatible, spawns positively buoyant egg-sperm bundles, and shows planktonic development. Yet, both species exhibited unexpected differentiation patterns in a spatial-autocorrelation analysis in the GBR (Miller and Ayre 2008b).

Some coral species often produce clonemates via fragmentation. This tendency is more pronounced in species that form the framework of ecosystems, (e.g., Baums et al. 2006, Fig. 8.5). However, clonal diversity differs widely among genera and species. Nakajima et al. (2015) developed 27 microsatellite markers for the broadcast-spawning species *Galaxea fascicularis*. These markers were amplifiable and polymorphic, of which 11 were cross-type microsatellite loci in the mt-Long and mt-Short colonies. These characteristics indicated that many clonemates were present in Zampa, Okinawa, for both mitochondrial DNA types (clonal diversity $R = 0.189$ in mt-L and $R = 0.159$ in mt-S). The octocoral species *Heliopora coerulea* also has clonal colonies; however, it only has one genotype in Oura Bay, Okinawa (Yasuda et al. 2012). In comparison, there are few or no clonemates in some *Acropora* species in the Nansei Islands (Nakajima et al. 2009b, 2010, 2012a; Zayasu et al. 2016). Increasing the number of clonal colonies is an effective strategy to maintain a population during adaptive circumstances because it helps preserve habitat efficiency. In contrast, effective population

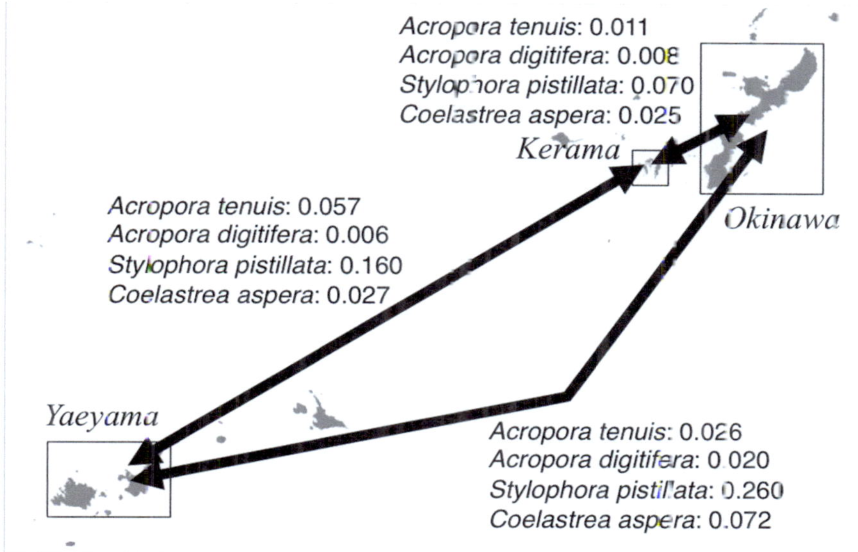


**Fig. 8.4** Genetic differentiation of corals, including *Acropora tenuis* (Nishikawa et al. 2003), *Acropora digitifera* (Nishikawa and Sakai 2005a), *Stylophora pistillata* (Nishikawa et al. 2003), and *Coelastrea aspera* (formerly *Goniastrea aspera*) (Nishikawa and Sakai 2003), using allozyme electrophoresis. Broadcast-spawning corals had relatively lower genetic differentiation, whereas brooding corals had relatively higher genetic differentiation, among regions

**Fig. 8.5** Several putative clonal colonies of *Galaxea fascicularis*, apparently produced by asexual reproduction, such as fragmentation (Tarama Island, Okinawa, Japan, 2015)

size will diminish if new genotypes are not recruited into the habitat from the outer reef (Muko et al. 2014). The decline in effective population size might be related to the extinction of a local population, especially in marginal habitats, after being exposed to extreme environmental changes. In such small populations, genetic drift will easily influence population size (Ouborg et al. 2010).

The occurrence of clonemates depends on physical and historical factors, as well as the species characteristics. Physical factors, like typhoons, storms, and strong waves, appear to promote fragmentation in coral species (e.g., Aranceta-Garza et al. 2012). In *Galaxea fascicularis*, populations in shallow waters may show an increase in clonal rate; however, further analyses with the standardization of sampling sites are needed, because previous studies did not account for the size of the collection site or the density of colonies at different sites (Nakajima et al. 2016). Because the number of alleles and loci is restricted in allozyme electrophoresis, microsatellite markers are more suitable than allozymes for estimating clonemates. However, problems have been reported in the scoring of multilocus genotypes. Arnaud-Haond et al. (2007) put forward the concept of clonal lineages, which takes somatic mutation and scoring errors into account. These indices envision further, precise clonality in analyses, due to the presence of large numbers of genetic loci in coral species, especially where fragmentation frequently occurs.

## 8.4 Genetic Diversity and Connectivity of Corals

Intraspecific genetic diversity is an important index for estimating the possibility of population extinction. The bottleneck effect causes population size and genetic diversity to decline. Previous population genetic analyses using microsatellite markers reported that *A. digitifera* seems to maintain genetic diversity, even in the high-latitude marginal habitats of the Nansei Islands, which cover an area of approximately 1000 km, and include Tanegashima Island (Nakajima et al. 2010). Furthermore, *Acropora solitaryensis* also maintains its genetic diversity in the range-edge habitats of Australia (Noreen et al. 2013). In contrast, the genetic diversity of the broadcast-spawning species *Pocillopora verrucosa* declined with increasing latitude in the coastal areas of east Africa (Ridgway et al. 2008). The genetic diversity of the brooding species *S. hystrix* also declined around the Lord Howe Island, which is a high-latitude area in Australia (Noreen et al. 2009).

In such populations, the restricted recruitment of larvae from external populations might cause low genetic diversity. In comparison with *Pocillopora* and *Seriatopora*, *Acropora* species tend to exhibit greater genetic connectivity and appear to maintain relatively high genetic diversity, even in marginal areas. The Kuroshio Current might also contribute toward maintaining the genetic diversity of coral populations in the high-latitude areas of Japan.

The Kuroshio Current flows northward along the Nansei Islands, with the current and its branch currents appearing to contribute to the dispersal of larvae among habitats. Marginal habitats in the area might also help maintain genetic diversity, providing corals with the ability to recover after disturbance events, including mass bleaching. In contrast, significant genetic differentiation was detected in *Acropora* sp. 1 aff. *digitifera* between Okinawa and Ogasawara, which are located at a similar latitude ($F_{ST}$ = 0.015–0.043, all $P < 0.05$, Nakajima et al. 2012a). These two regions are separated by a distance of approximately 1500 km, with no strong oceanographic current connecting them, leading to isolation; thus, larval dispersal cannot occur on a short-time scale between the two regions. On the local geographical scale, each region maintains its own genetic diversity, with Ogasawara rarely experiencing recruitment from other regions because it is an isolated island system.

*Acropora digitifera* maintains relatively high genetic diversity and connectivity near the northern limits of the species range. However, other coral species do not exhibit the same pattern. For example, one lineage of *Seriatopora* showed extremely low genetic diversity and strong genetic structure in the Ryukyu Archipelago (Nakajima et al. 2017). A similar pattern was also observed in previous studies in the GBR (see van Oppen et al. 2008; Noreen et al. 2009). Thus, local populations, especially those with low population size, might face extinction due to local disturbances in the near future. However, *S. hystrix* maintains connectivity by rare and long-distance dispersal migration via oceanographic currents in the GBR, despite being a brooding coral species (e.g., van Oppen et al. 2008). Nevertheless, gene flow varies among species, with significant genetic differentiation occasionally occurring within a region. Local recruitment contributes toward maintaining the regional population; however, multiple generations facilitate dispersion and habitat extension in the long term.

Geographic distance often influences genetic differentiation; however, the extent of differentiation is usually low in marine invertebrate species, especially for broadcast-spawning species (e.g., Palumbi 2003). Isolation-by-distance (IBD) is an index used to estimate the correlation between genetic differentiation and geographic distance (see Wright 1943; Kimura and Weiss 1964). On the local scale, there was no significant correlation in IBD for *A. digitifera*, whereas a significant correlation was found at large geographic distances (approximately 1000 km) in the Nansei Islands (Fig. 8.6). This result shows that geographic distance marginally influences the differentiation even in broadcast-spawning species. Extension of the sampling range would facilitate a comprehensive study of genetic isolation in corals. In addition, oceanographic distance should be used in place of geographic distances to estimate distance and the genetic index (see Sect. 8.5). Using the population genetic index for genetic diversity and differentiation, analyses using specific software for population genetics would facilitate the identification of genetic clusters when the extent of genetic differentiation is high (see Cornuet et al. 1999). For example, to determine genetic clustering, STRUCTURE (Pritchard et al. 2000), BAPS (Corander et al. 2003), InStruct (Gao et al. 2007), and ADMIXTURE (Alexander et al. 2009) are used for Bayesian clustering, whereas adegenet is used for DAPC (Discriminant Analysis of Principal Components; Jombart et al. 2010). To ascertain migration patterns, Migrate-n (Beerli and Felsenstein 1999, 2001; Beerli 2004) and IMa (Hey and Nielsen 2004) are often used. These population-based applications also contribute toward elucidating more specific genetic structures for inter- and intra-specific populations and migration patterns among habitats.

Patterns in genetic structure do not integrate well with the oceanographic current patterns, in many cases (e.g., Hellberg 2007). Ocean current patterns are not stable over longtime scales, with genetic structure reflecting historical fluctuations in genotypic patterns and population size (Zayasu et al. 2016). Apart from reproductive mode, larval duration, fertilization rates, habitat numbers, colony density, and mortality are factors that can influence successful long-distance dispersal and genetic differentiation among populations. Assuming 18 % larval mortality per day (Cowen et al. 2000), the pattern in larval dispersal differs from conventional simulation patterns. Mortality due to environmental factors during both larval period and colonization processes contributes to this inconsistency (see Cowen et al. 2000; Marshall et al. 2010), especially in population genetic samples collected from adult colonies during the growth process. Recent disturbances caused by rapid environmental changes might affect the pattern of gene flow between sites, by altering the population size and the recruitment habitat. In addition, increasing sea surface temperature might cause a decline in the survival rate of larvae, including the duration of survival, which, in turn, would affect the population persistence via reduced connectivity (Figueiredo et al. 2014).

## 8.5 Future of the Population Genetics of Corals in Japan

Predicted larval dispersal patterns based on geographic and oceanographic information do not reflect patterns in genetic differentiation, in some cases (see Sect. 8.4). This is because genetic data often reflect the historical background of popu-

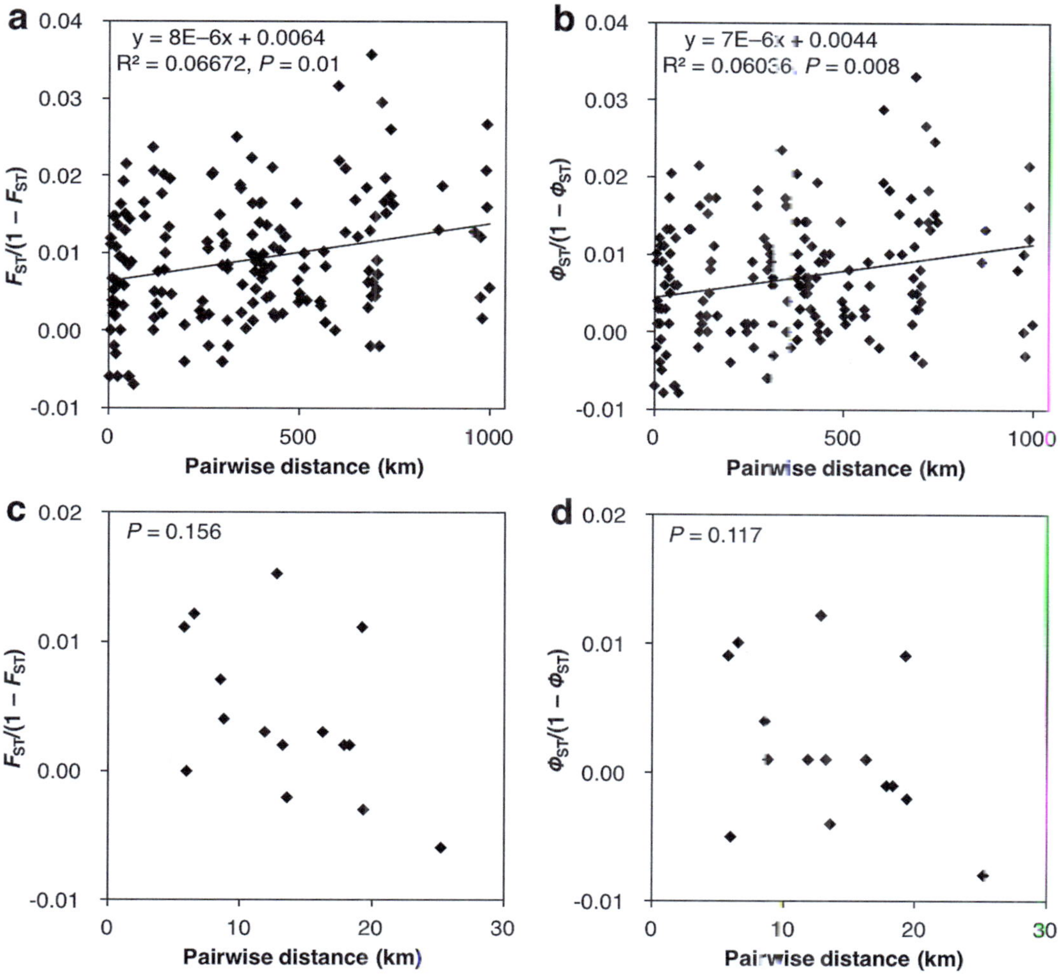


**Fig. 8.6** Isolation-by-distance results from the $F_{ST}/(1-F_{ST})$ and $\Phi_{ST}/(1-\Phi_{ST})$ values of *Acropora digitifera* at the Nansei Islands and Sekisei Reef. (**a**) $F_{ST}/(1-F_{ST})$ and (**b**) $\Phi_{ST}/(1-\Phi_{ST})$ at the Nansei Islands (original genetic data obtained by Nakajima et al. (2010)), (**c**) $F_{ST}/(1-F_{ST})$ and (**d**) $\Phi_{ST}/(1-\Phi_{ST})$ at Sekisei Reef (original genetic data was obtained by Nakajima et al. (2009b)). Distance is the minimum geographic distance in km. $F_{ST}$ and $\Phi_{ST}$ were calculated using GenAlEx ver. 6.2 (Peakall and Smouse 2006) and GenoDive ver. 2.0b17 (Meirmans and Van Tienderen 2004), respectively. The Mantel test was conducted using GenAlEx

lations on the evolutionary time scale. Nevertheless, multilocus genotypic data are used to estimate migration patterns on the ecological time scale. Assignment tests are used to estimate the recent migration and genetic background of each individual (Manel et al. 2005). However, the reliability of these methods is conditional, depending on large genetic differentiation among the analyzed populations and on a threshold number of genetic loci being studied (Cornuet et al. 1999). In broadcast-spawning species, including *Acropora* species, we need to study a larger geographic area to estimate the robustness of the data used for genetic differentiation. In contrast, brooding corals could be used to estimate migrants to the Nansei Islands, or more narrow geographic ranges, because these species have lower dispersal ranges, shorter larval duration, and smaller population sizes than broadcast-spawning species surveyed in previous studies (Nakajima et al. 2017). *Seriatopora hystrix* populations in Scott Reef, Western Australia, also exhibited large genetic differentiation, even though they only occur in a 65 km area and indicated migration patterns following fast recovery after recent disturbances (Underwood et al. 2007). In Japan, almost no present research provides detailed reports on the migration patterns of corals. As analytical techniques advance, such as the use of next-generation sequencing and the development of new software, it is expected that analyses of migration would be the focus of intensive research in the future.

In the population genetics of marine species, it is usually difficult to measure the actual distance, due to the influence of the direction and velocity of oceanographic currents. To solve this problem, oceanographic distances are being used (e.g., Alberto et al. 2011), following the method of seascape genetics. In addition, the coral habitats are affected not only by the oceanographic current and regional environmental changes but also by the ecological adaptations of the corals. Furthermore, factors such as sea temperature, coral cover, reef range, and genetic information on other competitive or coexisting marine species (see Galindo et al. 2006; Selkoe et al. 2008, 2010, 2016) need to be incorporated into analyses for more detailed seascape genetics. Thus, physical conditions should be continuously monitored to obtain environmental data. Along with genetic connectivity among populations, genetic differentiation within a local population is also informative for understanding the ecological equilibrium of coral populations. Information on the location and genotypic data of coral colonies is also essential when investigating local recruitment and fragmentation (Maier et al. 2009; Gorospe and Karl 2013; Gorospe et al. 2015).

The development of genetic markers using next-generation sequencing has accelerated the analytic process. Microsatellite markers based on next-generation sequencing have already been developed for some coral species (*Acropora*: Shinzato et al. 2014; *Galaxea*: Nakajima et al. 2015). Further research on other coral species is expected to help increase the number of these markers. Microsatellite markers are usually treated as neutral markers for population genetics. Thus, the frequency of genotypes in these loci is not affected by natural selection. Adaptation of populations also provides an important index of how populations are maintained. Non-neutral markers also help with estimating the possibility of adaptation. Lundgren et al. (2013) isolated single nucleotide polymorphisms (SNPs) in functional coding regions of coral species. SNPs help to specify the environmental factors that might influence the coral populations. For instance, Shinzato et al. (2015) detected four genetic clusters related with latitude in *A. digitifera*, based on principal components analysis using whole genome SNPs. Because corals have a symbiotic relationship with zooxanthella, population genetic studies on these symbionts have increased. Howells et al. (2013) used microsatellite markers to study the population genetics of zooxanthella in *A. millepora* and reported genetic differentiation on a small geographic scale. Although there is no clear genetic structure for *Pocillopora* in the eastern Pacific (Pinzón and LaJeunesse 2011), the biogeographic pattern of genetic structure in the symbiont clade D1 exhibits high-temperature stress tolerance (Pettay and LaJeunesse 2013). In addition to obtaining nuclear genetic information on corals, parallel information on symbionts is informative to estimate the extent of adaptation, especially in disturbed areas.

Population genetics has certainly developed in recent years; however, the results have not been applied toward conserving coral reefs. In Japan, a national park was recently formed in the Kerama Islands to conserve the seascape. Corals are considered as important frameworks for these seascapes and are critical to maintain biodiversity. Furthermore, species should be characterized in relation to biogeographic information. If a population undergoes local extinction with continuous inbreeding, inbreeding depression will occur, leading to a drop in genetic diversity. For instance, the genetic diversity of *Seriatopora* populations has declined at some sites in the Ryukyu Archipelago (Nakajima et al. 2017). Along with the geographical and environmental factors, biological effects, such as mortality, must be considered. Application of the advancing methods of population genetics should be used to further exploring the field of population genetics as it would be a valuable asset for the conservation of coral reef ecosystems.

## References


Abe M, Watanabe T, Hayakawa H, Hidaka M (2008) Breeding experiments of the hermatypic coral *Galaxea fascicularis*: partial reproductive isolation between colonies of different nematocyst types, and enhancement of fertilization success by the presence of parental colonies. Fish Sci 74:1342–1344

Adjeroud M, Tsuchiya M (1999) Genetic variation and clonal structure in the scleractinian coral *Pocillopora damicornis* in the Ryukyu Archipelago, Southern Japan. Mar Biol 134:753–760

Alberto F, Raimondi PT, Reed DC, Watson JR, Siegel DA, Mitarai S, Coelho N, Serrão EA (2011) Isolation by oceanographic distance explains genetic structure for *Macrocystis pyrifera* in the Santa Barbara Channel. Mol Ecol 20:2543–2554

Alexander DH, Novembre J, Lange K (2009) Fast model-based estimation of ancestry in unrelated individuals. Genome Res 19:1655–1664

Aranceta-Garza F, Balart EF, Reyes-Bonilla H, Cruz-Hernández P (2012) Effect of tropical storms on sexual and asexual reproduction in coral *Pocillopora verrucosa* subpopulations in the Gulf of California. Coral Reefs 31:1157–1167

Arnaud-Haond S, Duarte CM, Alberto F, Serrão EA (2007) Standardizing methods to address clonality in population studies. Mol Ecol 16:5115–5139

Ayre DJ, Dufty SL (1994) Evidence for restricted gene flow in the viviparous coral *Seriatopora hystrix* on Australia's Great Barrier Reef. Evolution 48:1183–1201

Ayre DJ, Hughes TP (2000) Genotypic diversity and gene flow in brooding and spawning corals along the Great Barrier Reef, Australia. Evolution 54:1590–1605

Ayre DJ, Hughes TP (2004) Climate change, genotypic diversity and gene flow in reef-building corals. Ecol Lett 7:273–278

Ayre DJ, Hughes TP, Standish RJ (1997) Genetic differentiation, reproductive mode, and gene flow in the brooding coral *Pocillopora damicornis* along the Great Barrier Reef, Australia. Mar Ecol Prog Ser 159:175–187

Baums IB, Hughes CR, Hellberg MH (2005) Mendelian microsatellite loci for the Caribbean coral *Acropora palmata*. Mar Ecol Prog Ser 288:115–127

Baums IB, Miller MW, Hellberg ME (2006) Geographic variation in clonal structure in a reef-building Caribbean coral, *Acropora palmata*. Ecol Monogr 76:503–519

Baums IB, Boulay JN, Polato NR, Hellberg ME (2012) No gene flow across the Eastern Pacific Barrier in the reef-building coral *Porites lobata*. Mol Ecol 21:5418–5433

Beerli P (2004) Effect of unsampled populations on the estimation of population sizes and migration rates between sampled populations. Mol Ecol 13:827–836

Beerli P, Felsenstein J (1999) Maximum-likelihood estimation of migration rates and effective population numbers in two populations using a coalescent approach. Genetics 152:763–773

Beerli P, Felsenstein J (2001) Maximum likelihood estimation of a migration matrix and effective population sizes in n subpopulations by using a coalescent approach. Proc Natl Acad Sci U S A 98:4563–4568

Bongaerts P, Riginos C, Ridgway T, Sampayo EM, van Oppen MJH, Englebert N, Vermeulen F, Hoegh-Guldberg O (2010) Genetic divergence across habitats in the widespread coral *Seriatopora hystrix* and its associated *Symbiodinium*. PLoS One 5:e10871

Chen C, Dai C-F, Plathong S, Chiou C-Y, Chen CA (2008) The complete mitochondrial genomes of needle corals, *Seriatopora* spp. (Scleractinia: Pocilloporidae): an idiosyncratic *atp8*, duplicated *trnW* gene, and hypervariable regions used to determine species phylogenies and recently diverged populations. Mol Phylogenet Evol 46:19–33

Corander J, Waldmann P, Sillanpaa MJ (2003) Bayesian analysis of genetic differentiation between populations. Genetics 163:367–374

Cornuet J-M, Piry S, Luikart G, Estoup A, Solignac M (1999) New methods employing multilocus genotypes to select or exclude populations as origins of individuals. Genetics 153:1989–2000

Cowen RK, Lwiza KMM, Sponaugle S, Paris CB (2000) Connectivity of marine populations: open or closed? Science 287:857–859

Figueiredo J, Baird AH, Harii S, Connolly SR (2014) Increased local retention of reef coral larvae as a result of ocean warming. Nat Clim Chang 4:498–502

Flot J-F, Tillier S (2007) The mitochondrial genome of *Pocillopora* (Cnidaria: Scleractinia) contains two variable regions: the putative D-loop and a novel ORF of unknown function. Gene 401:80–87

Flot J-F, Licuanan WY, Nakano Y, Payri C, Cruaud C, Tiller S (2008) Mitochondrial sequences of *Seriatopora* corals show little agreement with morphology and reveal the duplication of a tRNA gene near the control region. Coral Reefs 27:789–794

Fukami H, Omori M, Hatta M (2000) Phylogenetic relationships in the coral family Acroporidae, reassessed by inference from mitochondrial genes. Zool Sci 17:689–696

Galindo HM, Olson DB, Palumbi SR (2006) Seascape genetics: a coupled oceanographic-genetic model predicts population structure of Caribbean corals. Curr Biol 16:1622–1626

Gao H, Williamson S, Bustamante CD (2007) A Markov chain Monte Carlo approach for joint inference of population structure and inbreeding rates from multilocus genotype data. Genetics 176:1635–1651

Gorospe KD, Karl SA (2013) Genetic relatedness does not retain spatial pattern across multiple spatial scales: dispersal and colonization in the coral, *Pocillopora damicornis*. Mol Ecol 22:3721–3736

Gorospe KD, Donahue MJ, Karl SA (2015) The importance of sampling design: spatial patterns and clonality in estimating the genetic diversity of coral reefs. Mar Biol 162:917–928

Harrison PL, Wallace CC (1990) Reproduction, dispersal and recruitment of scleractinian corals. In: Dubinsky Z (ed) Coral reefs: ecosystems of the world, vol 25. Elsevier, Amsterdam, pp 133–207

Hayashibara T, Shimoike K (2002) Cryptic species of *Acropora digitifera*. Coral Reefs 21:224–225

Hedgecock D, Barber PH, Edmands SE (2007) Genetic approaches to measuring connectivity. Oceanography 20:70–79

Hellberg ME (2007) Footprints on water: the genetic wake of dispersal among reefs. Coral Reefs 26:463–473

Hey J, Nielsen R (2004) Multilocus methods for estimating population sizes, migration rates and divergence time, with applications to the divergence of *Drosophila pseudoobscura* and *D. persimilis*. Genetics 167:747–760

Heyward AJ, Babcock RC (1986) Self- and cross-fertilization in scleractinian corals. Mar Biol 90:191–195

Howells EJ, Willis BL, Bay LK, van Oppen MJH (2013) Spatial and temporal genetic structure of *Symbiodinium* populations within a common reef-building coral on the Great Barrier Reef. Mol Ecol 22:3693–3708

Isomura N, Iwao K, Morita M, Fukami H (2016) Spawning and fertility of $F_1$ hybrids of the coral genus *Acropora* in the Indo-Pacific. Coral Reefs 35:851–855

Jombart T, Devillard S, Balloux F (2010) Discriminant analysis of principal components: a new method for the analysis of genetically structured populations. BMC Genet 11:94

Kimura M, Weiss GH (1964) The stepping stone model of population structure and the decrease of genetic correlation with distance. Genetics 49:561–576

Kitanobo S, Isomura N, Fukami H, Iwao K, Morita M (2016) The reef-building coral *Acropora* conditionally hybridize under sperm limitation. Biol Lett 12:20160511

Leigh JW, Bryant D (2015) PopART: full-feature software for haplotype network construction. Methods Ecol Evol 6:1110–1116

Lowe WH, Allendorf FW (2010) What can genetics tell us about population connectivity? Mol Ecol 19:3038–3051

Lundgren P, Vera JC, Peplow L, Manel S, van Oppen MJH (2013) Genotype – environment correlations in corals from the Great Barrier Reef. BMC Genet 14:9

Maier E, Tollrian R, Nürnberger B (2009) Fine-scale analysis of genetic structure in the brooding coral *Seriatopora hystrix* from the Red Sea. Coral Reefs 28:751–756

Manel S, Gaggioti OE, Waples RS (2005) Assignment methods: matching biological questions with appropriate techniques. Trends Ecol Evol 20:136–142

Marshall DJ, Monro K, Bode M, Keough MJ, Swearer S (2010) Phenotype–environment mismatches reduce connectivity in the sea. Ecol Lett 13:128–140

Meirmans PG, Van Tienderen PH (2004) Genotype and Genodive: two programs for the analysis of genetic diversity of asexual organisms. Mol Ecol Notes 4:792–794

Miller KJ, Ayre DJ (2008a) Protection of genetic diversity and maintenance of connectivity among reef corals within marine protected areas. Conserv Biol 22:1245–1254

Miller KJ, Ayre DJ (2008b) Population structure is not a simple function of reproductive mode and larval type: insights from tropical corals. J Anim Ecol 77:713–724

Muko S, Arakaki S, Tamai R, Sakai K (2014) An individual-based model for population viability analysis of the brooding coral *Seriatopora hystrix*. Ecol Model 277:68–76

Nakajima Y, Nishikawa A, Iguchi A, Sakai K (2009a) Novel and cross-species amplifiable microsatellite markers in two *Acropora* species. Plankton Benthos Res 4:38–41

Nakajima Y, Nishikawa A, Isomura N, Iguchi A, Sakai K (2009b) Genetic connectivity in the broadcast-spawning coral *Acropora digitifera* analyzed by microsatellite markers on the Sekisei Reef, southwestern Japan. Zool Sci 26:209–215

Nakajima Y, Nishikawa A, Iguchi A, Sakai K (2010) Gene flow and genetic diversity of a broadcast-spawning coral in northern peripheral populations. PLoS One 5:e11149

Nakajima Y, Nishikawa A, Iguchi A, Sakai K (2012a) Regional genetic differentiation among northern high-latitude island populations of a broadcast-spawning coral. Coral Reefs 31:1125–1133

Nakajima Y, Nishikawa A, Iguchi A, Sakai K (2012b) The population genetic approach delineates the species boundary of reproductively isolated corymbose acroporid corals. Mol Phylogenet Evol 63:527–531

Nakajima Y, Shinzato C, Satoh N, Mitarai S (2015) Novel polymorphic microsatellite markers reveal genetic differentiation between two sympatric types of *Galaxea fascicularis*. PLoS One 10:e0130176

Nakajima Y, Zayasu Y, Shinzato C, Satoh N, Mitarai S (2016) Genetic differentiation and connectivity of morphological types of the broadcast-spawning coral *Galaxea fascicularis* in the Nansei Islands, Japan. Ecol Evol 6:1457–1469

Nakajima Y, Nishikawa A, Iguchi A, Nagata T, Uyeno D, Sakai K, Mitarai S (2017) Elucidating the multiple genetic lineages and population genetic structure of the brooding coral *Seriatopora* (Scleractinia: Pocilloporidae) in the Ryukyu Archipelago. Coral Reefs 36:415–426

Nishikawa A, Sakai K (2003) Genetic variation and gene flow of broadcast spawning and planula brooding coral, *Goniastrea aspera* (Scleractinia) in the Ryukyu Archipelago, southern Japan. Zool Sci 20:1031–1038

Nishikawa A, Sakai K (2005a) Settlement-competency period of planulae and genetic differentiation of the scleractinian coral *Acropora digitifera*. Zool Sci 22:391–399

Nishikawa A, Sakai K (2005b) Genetic connectivity of the scleractinian coral *Goniastrea aspera* around the Okinawa Islands. Coral Reefs 24:318–323

Nishikawa A, Katoh M, Sakai K (2003) Larval settlement rates and gene flow of broadcast-spawning (*Acropora tenuis*) and planula-brooding (*Stylophora pistillata*) corals (Scleractinia). Mar Ecol Prog Ser 256:87–97

Noreen AME, Harrison PL, van Oppen MJH (2009) Genetic diversity and connectivity in a brooding reef coral at the limit of its distribution. P R Soc B 276:3927–3935

Noreen AME, van Oppen MJH, Harrison PL (2013) Genetic diversity and differentiation among high-latitude broadcast-spawning coral populations disjunct from the core range. Mar Ecol Prog Ser 491:101–109

Ohki S, Kowalski RK, Kitanobo S, Morita M (2015) Changes in spawning time led to the speciation of the broadcast spawning corals *Acropora digitifera* and the cryptic species *Acropora* sp. 1 with similar gamete recognition systems. Coral Reefs 34:1189–1198

Ouborg NJ, Pertoldi C, Loeschcke V, Bijlsma RK, Hedrick PW (2010) Conservation genetics in transition to conservation genomics. Trends Genet 26:177–187

Palumbi SR (2003) Population genetics, demographic connectivity, and the design of marine reserves. Ecol Appl 13:S146–S158

Peakall R, Smouse PE (2006) GENALEX 6: genetic analysis in Excel. Population genetic software for teaching and research. Mol Ecol Notes 6:288–295

Pettay DT, LaJeunesse TC (2013) Long-range dispersal and high-latitude environments influence the population structure of a "stress-tolerant" dinoflagellate endosymbiont. PLoS One 8:e79208

Pinzón JH, LaJeunesse TC (2011) Species delimitation of common reef corals in the genus Pocillopora using nucleotide sequence phylogenies, population genetics and symbiosis ecology. Mol Ecol 20:311–325

Pinzón JH, Sampayo E, Cox E, Chauka LJ, Chen CA, Voolstra CR, LaJeunesse TC (2013) Blind to morphology: genetics identifies several widespread ecologically common species and few endemics among Indo-Pacific cauliflower corals (Pocillopora, Scleractinia). J Biogeogr 40:1595–1608

Pritchard JK, Stephens M, Donnelly P (2000) Inference of population structure using multilocus genotype data. Genetics 155:945–959

Richards ZT, van Oppen MJH (2012) Rarity and genetic diversity in Indo–Pacific *Acropora* corals. Ecol Evol 2:1867–1888

Richards ZT, Miller DJ, Wallace CC (2013) Molecular phylogenetics of geographically restricted Acropora species: implications for threatened species conservation. Mol Phylogenet Evol 69:837–851

Ridgway T, Riginos C, Davis J, Hoegh-Guldberg O (2008) Genetic connectivity patterns of *Pocillopora verrucosa* in southern African Marine Protected Areas. Mar Ecol Prog Ser 354:161–168

Rodriguez-Lanetty M, Hoegh-Guldberg O (2002) The phylogeography and connectivity of the latitudinally widespread scleractinian coral *Plesiastrea versipora* in the Western Pacific. Mol Ecol 11:1177–1189

Schmidt-Roach S, Lundgren P, Miller KJ, Gerlach G, Noreen AME, Andreakis N (2013) Assessing hidden species diversity in the coral *Pocillopora damicornis* from Eastern Australia. Coral Reefs 32:161–172

Schmidt-Roach S, Miller KJ, Lundgren P, Andreakis N (2014) With eyes wide open: a revision of species within and closely related to the *Pocillopora damicornis* species complex (Scleractinia; Pocilloporidae) using morphology and genetics. Zool J Linnean Soc 170:1–33

Selkoe KA, Henzler CM, Gaines SD (2008) Seascape genetics and the spatial ecology of marine populations. Fish Fish 9:363–377

Selkoe KA, Watson JR, White C, Horin TB, Iacchei M, Mitarai S, Siegel DA, Gaines SD, Toonen RJ (2010) Taking the chaos out of genetic patchiness: seascape genetics reveals ecological and oceanographic drivers of genetic patterns in three temperate reef species. Mol Ecol 19:3708–3726

Selkoe KA, Gaggiotti OE, Treml EA, Wren JL, Donovan MK, Toonen RJ, Hawai'i Reef Connectivity Consortium (2016) The DNA of coral reef biodiversity: predicting and protecting genetic diversity of reef assemblages. P R Soc B 283:1829

Shinzato C, Yasuoka Y, Mungpakdee S, Arakaki N, Fujie M, Nakajima Y, Satoh N (2014) Development of novel, cross-species microsatellite markers for *Acropora* corals using next-generation sequencing technology. Front Mar Sci 1:11

Shinzato C, Mungpakdee S, Arakaki N, Satoh N (2015) Genome-wide SNP analysis explains coral diversity and recovery in the Ryukyu Archipelago. Sci Rep 5:18211

Suzuki G, Keshavmurthy S, Hayashibara T, Wallace CC, Shirayama Y, Chen CA, Fukami H (2016) Genetic evidence of peripheral isolation and low diversity in marginal populations of the *Acropora hyacinthus* complex. Coral Reefs 35:1419–1432

Takabayashi M, Carter D, Lopez J, Hoegh-Guldberg O (2003) Genetic variation of the scleractinian coral *Stylophora pistillata*, from western Pacific reefs. Coral Reefs 22:17–22

Underwood JN, Smith LD, van Oppen MJH, Gilmour JP (2007) Multiple scales of genetic connectivity in a brooding coral on isolated reefs following catastrophic bleaching. Mol Ecol 16:771–784

van Oppen MJH, Willis BL, Miller DJ (1999) Atypically low rate of cytochrome b evolution in the scleractinian coral genus *Acropora*. P R Soc B 266:179–183

van Oppen MJH, McDonald BJ, Willis BL, Miller DJ (2001) The evolutionary history of the coral genus *Acropora* (Scleractinia, Cnidaria) based on a mitochondrial and a nuclear marker: reticulation, incomplete lineage sorting or morphological convergence? Mol Biol Evol 18:1315–1329

van Oppen MJH, Underwood JN, Muirhead AN, Peplow L (2007) Ten microsatellite loci for the reef-building coral *Acropora millepora* (Cnidaria, Scleractinia) from the Great Barrier Reef, Australia. Mol Ecol Notes 7:436–438

van Oppen MJH, Lutz A, De'ath G, Peplow L, Kininmonth S (2008) Genetic traces of recent long-distance dispersal in a predominantly self-recruiting coral. PLoS One 3:e3401

van Oppen MJH, Peplow LM, Kininmonth S, Berkelmans R (2011) Historical and contemporary factors shape the population genetic structure of the broadcast spawning coral, *Acropora millepora*, on the Great Barrier Reef. Mol Ecol 20:4899–4914

Villanueva RD (2015) Cryptic speciation in the stony octocoral *Heliopora coerulea*: temporal reproductive isolation between two growth forms. Mar Biodivers 46:503–507

Warner PA, van Oppen MJH, Willis BL (2015) Unexpected cryptic species diversity in the widespread coral *Seriatopora hystrix* masks spatial-genetic patterns of connectivity. Mol Ecol 24:2993–3008

Watanabe T, Nishida M, Watanabe K, Wewengkang DS, Hidaka M (2005) Polymorphism in the nucleotide sequence of a mitochondrial intergenic region in the scleractinian coral *Galaxea fascicularis*. Mar Biotechnol 7:33–39

Wright S (1943) Isolation by distance. Genetics 28:114–138

Yamano H (2008) Distribution of coral reefs in Japan. Bull Coast Oceanogr 46:3–9

Yasuda N, Nagai S, Lian C, Hamaguchi M, Hayashibara T, Nadaoka K (2012) Identification and characterization of microsatellite loci in the blue coral *Heliopora coerulea* (Alcynonaria: Coenothecalia). Conserv Genet 9:1011–1013

Yasuda N, Taquet C, Nagai S, Fortes M, Fan TY, Phongsuwan N, Nadaoka K (2014) Genetic structure and cryptic speciation in the threatened reef-building coral *Heliopora coerulea* along Kuroshio Current. Bull Mar Sci 90:233–255

Yasuda N, Taquet C, Nagai S, Fortes M, Fan TY, Harii S, Yoshida T, Sito Y, Nadaoka K (2015) Genetic diversity, paraphyly and incomplete lineage sorting of mtDNA, ITS2 and microsatellite flanking region in closely related *Heliopora* species (Octocorallia). Mol Phylogenet Evol 93:161–171

Zayasu Y, Nakajima Y, Sakai K, Suzuki G, Satoh N, Shinzato C (2016) Unexpectedly complex gradation of coral population structure in the Nansei Islands, Japan. Ecol Evol 6:5491–5505

# Distribution Expansion and Historical Population Outbreak Patterns of Crown-of-Thorns Starfish, *Acanthaster planci* sensu lato, in Japan from 1912 to 2015

9

Nina Yasuda

**Abstract**

The present chapter reviews the distribution and population outbreak records of the crown-of-thorns starfish, *Acanthaster planci* sensu lato, in Japan from 1912 to 2015. The literature survey suggests that *A. planci* sensu lato distribution has been extending northward since 1945 from Amami Oshima (its previous northernmost distribution) to Miyake Island and Goto Island. Genetic homogeneity within Japanese *A. planci* sensu lato populations indicates that larval dispersal has likely caused this poleward migration. Water temperatures have significantly increased in the temperate area of Japan, implying that global warming is partly responsible for this poleward migration. More frequent and intense population outbreaks in temperate areas were also observed, possibly in relation to increased water temperatures and successive larval dispersal from the south. Overall, complex and persistent patterns were observed for two major successive population outbreaks in Japan: from 1969 to 1991 and from 1995 to now. The evidence suggests that the western Okinawa populations are the most likely origin for secondary outbreaks within Japan. The Amami population is also likely to be an important source for outbreaks in temperate regions. However, no records of population outbreaks were found for least in two regions: Ogasawara and the Osumi Islands. Ogasawara is located approximately 1000 km south of the Kuroshio Current, so infestation via larval dispersal from other populations is more limited than in other Kuroshio regions. The Osumi Islands are, however, located in the middle of the Kuroshio Current, implying that insufficient corals are available for the growth of *A. planci* sensu lato or that unknown environment factors such as abundant predators of juvenile starfish suppress recruitment and juvenile survival.

**Keywords**

Crown-of-thorns starfish • Population outbreak • Genetic analysis • Mitochondrial DNA • Early life history • Microsatellite loci • Climate change • Larval dispersal • Global warming • Poleward migration

## 9.1 Introduction

### 9.1.1 *Acanthaster planci* sensu lato

Many detailed reviews are available about the biology and ecology of the crown-of-thorns starfish, *Acanthaster planci* sensu lato (e.g., Moran 1986; Birkeland and Lucas 1990; Pratchett et al. 2014). Therefore, I focus in the present chapter on reviewing key aspects of *A. planci* sensu lato

The original version of this chapter was revised. An erratum to this chapter can be found at https://doi.org/10.1007/978-981-10-6473-9_13

N. Yasuda (✉)
Organization for Promotion of Tenure Track, University of Miyazaki, 1-1 Gakuenkibanadai Nishi, Miyazaki, Japan
e-mail: nina27@cc.miyazaki-u.ac.jp


A. Iguchi, C. Hongo (eds.), *Coral Reef Studies of Japan*, Coral Reefs of the World 13,
https://doi.org/10.1007/978-981-10-6473-9_9

associated with larval dispersal, drawing on information from Japanese-based literature and unpublished data to improve the understanding of previous population outbreak patterns in Japan. I was interested in the following three questions: (1) Is *A. planci* sensu lato migrating toward the north like other coral species? (2) Has the intensity of population outbreaks increased in recent years? (3) Is the western Okinawan population really the source of other population outbreaks in Japan?

### 9.1.2 Species Status

Currently, the genus *Acanthaster* contains five species in the Indo-Pacific Ocean, but only two official species names exist: (1) *A. brevispinus*, which is a single species, and (2) *A. planci* (East Indian Ocean species) and the others called *A. planci* sensu lato, which is a complex of three cryptic species (Haszprunar and Spies 2014).

***Acanthaster planci* sensu lato**: *A. planci* is well-documented as a cause of coral-devastating population outbreaks in the Indian and Pacific Oceans (Pratchett et al. 2014). Vogler et al. (2008) showed that there are actually four closely related *A. planci* species complex. These four species include three genetically distinct species in the Indian Ocean and a single species in the Pacific Ocean, as determined by phylogenetic analysis using the mitochondrial CO1 region. One Indian Ocean *A. planci* lineage is distributed mainly in the western Indian Ocean, another lineage is mainly distributed in the Red Sea, and the last lineage is distributed in the eastern Indian Ocean. On Pari Island in northern Jakarta, the Pacific *A. planci* lineage (orange color morph) and eastern Indian Ocean *A. planci* lineage (purple color morph) are co-distributed (Yasuda et al. 2010). Although nuclear microsatellite analysis indicated distinct genotypes between the two *A. planci* lineages (*A. planci* and *A. planci* sensu lato) found on Pari Island and no intermediate color morphs were observed, one specimen had a mitochondrial haplotype incongruent with its color morph, implying that natural hybridization may have occurred between the two lineages in the past.

***Acanthaster brevispinus***: In contrast with *A. planci* species complex, *A. brevispinus* is relatively rare. The holotype specimen of *A. brevispinus* was collected at Sirun Island in the Philippines, with no subsequent documentation (Birkeland and Lucas 1990). In Japan, *A. brevispinus* was occasionally caught in lobster gill nets south of Wakayama between 2003 and 2005 (Saba and Iriyama 2002), and there was an initial report of *A. brevispinus* being detected in Sukumo (western Shikoku) during the removal of *A. planci* sensu lato in 2006 (Nakachi 2007), but there have been no subsequent reports. *A. brevispinus* was also collected in the Seychelles by Jangoux and Aziz (1984) in the western Indian Ocean. These reports imply that *A. brevispinus* is distributed in both the Indian (Seychelles) and Pacific Oceans (Japan, Philippines, and eastern Australia) but are seldom found (Birkeland and Lucas 1990). This huge difference in population density between *A. planci* species complex and *A. brevispinus* is surprising, given that *A. brevispinus* has a life cycle, fecundity, and spawning peak (Lucas and Jones 1976) similar to that of *A. planci* species complex. Unlike in the Great Barrier Reef (GBR, Australia), where *A. planci* sensu lato mainly appears around the reefs, while *A. brevispinus* is found in deeper lagoons, the habitat use of the two species partly overlaps to the south of Wakayama and in western Shikoku. Here, the total area is much smaller than that of the GBR, and the sea is shallow for some distance with a gentle slope from the shore. No real coral "reefs" exist in the temperate coral communities of Wakayama and Shikoku, but corals can be found in relatively shallow rocky areas near the shore. Corals, *A. planci* sensu lato, and *A. brevispinus* can thus be found in the same areas. Although *A. brevispinus* has never eaten corals in Australia (Lucas and Jones 1976), volunteer divers in Wakayama engaged in the control of *A. planci* sensu lato observed *A. brevispinus* digesting corals by baring its stomach, much like *A. planci* sensu lato. If *A. planci* and *A. brevispinus* with similar reproductive characteristics live in the same region and feed on the similar diet, differences between the two species (such as fertilization rates, larval behavior patterns, juvenile habitat use, and predators) should be examined to explain the difference in population density.

***Acanthaster ellisii***: Historically, the genus *Acanthaster* has contained three species separated by morphology: (1) *A. planci* species complex, found in the Indian Ocean and western and central Pacific Ocean (Pacific and Indian Ocean *A. planci* species complex was considered to be a single species), (2) *A. brevispinus*, and (3) *A. ellisii*, found only in the eastern Pacific Ocean (e.g., the Gulf of California).

The crown-of-thorns starfish in the Gulf of California, where it has short spines with short arms, was called *A. ellisii* and regarded as a different species from other Pacific crown-of-thorns starfish. However, population genetic analysis of *A. ellisii* and several Pacific *A. planci* sensu lato populations by Nishida and Lucas (1988) using allozyme markers showed that *A. ellisii* is genetically closer to other western Pacific crown-of-thorns starfish populations than to that in Hawaii. The result was unexpected because crown-of-thorns starfish in Hawaii has the same morphological features as its western Pacific counterpart. Given that the crown-of-thorns starfish in Hawaii is *A. planci*, *A. ellisii*, which is genetically closer to other Pacific *A. planci* than to *A. planci* in Hawaii, became synonymous with *A. planci*.

Two *Acanthaster* species are present in Japan: Pacific *A. planci* sensu lato and *A. brevispinus* (Fig. 9.1). *A. planci* sensu lato occurs in the Nansei Islands (all of the islands

**Fig. 9.1** Two *Acanthaster* species from Japan. *Left*, *Acanthaster brevispinus* collected in Susami, Wakayama, in 2004. *Right*, *A. planci* sensu lato collected from western Okinawa

**Table 9.1** Summary of peak spawning periods of *Acanthaster planci* in Japan

| Latitude (N) | Locality | Year | Apr | May | Jun | Jul | Aug | Sep | Oct | Source |
|---|---|---|---|---|---|---|---|---|---|---|
| 33.5 | Mainland Japan | 1973 | | | | | | | | The Environmental Agancy (1974) |
| 31.5 | Kyushu South | 2014 and 2015 | | | | | | | | Yasuda observation |
| 28 | Amami-Ohshima | 2004 | | | | | | | | Yasuda et al. (2010) |
| 26.5 | Okinawa Mainland | 2004 | | | | | | | | Yasuda et al. (2010) |
| 26.5 | Okinawa Mainland | 2005 | | | | | | | | Yasuda unpublished data |
| 26 | Kerama Islands | 2004 | | | | | | | | Yasuda et al. (2010) |
| 24.5 | Miyako Island | 2004 | | | | | | | | Yasuda et al. (2010) |
| 24 | Sekisei-Lagoon | 2004 | | | | | | | | Yasuda et al. (2010) |
| 24 | Iriomote Island | 1984 and 1985 | | | | | | | | Yokochi and Ogura (1987) |

between Kyushu and Taiwan) and the southern parts of Kyushu, Shikoku, and Honshu islands. *A. brevispinus* is much rarer than *A. planci* sensu lato and can be found in Shikoku and Wakayama, as noted above.

### 9.1.3 Life Cycle and Early Life Ecology

*A. planci* sensu lato is dioecious (i.e., contains both males and females) with high fecundity. *A. planci* sensu lato spawning in Japan is summarized in Table 9.1. Spawning starts from late May to early June in the southern part of its range, i.e., the Yaeyama region (24.31°N) of Japan. As spawning shifts to higher latitudes (e.g., Kushimoto, 33.48°N), the spawning period becomes shorter and occurs later. Because *A. planci* sensu lato spawning in Japan peaks when the water temperature exceeds approximately 28 °C in Japan (Yokochi and Ogura 1987; Yasuda et al. 2010), the timing and duration of spawning may change slightly at the same location across years. Such variations likely exist because 28 °C is the optimum temperature for *A. planci* sensu lato larvae, which have a relatively narrow temperature tolerance (26–31 °C; Lucas 1973). For example, on Okinawa Island (26.61°N), peak spawning was slightly delayed in 2005 compared to 2004 due to lower water temperatures (Table 9.1). While *A. planci* sensu lato spawning is not necessarily associated with the phase of the moon (Birkeland and Lucas 1990; Pratchett et al. 2014), *A. planci* sensu lato in a tank did spawn during the full moon in July and August near the northernmost part of their distribution range, in Wakayama (33.48°N) (Nature Conservation Bureau, Environment Agency 1974).

When the spawning peak starts, over half of both males and females have partially spawned or spent gonads, which are easily detected by dissecting the roots of their arms (Yasuda et al. 2010, Fig. 9.2). Each individual appears to spawn several times during the spawning period, because during the middle of a spawning period, a population contains individuals with full gonads, gonads at approximately

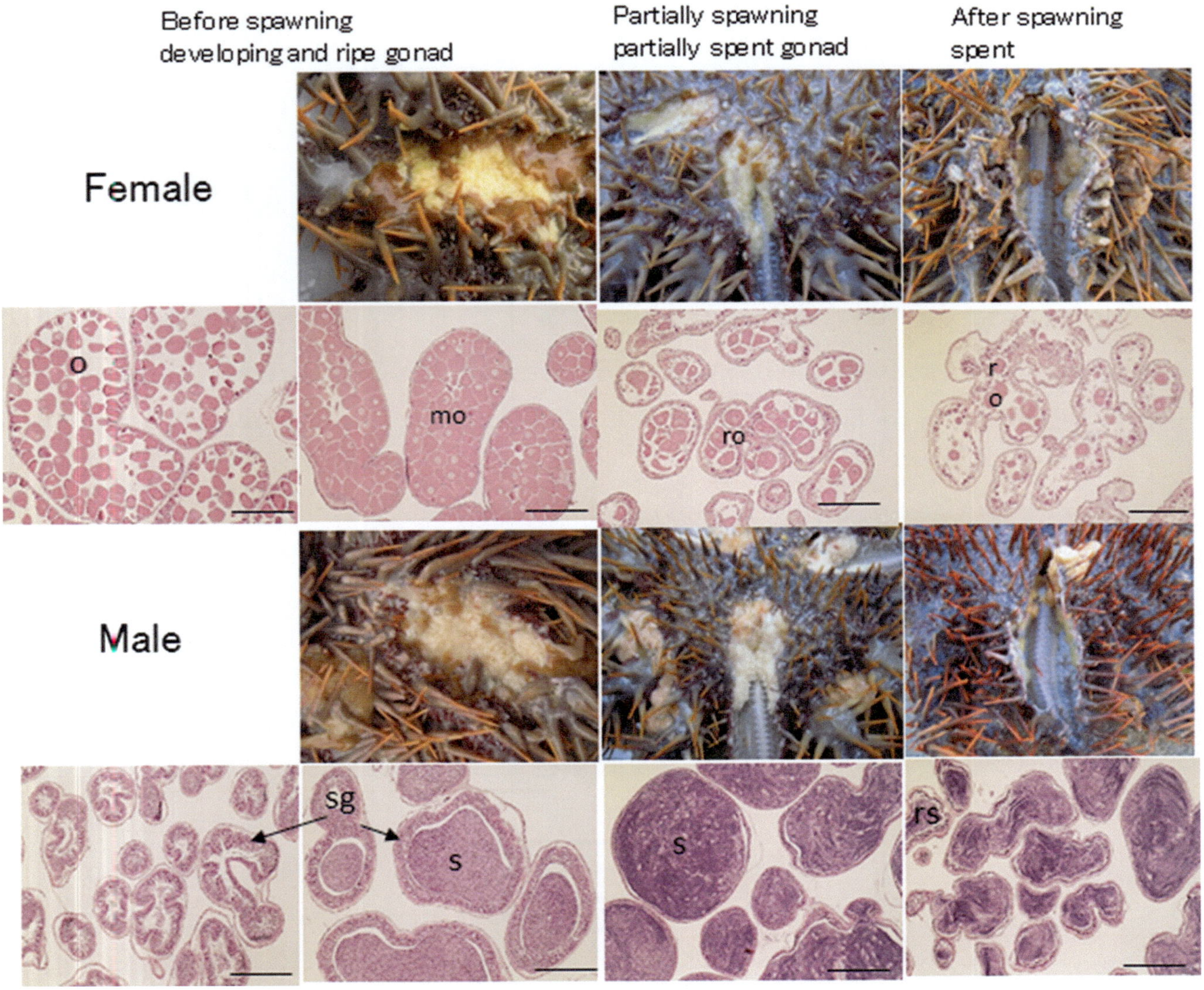


**Fig. 9.2** Comparison of simple dissection and histological examinations during Japanese *Acanthaster planci* sensu lato gonad development and spawning (dissection pictures from Yasuda et al. 2010). *o* oocytes, *mo* mature oocytes, *ro* residual oocytes, *s* spermatozoa, *sg* spermatogenic layer, *rs* residual spermatozoa. Scale bars = 500 μm

60 % volume, gonads at approximately 30 %, and no gonads; all individuals eventually become spent (Yasuda et al. 2010). Surprisingly, while Environment Agency researchers observed the spawning activities during the summer, artificial fertilization was achieved in December (the middle of winter in Japan) but failed in May, even though the gonads appeared to be more mature at that point than in December (Nature Conservation Bureau, Environment Agency 1974). Mature gonads may have remained in the starfish until December from the preceding summer, whereas only immature eggs and sperm were present in May, as suggested by the histology images shown in Fig. 9.2 (left side of slide). Yamazato and Kiyan (1973) implied possible reproduction during September and October on Okinawa Island.

Eggs are fertilized in the water column, and the resultant larvae disperse over a 3–4-week period, depending primarily on food availability and temperature, before settling on the ocean floor (Yamaguchi 1973). Juvenile *A. planci* sensu lato start to eat coral from approximately 4 months in age (8 mm in diameter; Yamaguchi 1974). Before this stage, larval and juvenile *A. planci* sensu lato are often eaten or killed by corals (Yamaguchi 1981). Interestingly, Yamaguchi (1981) observed cauliflower corals, *Pocillopora* spp., catching and killing *A. planci* sensu lato larvae but not eating or digesting them.

In the northern regions (>31°N), at the periphery of their distribution range (e.g., south of Honshu, Shikoku, and Kyushu in Japan), *A. planci* sensu lato are subject to low temperatures during the winter (e.g., <18 °C). However, *A.*

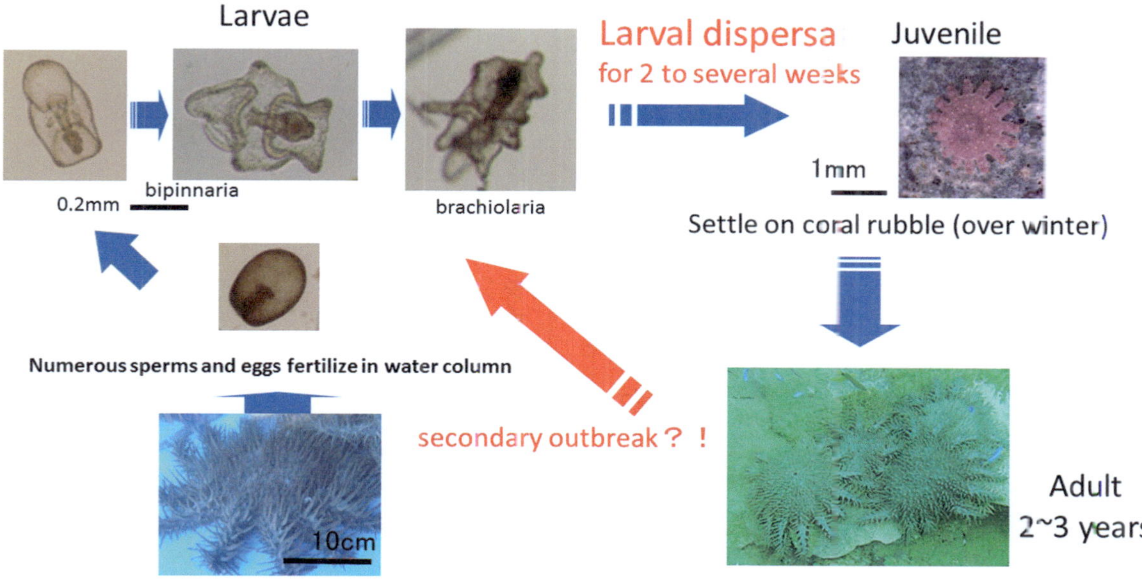


**Fig. 9.3** Life cycle of *Acanthaster planci* sensu lato. *A. planci* sensu lato releases numerous eggs and sperm. After fertilization in the water column, larvae develop from bipinnaria to brachiolaria before settlement. When the juvenile becomes big enough to eat corals, *A. planci* sensu lato grows fast and becomes a mature adult in 2–3 years

*planci* sensu lato juveniles appear to have relatively wide temperature tolerance. Researchers in the 1970s showed that most adult *A. planci* sensu lato collected from Kushimoto, south of Honshu (33.48° N), survived when kept in a tank at 15 °C, the coldest water temperature observed around Kushimoto, suggesting that *A. planci* sensu lato can overwinter in this area (Nature Conservation Bureau, Environment Agency 1974). In Okinawa, however, Takahasi (1986) and Yamaguchi (1987) found that juvenile and subadult *A. planci* sensu lato behaved normally at 18 °C but stopped eating at approximately 16 °C. The starfish survived only for several days at 14 °C and lost their ability to hang on to the substrate. Generally, most adult *A. planci* sensu lato die at temperatures below 13–14 °C (Okaji personal communication). These results suggest that 17–18 °C is the cold-tolerance threshold for juvenile and subadult *A. planci* sensu lato, and 15 °C is the threshold for adult *A. planci* sensu lato.

### 9.1.4 Genetic Structure of *A. planci* sensu lato in Japan

When fertilization is successful and more larvae and juveniles survive than usual, population outbreaks occur (Fig. 9.3). If no prior population outbreaks have occurred around that area and the population outbreak appears suddenly, it is designated as a primary population outbreak. Once a population outbreak has started in an area, the population will likely continue to produce large numbers of larvae. Successive population outbreaks will be caused by the recruitment of larvae produced by upstream population outbreaks, and these are designated as secondary outbreaks (Endean 1974). Secondary population outbreaks are suspected to have occurred in the GBR, Japan, and French Polynesia.

To test the secondary population outbreak hypothesis and estimate larval dispersal, many population genetic analyses have been conducted. Indeed, the genetic structure of *A. planci* sensu lato has been the most intensively and widely studied of all coral reef invertebrate species (Benzie 1999; Volger et al. 2008; Yasuda et al. 2009, 2011, 2013, 2014; Timmers et al. 2011, 2012). Significant genetic isolation by distance across remote Pacific Islands (Yasuda et al. 2009; Vogler et al. 2013) and North Indian Ocean species (Vogler et al. 2012) has been found using mitochondrial and nuclear microsatellite DNA.

In Japan, genetic homogeneity between different populations has been reported for different times and sites. Interestingly, genetic similarity between two outbreak populations 15 years apart examined using allozyme loci in Okinawa implied that relatively large populations are maintained during non-population outbreak (Katoh and Hashimoto 2003). Population genetic analysis using seven highly polymorphic nuclear microsatellite markers (Yasuda et al. 2006) from ten *A. planci* sensu lato populations extending from Sekisei Lagoon (24.31°N, near Taiwan; see Fig. 9.5) to the southern part of Wakayama (33.48°N, south of Honshu) showed no population differentiation based on traditional *F* statistics (Yasuda et al. 2009). Subsequent analysis using 14 microsatellite markers and mitochondrial control

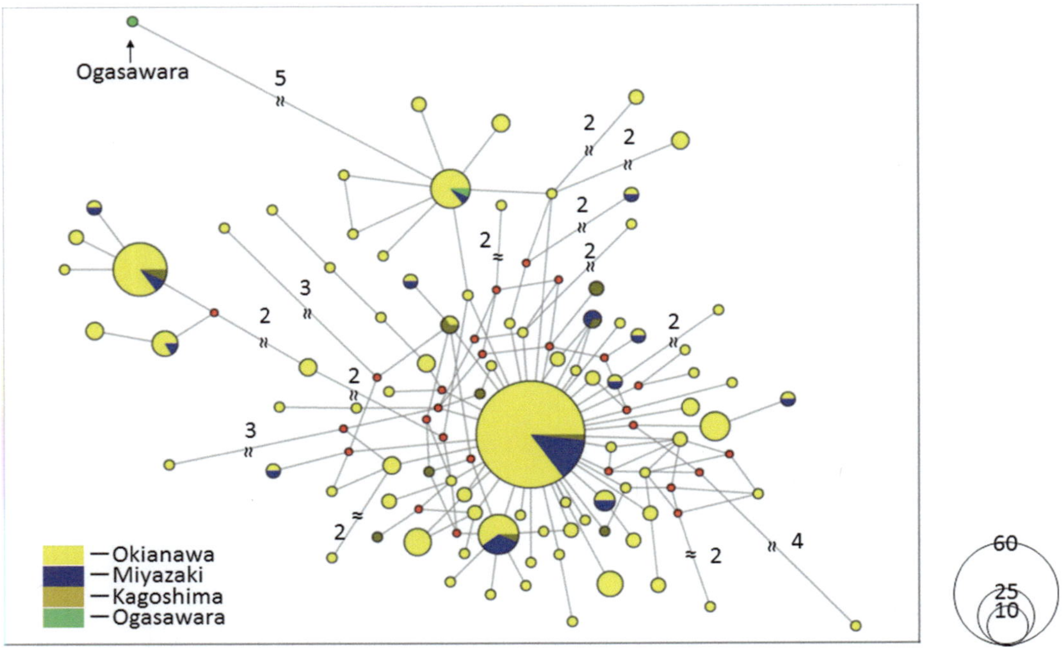


**Fig. 9.4** Mitochondrial DNA haplotype network in Japan. Each *circle* represents a different haplotype, and size is proportional to haplotype frequency. Node *numbers* indicate the number of substitutions between haplotypes

region analysis also showed genetic homogeneity among four of the populations (Miyazaki, Okinawa mainland, Miyako, and Yaeyama). Haplotype network analysis of two individuals from Ogasawara (approximately 1400 km east from Okinawa) suggested that one shared the same haplotype with the Okinawan population while the other differed (Fig. 9.4). Therefore, *A. planci* sensu lato in Ogasawara may originate from both the Kuroshio region (which is equivalent to all of the Japanese population) and another region. Although it is difficult to find many *A. planci* sensu lato from Ogasawara, analyzing further samples will be interesting to trace their origin. While just five individuals were obtained from Miyake Island, population differentiation and microsatellite private alleles were not observed when they were compared with other Japanese *A. planci* sensu lato populations along the Kuroshio Current (Yasuda unpublished data). This observation implies that the Miyake Island population is genetically similar to the Okinawa populations. Additional samples are required to statistically prove a genetic relationship between the Miyake Island and other Japanese populations.

### 9.1.5 Definition of Population Outbreaks

*A. planci* sensu lato population outbreaks are not consistently defined. For example, Moran and De'ath (1992) defined population outbreaks of *A. planci* sensu lato as occurring at a density of 1500 individuals per km$^2$. Coral coverage is expected to decline above that density because the starfish's eating rate exceeds coral growth in the GBR. However, the rate of coral consumption may be different in Japan, because the average size of *A. planci* sensu lato is larger (approximately 40 cm) in the GBR than in Japan (<35 cm), and Japanese *A. planci* sensu lato are thought to eat less coral per individual. The growth rate and composition of coral species also vary across reefs in Japan; consequently, the tolerance threshold of *Acanthaster* density for corals also varies depending on sites. In addition, the water temperature in northern temperate Japan during winter is lower than that in the GBR. For conveniences, a density of 4,000 individuals per km$^2$ has been defined as the tolerance threshold in Japan, as calculated from a survey conducted by Kushimoto Marine Park (Nomura et al. 2001). Thus, the health of coral reefs is conveniently examined by monitoring *A. planci* sensu lato impact based on the criteria in the spot-check manual (Nomura et al. 2001; Ministry of the Environment, Natural Environmental Bureau 2005). The expected area for this spot check is approximately 2500 m$^2$ (50 × 50 m), covered by 15 min of swimming. The manual defines spotting fewer than one *A. planci* sensu lato per 15-min swim per person as normal density, two to four individuals as relatively high density requiring caution, five to nine individuals as a semi-population outbreak, and >10 individuals as an outbreak.

A variety of sources exist regarding the information for *A. planci* sensu lato population outbreaks in Japan. These include anecdotal reports by old fishermen or locals, observation records from fishermen and coral reef researchers, and monitoring data obtained by the Ministry of the Environment

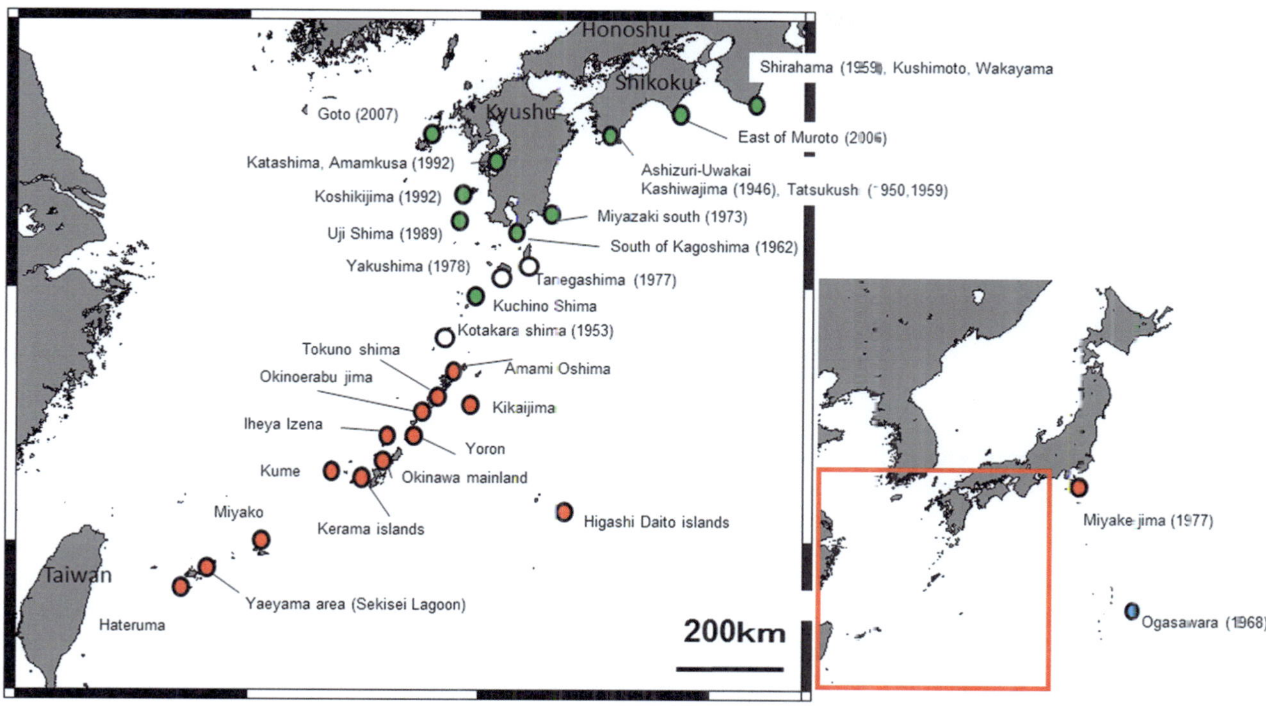


**Fig. 9.5** Places in Japan where *Acanthaster planci* sensu lato was observed before WWII (*red*), observed after WWII (*green*), and where population outbreaks have never been reported (*white*). *Numbers in parentheses* indicate the years in which *A. planci* sensu lato was first observed

and other local government agencies. Intensive control efforts have been led by national and local government agencies, local volunteer divers, and fishermen over seven decades. However, the success rate of removal depends on the number of participants, required effort, detectability of *A. planci* sensu lato individuals, and target area. In particular, it is often difficult to estimate the intensity of population outbreaks based on *A. planci* sensu lato numbers alone.

To clarify the patterns of population outbreaks from the past to the present and to see if the intensity of population outbreaks has increased in recent years, I visualized the intensity of previous population outbreaks in Japan. I used the spot-check manual criteria and an overview of available outbreak information to define the intensity of *A. planci* sensu lato population outbreaks, as shown in Table 9.2.

## 9.2 Review of A. planci sensu lato Occurrence and Population Outbreaks in Japan from 1912 to 2015

Here, I have accumulated and summarized all existing records I could find about this phenomenon. Reports about *A. planci* sensu lato population outbreaks were based on published reports (e.g., Yamaguchi 1986, 1987), survey reports (e.g., Ministry of the Environment), control program reports (e.g., Okinawa Prefectural Government), newspapers, information from local fishermen and professional divers, and other anecdotal information such as blogs with photographs. The oldest written record of *A. planci* sensu lato in Japan was published in 1903: the species was on Amami Oshima by Mitsukuri (1903), who named it Onihitode (demon starfish in English). The oldest anecdotal record of a population outbreak in Japan is sometime between 1912 and 1926 in Yoron (Marine Parks Center of Japan 1987). All of these sources are integrated in Fig. 9.5 and Table 9.3.

### 9.2.1 Before 1960: Population Outbreaks Around the Ryukyu Islands with Relatively Low Human Impact

It was challenging to find official written information on population outbreaks of *A. planci* sensu lato before World War II (WWII). However, because of the starfish's toxic spine and unusual appearance in the sea, local people, especially fishermen, often remember if a population outbreak of *A. planci* sensu lato occurred at least once, and each island often has a local name for *A. planci* sensu lato (Birkeland 1982; Marine Parks Center of Japan 1987). Inquiry surveys were conducted from 1984 to 1987 by the Marine Parks Center of Japan (1987), asking local elders (fishermen's

**Table 9.2** Criteria for the intensity of population outbreak used in this study

| | Intensity | 15 min observation by a diver | per control effort by 3-10 divers with different detactability | Control effort per year per region (e.g. 20 times at 1-25 sites by 10 divers per effort) |
|---|---|---|---|---|
| Not distributed | 0 | Not distributed | | |
| Very low | 1 | 0 - 1 | < 5 | < 100 |
| Sign of increasing | 2 | 2 - 4 | 6 - 49 | 100 - 2,000 |
| Semi-outbreak | 3 | 5 -9 | 50 -100 | 2,000 - 10,000 |
| Outbreak | 4 | 10 - 49 | 100 - 500 | 10,000 - 100,000 |
| Massive outbreak | 5 | 50 < | 500 < | 100,000 < |

unions, fisheries divisions, and directors of community centers) about local names and known population outbreaks of *A. planci* sensu lato on 11 Okinawan Islands, five Amami Islands, and two sites south of Shikoku. This survey clearly showed the presence of local names for *A. planci* sensu lato on tropical islands such as the Okinawan and Amami Islands; however, this was not the case in Shikoku, one of Japan's temperate areas. The oldest anecdotal evidence of *A. planci* sensu lato population outbreaks was observed by local fishermen sometime between 1912 and 1926 on Yoron Island (27.05°N), one of the Nansei Islands in Japan (Marine Parks Center of Japan 1987). Based on the talk from the head of the Yoron fisherman's cooperative association, Mr. Tokuzo Sako, at least three *A. planci* sensu lato population outbreaks were reported to occur around Yoron Island: sometime between 1912 and 1926, 1939–1940, and 1950–1951 (Marine Parks Center of Japan 1987). Mr. Sako remembered that removed starfish were piled up on a sabani, a traditional small fishing boat from Okinawa Island, within an hour during the severest population outbreak in 1950–1951. In Yoron, the removed *A. planci* sensu lato was used as fertilizer for sugarcane farming. A localized *A. planci* sensu lato population outbreak (inferred intensity 4 in Table 9.2) was detected at Amami Oshima (28.48°N) in 1955 (Shirai 1956; see also the oldest *Acanthaster* picture on record, Fig. 9.6). In a questionnaire, Mr. Katsuki Oki, an old local fisherman from Amami Oshima, mentioned a large number of dead *A. planci* sensu lato being washed ashore after a typhoon in 1955, which was then used as fertilizer for pumpkins, supporting the suggestion that a severe outbreak occurred at that time (Oki 2014). Local people from the Amami Islands called *A. planci* sensu lato "America Yui (Friend)" at that time because they believed that the starfish were being transported by the many boats traveling from Okinawa to the Amami Islands during the US administration after WWII. Another anecdotal record of an *A. planci* sensu lato population outbreak was reported from Sesoko (26.63° N), near the Okinawa mainland, in 1942 (Okinawa Prefectural Tourism Development Corporation 1976, Marine Parks Center of Japan 1987). The intensity of this population outbreak was almost the same as that in the 1970s. Anecdotal reports of population outbreaks were also obtained from Hatoma (24.47°N, in 1952 and 1953) and Kabira Bay (24.45°N, in 1958) in the Yaeyama region (see Fig. 9.5, Okinawa Prefectural Tourism Development Bureau 1976; Nature Conservation Division, Department of Cultural and Environmental Affairs, Okinawa Prefectural Government 2006). During the 1953 population outbreak at Hatoma Island, a truckload of starfish was seen almost every day (Nature Conservation Bureau, Environment Agency 1973, 1974). Almost at the same time as the events documented in the Yaeyama region and a few years after the Yoron (27.05°N) population outbreak, an *A. planci* sensu lato population outbreak was reported on Miyako Island (24.90°N, Fig. 9.5) from 1958 to 1959, with 212,700 individuals being removed in 1957 (Yamazato 1969; Nature Conservation Bureau, Environment Agency 1973). Reports of *A. planci* sensu lato population outbreaks were limited before World War II, partly because scuba diving was not as common as it is today, and no one knew that *A. planci* sensu lato ate corals at that time. However, outbreaks of *A. planci* sensu lato populations have been historically reported in Japan (as documented on the Nansei Islands) before the anthropogenic impact became as high as it is today.

Notably, the possible first population outbreak of *A. planci* sensu lato in a temperate region (Kashiwa Island, 32.76°N) occurred in 1946 (Tada 1982). No one in the region, including old fishermen, had seen that animal before. The inquiry survey conducted in 1986 indicated that *A. planci* sensu lato was locally called Genbaku (atomic bomb in English) at that time because skin injuries due to stings by the crown-of-thorns starfish look like radiation sickness (Marine Parks Center of Japan 1987).

**Table 9.3** Population outbreak of *Acanthaster planci* in Japan

| | Region | Latitude | Longtitude | 1912 | 1913 | 1914 | 1915 | 1916 | 1917 | 1918 | 1919 | 1920 | 1921 | 1922 | 1923 | 1924 | 1925 | 1926 | 1927 | 1928 | 1929 | 1930 | 1931 | 1932 | 1933 | 1934 | 1935 | 1936 | 1937 |
|---|---|---|---|---|---|---|---|---|---|---|---|---|---|---|---|---|---|---|---|---|---|---|---|---|---|---|---|---|---|
| Honshu | Choshi | 35.70433 | 140.87738 | 0 | 0 | 0 | 0 | 0 | 0 | 0 | 0 | 0 | 0 | 0 | 0 | 0 | 0 | 0 | 0 | 0 | 0 | 0 | 0 | 0 | 0 | 0 | 0 | 0 | 0 |
| Honshu | Tateyama | 34.98333 | 139.81667 | 0 | 0 | 0 | 0 | 0 | 0 | 0 | 0 | 0 | 0 | 0 | 0 | 0 | 0 | 0 | 0 | 0 | 0 | 0 | 0 | 0 | 0 | 0 | 0 | 0 | 0 |
| Kyusuh NW | Tsushima | 34.26697 | 129.32529 | 0 | 0 | 0 | 0 | 0 | 0 | 0 | 0 | 0 | 0 | 0 | 0 | 0 | 0 | 0 | 0 | 0 | 0 | 0 | 0 | 0 | 0 | 0 | 0 | 0 | 0 |
| Kyushu NE | Nobeoka | 32.58912 | 131.72087 | 0 | 0 | 0 | 0 | 0 | 0 | 0 | 0 | 0 | 0 | 0 | 0 | 0 | 0 | 0 | 0 | 0 | 0 | 0 | 0 | 0 | 0 | 0 | 0 | 0 | 0 |
| Oagasawara | Chichijima | 27.04727 | 142.18268 | 0 | 0 | 0 | 0 | 0 | 0 | 0 | 0 | 0 | 0 | 0 | 0 | 0 | 0 | 0 | 0 | 0 | 0 | 0 | 0 | 0 | 0 | 0 | 0 | 0 | 0 |
| Izu | Niijima | 34.35587 | 139.24278 | 0 | 0 | 0 | 0 | 0 | 0 | 0 | 0 | 0 | 0 | 0 | 0 | 0 | 0 | 0 | 0 | 0 | 0 | 0 | 0 | 0 | 0 | 0 | 0 | 0 | 0 |
| Izu | Miyake | 34.04719 | 139.49619 | 0 | 0 | 0 | 0 | 0 | 0 | 0 | 0 | 0 | 0 | 0 | 0 | 0 | 0 | 0 | 0 | 0 | 0 | 0 | 0 | 0 | 0 | 0 | 0 | 0 | 0 |
| Izu | Hachijyo Jima | 33.14901 | 139.74068 | 0 | 0 | 0 | 0 | 0 | 0 | 0 | 0 | 0 | 0 | 0 | 0 | 0 | 0 | 0 | 0 | 0 | 0 | 0 | 0 | 0 | 0 | 0 | 0 | 0 | 0 |
| Honshu | Kushimoto | 33.47767 | 135.72431 | 0 | 0 | 0 | 0 | 0 | 0 | 0 | 0 | 0 | 0 | 0 | 0 | 0 | 0 | 0 | 0 | 0 | 0 | 0 | 0 | 0 | 0 | 0 | 0 | 0 | 0 |
| Shikoku E | Muroto | 33.27030 | 134.15907 | 0 | 0 | 0 | 0 | 0 | 0 | 0 | 0 | 0 | 0 | 0 | 0 | 0 | 0 | 0 | 0 | 0 | 0 | 0 | 0 | 0 | 0 | 0 | 0 | 0 | 0 |
| Shikoku W | Ashizuri-uwakai | 32.76247 | 132.86831 | 0 | 0 | 0 | 0 | 0 | 0 | 0 | 0 | 0 | 0 | 0 | 0 | 0 | 0 | 0 | 0 | 0 | 0 | 0 | 0 | 0 | 0 | 0 | 0 | 0 | 0 |
| Kyushu W | Goto(Fukue) | 32.74684 | 128.86777 | 0 | 0 | 0 | 0 | 0 | 0 | 0 | 0 | 0 | 0 | 0 | 0 | 0 | 0 | 0 | 0 | 0 | 0 | 0 | 0 | 0 | 0 | 0 | 0 | 0 | 0 |
| Kyushu W | Amakusa | 32.19311 | 129.99381 | 0 | 0 | 0 | 0 | 0 | 0 | 0 | 0 | 0 | 0 | 0 | 0 | 0 | 0 | 0 | 0 | 0 | 0 | 0 | 0 | 0 | 0 | 0 | 0 | 0 | 0 |
| Kyushu W | Koshikijima | 31.84601 | 129.90373 | 0 | 0 | 0 | 0 | 0 | 0 | 0 | 0 | 0 | 0 | 0 | 0 | 0 | 0 | 0 | 0 | 0 | 0 | 0 | 0 | 0 | 0 | 0 | 0 | 0 | 0 |
| Kyushu W | Ujishima | 31.20592 | 129.47001 | 0 | 0 | 0 | 0 | 0 | 0 | 0 | 0 | 0 | 0 | 0 | 0 | 0 | 0 | 0 | 0 | 0 | 0 | 0 | 0 | 0 | 0 | 0 | 0 | 0 | 0 |
| Kyushu E | Miyazaki south | 31.46933 | 131.39251 | 0 | 0 | 0 | 0 | 0 | 0 | 0 | 0 | 0 | 0 | 0 | 0 | 0 | 0 | 0 | 0 | 0 | 0 | 0 | 0 | 0 | 0 | 0 | 0 | 0 | 0 |
| Kyushu S | South Kagoshima | 30.99628 | 130.67078 | 0 | 0 | 0 | 0 | 0 | 0 | 0 | 0 | 0 | 0 | 0 | 0 | 0 | 0 | 0 | 0 | 0 | 0 | 0 | 0 | 0 | 0 | 0 | 0 | 0 | 0 |
| Ohsumi | Tanegashima | 30.39915 | 130.98550 | 0 | 0 | 0 | 0 | 0 | 0 | 0 | 0 | 0 | 0 | 0 | 0 | 0 | 0 | 0 | 0 | 0 | 0 | 0 | 0 | 0 | 0 | 0 | 0 | 0 | 0 |
| Ohsumi | Yakushima | 30.46281 | 130.49343 | 0 | 0 | 0 | 0 | 0 | 0 | 0 | 0 | 0 | 0 | 0 | 0 | 0 | 0 | 0 | 0 | 0 | 0 | 0 | 0 | 0 | 0 | 0 | 0 | 0 | 0 |
| Tokara | Kodakara | 29.22063 | 129.32215 | 0 | 0 | 0 | 0 | 0 | 0 | 0 | 0 | 0 | 0 | 0 | 0 | 0 | 0 | 0 | 0 | 0 | 0 | 0 | 0 | 0 | 0 | 0 | 0 | 0 | 0 |
| Amami | Kikaijima | 28.33752 | 129.96209 | 1 | 1 | 1 | 1 | 1 | 1 | 1 | 1 | 1 | 1 | 1 | 1 | 1 | 1 | 1 | 1 | 1 | 1 | 1 | 1 | 1 | 1 | 1 | 1 | 1 | 1 |
| Amami | Amami | 28.47767 | 129.60981 | 1 | 1 | 1 | 1 | 1 | 1 | 1 | 1 | 1 | 1 | 1 | 1 | 1 | 1 | 1 | 1 | 1 | 1 | 1 | 1 | 1 | 1 | 1 | 1 | 1 | 1 |
| Amami | Tokunoshima | 27.86615 | 128.96590 | | | | | | | | | | | | | | | | | | | | | | | | | | |
| Amami | Okinoerabu | 27.41105 | 128.63056 | | | | | | | | | | | | | | | | | | | | | | | | | | |
| Amami | Yoron | 27.05149 | 128.45914 | 3 | 3 | 3 | 3 | 3 | 3 | 3 | 3 | 3 | 3 | 3 | 3 | 3 | 3 | 3 | 1 | 1 | 1 | 1 | 1 | 1 | 1 | 1 | 1 | 1 | 1 |
| Okinawa | Izena/Iheyajima | 26.96472 | 127.92218 | | | | | | | | | | | | | | | | | | | | | | | | | | |
| Okinawa | Daito-shoto | 25.87377 | 131.24963 | | | | | | | | | | | | | | | | | | | | | | | | | | |
| Okinawa | Okinawa E | 26.29749 | 127.82141 | | | | | | | | | | | | | | | | | | | | | | | | | | |
| Okinawa | Okinawa W | 26.50742 | 127.85324 | | | | | | | | | | | | | | | | | | | | | | | | | | |
| Okinawa | Kerama | 26.20314 | 127.26717 | | | | | | | | | | | | | | | | | | | | | | | | | | |
| Okinawa | Kume | 26.33156 | 126.82681 | | | | | | | | | | | | | | | | | | | | | | | | | | |
| Miyako | Miyako | 24.99089 | 125.28617 | | | | | | | | | | | | | | | | | | | | | | | | | | |
| Yaeyama | Yaeyama (Hatoma, Ishigaki, Sekisei) | 24.31366 | 124.02195 | | | | | | | | | | | | | | | | | | | | | | | | | | |
| Yaeyama | Hateruma | 24.07135 | 123.77215 | | | | | | | | | | | | | | | | | | | | | | | | | | |

(continued)

**Table 9.3** (continued)

| | Region | Latitude | Longtitude | 1938 | 1939 | 1940 | 1941 | 1942 | 1943 | 1944 | 1945 | 1946 | 1947 | 1948 | 1949 | 1950 | 1951 | 1952 | 1953 | 1954 | 1955 | 1956 | 1957 | 1958 | 1959 | 1960 | 1961 | 1962 | 1963 |
|---|---|---|---|---|---|---|---|---|---|---|---|---|---|---|---|---|---|---|---|---|---|---|---|---|---|---|---|---|---|
| Honshu | Choshi | 35.70433 | 140.87738 | 0 | 0 | 0 | 0 | 0 | 0 | 0 | 0 | 0 | 0 | 0 | 0 | 0 | 0 | 0 | 0 | 0 | 0 | 0 | 0 | 0 | 0 | 0 | 0 | 0 | 0 |
| Honshu | Tateyama | 34.98333 | 139.81667 | 0 | 0 | 0 | 0 | 0 | 0 | 0 | 0 | 0 | 0 | 0 | 0 | 0 | 0 | 0 | 0 | 0 | 0 | 0 | 0 | 0 | 0 | 0 | 0 | 0 | 0 |
| Kyusuh NW | Tsushima | 34.26697 | 129.32529 | 0 | 0 | 0 | 0 | 0 | 0 | 0 | 0 | 0 | 0 | 0 | 0 | 0 | 0 | 0 | 0 | 0 | 0 | 0 | 0 | 0 | 0 | 0 | 0 | 0 | 0 |
| Kyushu NE | Nobeoka | 32.58912 | 131.72087 | 0 | 0 | 0 | 0 | 0 | 0 | 0 | 0 | 0 | 0 | 0 | 0 | 0 | 0 | 0 | 0 | 0 | 0 | 0 | 0 | 0 | 0 | | | | |
| Oagasawara | Chichijima | 27.04727 | 142.18268 | 0 | 0 | 0 | 0 | 0 | 0 | 0 | 0 | | | | | | | | | | | | | | | | | | |
| Izu | Niijima | 34.35587 | 139.24278 | 0 | 0 | 0 | 0 | 0 | 0 | 0 | 0 | 0 | 0 | 0 | 0 | 0 | 0 | 0 | 0 | 0 | 0 | 0 | 0 | 0 | 0 | 0 | 0 | 0 | 0 |
| Izu | Miyake | 34.04719 | 139.49619 | 0 | 0 | 0 | 0 | 0 | 0 | 0 | 0 | 0 | 0 | 0 | 0 | 0 | 0 | 0 | 0 | 0 | 0 | 0 | 0 | 0 | 0 | 0 | 0 | 0 | 0 |
| Izu | Hachijyo Jima | 33.14901 | 139.74068 | 0 | 0 | 0 | 0 | 0 | 0 | 0 | 0 | 0 | 0 | 0 | 0 | 0 | 0 | 0 | 0 | 0 | 0 | 0 | 0 | 0 | 0 | 0 | 0 | 0 | 0 |
| Honshu | Kushimoto | 33.47767 | 135.72431 | 0 | 0 | 0 | 0 | 0 | 0 | 0 | 0 | 0 | 0 | 0 | 0 | 0 | 0 | 0 | 0 | 0 | 0 | 0 | 0 | 0 | 1 | 1 | 1 | 1 | 1 |
| Shikoku E | Muroto | 33.27030 | 134.15907 | 0 | 0 | 0 | 0 | 0 | 0 | 0 | 0 | 0 | 0 | 0 | 0 | 0 | 0 | 0 | 0 | 0 | 0 | 0 | 0 | 0 | | | | | |
| Shikoku W | Ashizuri-uwakai | 32.76247 | 132.86831 | 0 | 0 | 0 | 0 | 0 | 0 | 0 | 0 | 3 | 3 | 1 | 1 | 1 | 1 | 1 | 1 | 1 | 1 | 1 | 1 | 1 | 1 | 1 | 1 | 1 | 1 |
| Kyushu W | Goto(Fukue) | 32.74684 | 128.86777 | 0 | 0 | 0 | 0 | 0 | 0 | 0 | 0 | 0 | 0 | 0 | 0 | 0 | 0 | 0 | 0 | 0 | 0 | | | | 0 | 0 | 0 | 0 | 0 |
| Kyushu W | Amakusa | 32.19311 | 129.99381 | 0 | 0 | 0 | 0 | 0 | 0 | 0 | 0 | 0 | 0 | 0 | 0 | 0 | 0 | 0 | 0 | 0 | 0 | | | | 0 | 0 | 0 | 0 | 0 |
| Kyushu W | Koshikijima | 31.84601 | 129.90373 | 0 | 0 | 0 | 0 | 0 | 0 | 0 | 0 | 0 | 0 | 0 | 0 | 0 | 0 | 0 | 0 | 0 | 0 | | | | | | | | |
| Kyushu W | Ujishima | 31.20592 | 129.47001 | 0 | 0 | 0 | 0 | 0 | 0 | 0 | 0 | 0 | 0 | 0 | 0 | 0 | 0 | 0 | 0 | 0 | 0 | | | | | | | | |
| Kyushu E | Miyazaki south | 31.46933 | 131.39251 | 0 | 0 | 0 | 0 | 0 | 0 | 0 | 0 | 0 | 0 | 0 | 0 | 0 | 0 | 0 | 0 | 0 | 0 | | | | | | | | |
| Kyushu S | South Kagoshima | 30.99628 | 130.67078 | 0 | 0 | 0 | 0 | 0 | 0 | 0 | 0 | 0 | 0 | 0 | 0 | 0 | 0 | 0 | 0 | 0 | 0 | | | | | | | 1 | 1 |
| Ohsumi | Tanegashima | 30.39915 | 130.98550 | 0 | 0 | 0 | 0 | 0 | 0 | 0 | 0 | 0 | 0 | 0 | 0 | 0 | 0 | 0 | 0 | 0 | 0 | | | | | | | 1 | 1 |
| Ohsumi | Yakushima | 30.46281 | 130.49343 | 0 | 0 | 0 | 0 | 0 | 0 | 0 | 0 | 0 | 0 | 0 | 0 | 0 | 0 | 0 | 0 | 0 | 0 | | | | | | | 1 | 1 |
| Tokara | Kodakara | 29.22063 | 129.32215 | 0 | 0 | 0 | 0 | 0 | 0 | 0 | 0 | 1 | 1 | 1 | 1 | 1 | 1 | 1 | 1 | 1 | 1 | 1 | 1 | 1 | 1 | 1 | 1 | 1 | 1 |
| Amami | Kikaijima | 28.33752 | 129.96209 | 1 | 1 | 1 | 1 | 1 | 1 | 1 | 3 | 1 | 1 | 1 | 1 | 1 | 1 | 1 | 1 | 1 | 1 | 1 | 1 | 1 | 1 | 1 | 1 | 1 | 1 |
| Amami | Amami | 28.47767 | 129.60981 | 1 | 1 | 1 | 1 | 1 | 1 | 1 | 1 | 1 | 1 | 1 | 1 | 1 | 1 | 1 | 1 | 1 | 4 | 1 | 1 | 1 | 1 | 1 | 1 | 1 | 1 |
| Amami | Tokunoshima | 27.86615 | 128.96590 | | | | | | | | | | | | | | | | | | | | | | | | | | |
| Amami | Okinoerabu | 27.41105 | 128.63056 | | | | | | | | | | | | | | | | | | | | | | | | | | |
| Amami | Yoron | 27.05149 | 128.45914 | 1 | 3 | 3 | 1 | 1 | 1 | 1 | 1 | 1 | 1 | 1 | 1 | 4 | 4 | 1 | 1 | 1 | 1 | 1 | 1 | 1 | 1 | 1 | 1 | 1 | 1 |
| Okinawa | Izena/Iheyajima | 26.96472 | 127.92218 | | | | | | | | | | | | | 2 | 3 | 3 | 3 | 3 | 2 | | | | | | | | |
| Okinawa | Daito-shoto | 25.87377 | 131.24963 | | | | | | | | | | | | | | | | | | | | | | 1 | | | | |
| Okinawa | Okinawa E | 26.29749 | 127.82141 | | | | | | | | | | | | | | | | | | | | | 1 | 1 | 1 | 1 | 1 | 1 |
| Okinawa | Okinawa W | 26.50742 | 127.85324 | | | | | 5 | | | | | | | | | | | | | | 1 | 5 | 1 | 1 | 1 | 1 | 1 | 1 |
| Okinawa | Kerama | 26.20314 | 127.26717 | | | | | | | | | | | | | | | | | | | | 1 | 1 | 1 | 1 | 1 | 1 | 1 |
| Okinawa | Kume | 26.33156 | 126.82681 | | | | | | | | | | | | | | | | | | | | | 4 | 1 | 1 | 1 | 1 | 1 |
| Miyako | Miyako | 24.99089 | 125.28617 | | | | | | | | | | | | | | | | | | | | 5 | 5 | 1 | 1 | 1 | 1 | 1 |
| Yaeyama | Yaeyama (Hatoma, Ishigaki, Sekisei) | 24.31366 | 124.02195 | | | | | | | | | | | 4 | | | | 4 | | | | | | 4 | 1 | 1 | 1 | 1 | 1 |
| Yaeyama | Hateruma | 24.07135 | 123.77215 | | | | | | | | | | | | | | | | | | | | | | | | | | |

| | Region | Latitude | Longtitude | 1964 | 1965 | 1966 | 1967 | 1968 | 1969 | 1970 | 1971 | 1972 | 1973 | 1974 | 1975 | 1976 | 1977 | 1978 | 1979 | 1980 | 1981 | 1982 | 1983 | 1984 | 1985 | 1986 | 1987 | 1988 | 1989 |
|---|---|---|---|---|---|---|---|---|---|---|---|---|---|---|---|---|---|---|---|---|---|---|---|---|---|---|---|---|---|
| Honshu | Choshi | 35.70433 | 140.87738 | 0 | 0 | 0 | 0 | 0 | 0 | 0 | 0 | 0 | 0 | 0 | 0 | 0 | 0 | 0 | 0 | 0 | 0 | 0 | 0 | 0 | 0 | 0 | 0 | 0 | 0 |
| Honshu | Tateyama | 34.98333 | 139.81667 | 0 | 0 | 0 | 0 | 0 | 0 | 0 | 0 | 0 | 0 | 0 | 0 | 0 | 0 | 0 | 0 | 0 | 0 | 0 | 0 | 0 | 0 | 0 | 0 | 0 | 0 |
| Kyusuh NW | Tsushima | 34.26697 | 129.32529 | 0 | 0 | 0 | 0 | 0 | 0 | 0 | 0 | 0 | 0 | 0 | 0 | 0 | 0 | 0 | 0 | 0 | 0 | 0 | 0 | 0 | 0 | 0 | 0 | 0 | 0 |
| Kyushu NE | Nobeoka | 32.58912 | 131.72087 | | | | | | | | | | 1 | 1 | 1 | 1 | 1 | 1 | 1 | 1 | 1 | 1 | 1 | 1 | 1 | 1 | 1 | 1 | 1 |
| Oagasawara | Chichijima | 27.04727 | 142.18268 | | | | | 1 | 1 | 1 | 1 | 1 | 1 | 1 | 1 | 1 | 1 | 1 | 2 | 1 | 1 | 1 | 1 | 1 | 1 | 1 | 1 | 1 | 1 |
| Izu | Niijima | 34.35587 | 139.24278 | 0 | 0 | 0 | 0 | 0 | 0 | 0 | 0 | 0 | 0 | 0 | 0 | 0 | 0 | 0 | 1 | 1 | 1 | 1 | 1 | 1 | 1 | 1 | 1 | 1 | 1 |
| Izu | Miyake | 34.04719 | 139.49619 | 0 | 0 | 0 | 0 | 0 | 0 | 0 | 0 | 0 | 0 | 0 | 0 | 0 | 1 | 3 | 3 | 3 | 1 | 1 | 1 | 1 | 1 | 1 | 1 | 1 | 1 |
| Izu | Hachijyo Jima | 33.14901 | 139.74068 | 0 | 0 | 0 | 0 | 0 | 0 | 0 | 0 | 0 | 0 | 0 | 1 | 1 | 1 | 1 | 1 | 1 | 1 | 1 | 1 | 1 | 1 | 1 | 1 | 1 | 1 |
| Honshu | Kushimoto | 33.47767 | 135.72431 | 1 | 1 | 1 | 1 | 1 | 1 | 1 | 1 | 1 | 2 | 2 | 2 | 2 | 1 | 1 | 1 | 1 | 1 | 1 | 1 | 1 | 1 | 1 | 1 | 1 | 1 |
| Shikoku E | Muroto | 33.27030 | 134.15907 | | | | | | | | | | | | | | | | | 1 | 1 | 1 | 1 | 1 | 1 | 1 | 1 | 1 | 1 |
| Shikoku W | Ashizuri-uwakai | 32.76247 | 132.86831 | 1 | 1 | 1 | 1 | 1 | 1 | 1 | 1 | 1 | 4 | 3 | 3 | 3 | 4 | 4 | 4 | 3 | 3 | 2 | 1 | 1 | 1 | 1 | 1 | 1 | 1 |
| Kyushu W | Goto(Fukue) | 32.74684 | 128.86777 | 0 | 0 | 0 | 0 | 0 | 0 | 0 | 0 | 0 | 0 | 0 | 0 | 0 | 0 | 0 | 0 | 0 | 0 | 0 | 0 | 0 | 0 | 0 | 0 | 0 | 0 |
| Kyushu W | Amakusa | 32.19311 | 129.99381 | 0 | 0 | 0 | 0 | 0 | 0 | 0 | 0 | 0 | 0 | 0 | 0 | 0 | 0 | 0 | 0 | 0 | 0 | 0 | 0 | 0 | 0 | 0 | 0 | 0 | 0 |
| Kyushu W | Koshikijima | 31.84601 | 129.90373 | | | | | | | | | | | | | | | | | | | | | | | | | | |
| Kyushu W | Ujishima | 31.20592 | 129.47001 | | | | | | | | | | | | | | | | | | | | | | | | 2 | 1 | 1 |
| Kyushu E | Miyazaki south | 31.46933 | 131.39251 | | | | | | | | | | 2 | 2 | 2 | 2 | 2 | 3 | 2 | 2 | 2 | 2 | 2 | 2 | 2 | 1 | 1 | 1 | 1 |
| Kyushu S | South Kagoshima | 30.99628 | 130.67078 | 1 | 1 | 1 | 1 | 1 | 1 | 1 | 1 | 1 | 1 | 1 | 1 | 2 | 2 | 3 | 2 | 2 | 2 | 2 | 2 | 2 | 2 | 1 | 1 | 1 | 1 |
| Ohsumi | Tanegashima | 30.39915 | 130.98550 | 1 | 1 | 1 | 1 | 1 | 1 | 1 | 1 | 1 | 1 | 1 | 1 | 1 | 1 | 1 | 1 | 1 | 1 | 1 | 1 | 1 | 1 | 1 | 1 | 1 | 1 |
| Ohsumi | Yakushima | 30.46281 | 130.49343 | 1 | 1 | 1 | 1 | 1 | 1 | 1 | 1 | 1 | 1 | 1 | 1 | 1 | 1 | 1 | 1 | 1 | 1 | 1 | 1 | 1 | 1 | 1 | 1 | 1 | 1 |
| Tokara | Kodakara | 29.22063 | 129.32215 | 1 | 1 | 1 | 1 | 1 | 1 | 1 | 1 | 1 | 1 | 1 | 1 | 1 | 1 | 1 | 1 | 1 | 1 | 1 | 1 | 1 | 1 | 1 | 1 | 1 | 1 |
| Amami | Kikaijima | 28.33752 | 129.96209 | 1 | 1 | 1 | 1 | 1 | 1 | 1 | 1 | 1 | 1 | 1 | 1 | 1 | 1 | 1 | 1 | 1 | 1 | 1 | 1 | 1 | 1 | 1 | 1 | 1 | 1 |
| Amami | Amami | 28.47767 | 129.60981 | 1 | 1 | 1 | 1 | 1 | 1 | 1 | 1 | 1 | 4 | 4 | 4 | 4 | 4 | 4 | 4 | 5 | 4 | 4 | 4 | 4 | 4 | 4 | 3 | 3 | 3 |
| Amami | Tokunoshima | 27.86615 | 128.96590 | | | | | | | | | | 1 | 1 | 1 | 3 | 3 | 2 | 3 | 3 | 2 | 3 | 3 | 2 | 3 | 3 | 3 | 3 | 3 |
| Amami | Okinoerabu | 27.41105 | 128.63056 | | | | | | | | | | | | | | | | | | | 4 | 4 | 4 | | | | | |
| Amami | Yoron | 27.05149 | 128.45914 | 1 | 1 | 1 | 1 | 1 | 1 | 1 | 1 | | 5 | 5 | 4 | 4 | 4 | 4 | 4 | 4 | 4 | 4 | 4 | 3 | 3 | 2 | 2 | 2 | 2 |
| Okinawa | Izena/Iheyajima | 26.96472 | 127.92218 | | | | | | | | | | 4 | 4 | 5 | 5 | 5 | 5 | 5 | | | | | | | | | | |
| Okinawa | Daito-shoto | 25.87377 | 131.24963 | | | | | | | | | | | | | | | | | 1 | 1 | 1 | 1 | 1 | 1 | 1 | 1 | 1 | 1 |
| Okinawa | Okinawa E | 26.29749 | 127.82141 | 1 | 1 | 1 | 1 | 1 | 1 | 1 | 1 | 1 | 1 | 1 | 4 | 4 | 5 | 5 | 5 | 4 | 4 | 4 | 4 | | | | | | |
| Okinawa | Okinawa W | 26.50742 | 127.85324 | 1 | 1 | 1 | 1 | 1 | 3 | 4 | 5 | 5 | 5 | 5 | 5 | 5 | 5 | | | | | | 3 | 4 | 3 | 3 | 2 | 2 | 2 |
| Okinawa | Kerama | 26.20314 | 127.26717 | 1 | 1 | 1 | 1 | 1 | 1 | 1 | 1 | 3 | 3 | 4 | 3 | 3 | 3 | 3 | 3 | 3 | 3 | 3 | 3 | 3 | 3 | 3 | 3 | 4 | 4 |
| Okinawa | Kume | 26.33156 | 126.82681 | 1 | 1 | 1 | 1 | 1 | 2 | 3 | 3 | 5 | 5 | 3 | 3 | | | | | | | | | | | | | | |
| Miyako | Miyako | 24.99089 | 125.28617 | 1 | 1 | 1 | 1 | 1 | 1 | 1 | 1 | 1 | 4 | | | | | | 5 | 5 | 5 | 5 | 5 | 4 | | | | | |
| Yaeyama | Yaeyama (Hatoma, Ishigaki, Sekisei) | 24.31366 | 124.02195 | 1 | 1 | 1 | 1 | 1 | 2 | 3 | 4 | 4 | 4 | 4 | 4 | 4 | 4 | 4 | 4 | 5 | 5 | 5 | 5 | 5 | 4 | 3 | 3 | 3 | 3 |
| Yaeyama | Hateruma | 24.07135 | 123.77215 | | | | | | | | | | | | | | | | | | | | | 5 | 4 | 3 | 2 | 1 | |

(continued)

**Table 9.3** (continued)

| | Region | Latitude | Longtitude | 1990 | 1991 | 1992 | 1993 | 1994 | 1995 | 1996 | 1997 | 1998 | 1999 | 2000 | 2001 | 2002 | 2003 | 2004 | 2005 | 2006 | 2007 | 2008 | 2009 | 2010 | 2011 | 2012 | 2013 | 2014 | 2015 |
|---|---|---|---|---|---|---|---|---|---|---|---|---|---|---|---|---|---|---|---|---|---|---|---|---|---|---|---|---|---|
| Honshu | Choshi | 35.70433 | 140.87738 | 0 | 0 | 0 | 0 | 0 | 0 | 0 | 0 | 0 | 0 | 0 | 0 | 0 | 0 | 0 | 0 | 0 | 0 | 0 | 0 | 0 | 0 | 0 | 0 | 0 | 0 |
| Honshu | Tateyama | 34.98333 | 139.81667 | 0 | 0 | 0 | 0 | 0 | 0 | 0 | 0 | 0 | 0 | 0 | 0 | 0 | 0 | 0 | 0 | 0 | 0 | 0 | 0 | 0 | 0 | 0 | 0 | 0 | 0 |
| Kyusuh NW | Tsushima | 34.26697 | 129.32529 | 0 | 0 | 0 | 0 | 0 | 0 | 0 | 0 | 0 | 0 | 0 | 0 | 0 | 0 | 0 | 0 | 0 | 0 | 0 | 0 | 0 | 0 | 0 | 0 | 0 | 0 |
| Kyushu NE | Nobeoka | 32.58912 | 131.72087 | 1 | 1 | 1 | 1 | 1 | 1 | 1 | 1 | 1 | 1 | 1 | 1 | 1 | 1 | 1 | 1 | 1 | 1 | 1 | 1 | 1 | 1 | 1 | 1 | 1 | 1 |
| Oagasawara | Chichijima | 27.04727 | 142.18268 | 1 | 1 | 1 | 1 | 1 | 1 | 1 | 1 | 1 | 1 | 1 | 1 | 1 | 1 | 1 | 1 | 1 | 1 | 1 | 1 | 1 | 1 | 1 | 1 | 1 | 1 |
| Izu | Niijima | 34.35587 | 139.24278 | 1 | 1 | 1 | 1 | 1 | 1 | 1 | 1 | 1 | 1 | 1 | 1 | 1 | 1 | 1 | 1 | 1 | 1 | 1 | 1 | 1 | 1 | 1 | 1 | 1 | 1 |
| Izu | Miyake | 34.04719 | 139.49619 | 1 | 1 | 1 | 1 | 1 | 1 | 1 | 1 | 1 | 1 | 1 | 1 | 1 | 1 | 1 | 1 | 1 | 1 | 2 | 1 | 1 | 1 | 1 | 1 | 1 | 1 |
| Izu | Hachijyo Jima | 33.14901 | 139.74068 | 1 | 1 | 1 | 1 | 1 | 1 | 1 | 1 | 1 | 1 | 1 | 1 | 1 | 1 | 1 | 1 | 1 | 1 | 1 | 1 | 1 | 1 | 1 | 1 | 1 | 1 |
| Honshu | Kushimoto | 33.47767 | 135.72431 | 1 | 1 | 1 | 1 | 1 | 1 | 1 | 1 | 1 | 1 | 1 | 1 | 2 | 1 | 4 | 4 | 4 | 3 | 4 | 2 | 1 | 2 | 2 | 2 | 2 | 2 |
| Shikoku E | Muroto | 33.27030 | 134.15907 | 1 | 1 | 1 | 1 | 1 | 1 | 1 | 1 | 1 | 1 | 1 | 1 | 1 | 1 | 1 | 1 | 2 | 2 | 2 | 1 | 1 | 1 | 1 | 1 | 1 | |
| Shikoku W | Ashizuri-uwakai | 32.76247 | 132.86831 | 1 | 1 | 1 | 1 | 1 | 1 | 1 | 1 | 1 | 1 | 2 | 2 | 2 | 2 | 3 | 3 | 3 | 3 | 3 | 3 | 4 | 5 | 4 | 3 | 3 | |
| Kyushu W | Goto(Fukue) | 32.74684 | 128.86777 | 0 | 0 | 0 | 0 | 0 | 0 | 0 | 0 | 0 | 0 | 0 | 0 | 0 | 0 | 0 | 0 | 0 | 1 | 1 | 1 | 1 | 2 | 1 | 1 | 1 | 1 |
| Kyushu W | Amakusa | 32.19311 | 129.99381 | 0 | 0 | 0 | 0 | 1 | 1 | 1 | 1 | 1 | 1 | 1 | 1 | 2 | 2 | 1 | 1 | 1 | 1 | 2 | 3 | 3 | 2 | 2 | 1 | 1 | 1 |
| Kyushu W | Koshikijima | 31.84601 | 129.90373 | 2 | 2 | 1 | 1 | 1 | 1 | 1 | 1 | 1 | 1 | 1 | 1 | 1 | 1 | 1 | 1 | 1 | 1 | 1 | 1 | 4 | 3 | 2 | 2 | 2 | 1 |
| Kyushu W | Ujishima | 31.20592 | 129.47001 | 1 | 1 | 1 | 1 | 1 | 1 | 1 | 1 | 1 | 1 | 1 | 1 | 1 | 1 | 1 | 1 | 1 | 1 | 1 | 1 | 1 | 1 | 1 | 1 | 1 | 1 |
| Kyushu E | Miyazaki south | 31.46933 | 131.39251 | 1 | 1 | 1 | 1 | 1 | 1 | 1 | 1 | 1 | 1 | 1 | 1 | 1 | 1 | 1 | 1 | 1 | 1 | 1 | 1 | 1 | 2 | 3 | 2 | 2 | 2 |
| Kyushu S | South Kagoshima | 30.99628 | 130.67078 | 1 | 1 | 1 | 1 | 1 | 1 | 1 | 1 | 1 | 1 | 1 | 1 | 1 | 2 | 2 | 2 | 1 | 4 | 3 | 1 | 2 | 5 | 4 | 3 | 3 | |
| Ohsumi | Tanegashima | 30.39915 | 130.98550 | 1 | 1 | 1 | 1 | 1 | 1 | 1 | 1 | 1 | 1 | 1 | 1 | 1 | 1 | 1 | 1 | 1 | 1 | 1 | 1 | 1 | 1 | 1 | 1 | 1 | 1 |
| Ohsumi | Yakushima | 30.46281 | 130.49343 | 1 | 1 | 1 | 1 | 1 | 1 | 1 | 1 | 1 | 1 | 1 | 1 | 1 | 1 | 1 | 1 | 1 | 1 | 1 | 1 | 1 | 1 | 1 | 1 | 1 | 1 |
| Tokara | Kodakara | 29.22063 | 129.32215 | 1 | 1 | 1 | 1 | 1 | 1 | 1 | 1 | 1 | 1 | 1 | 1 | 1 | 1 | 1 | 1 | 1 | 1 | 1 | 1 | 1 | 1 | 1 | 1 | 1 | 1 |
| Amami | Kikaijima | 28.33752 | 129.96209 | 1 | 1 | 1 | 1 | 1 | 1 | 1 | 1 | 1 | 1 | 1 | 1 | 1 | 1 | 1 | 1 | 1 | 1 | 1 | 1 | 1 | 1 | 1 | 1 | 1 | 1 |
| Amami | Amami | 28.47767 | 129.60981 | 3 | 3 | 3 | 3 | 3 | 3 | 3 | 3 | 3 | 3 | 4 | 5 | 5 | 4 | 2 | 3 | 4 | 3 | 1 | 1 | 1 | 1 | 1 | 1 | 1 | 1 |
| Amami | Tokunoshima | 27.86615 | 128.96590 | 3 | 3 | 3 | 3 | 3 | 2 | 2 | 2 | 2 | 2 | 2 | 2 | 3 | 3 | 3 | 2 | 3 | 3 | 3 | 1 | 1 | 1 | 1 | 1 | 1 | 1 |
| Amami | Okinoerabu | 27.41105 | 128.63056 | | | | | | | | | | | | | | | | 2 | 2 | 2 | 2 | 1 | 1 | 1 | 1 | 1 | 1 | 1 |
| Amami | Yoron | 27.05149 | 128.45914 | 2 | 2 | 2 | 2 | 2 | 2 | 2 | 2 | 2 | 2 | 1 | 1 | 1 | 1 | 1 | 2 | 1 | 1 | 1 | 2 | 1 | 2 | 2 | 2 | | |
| Okinawa | Izena/Iheyajima | 26.96472 | 127.92218 | 3 | 3 | | | | | | | | | | | 3 | 3 | | | 1 | 1 | 1 | 1 | 1 | 1 | 1 | 1 | 1 | 1 |
| Okinawa | Daito-shoto | 25.87377 | 131.24963 | 1 | 1 | 1 | 1 | 1 | 1 | 1 | 1 | 1 | 1 | 1 | | | | 3 | 3 | 3 | 3 | | | | 1 | | | | |
| Okinawa | Okinawa E | 26.29749 | 127.82141 | | 4 | | | | | | 5 | | | | | 4 | 4 | 3 | 3 | 1 | 1 | 1 | 1 | 1 | 1 | 1 | | | |
| Okinawa | Okinawa W | 26.50742 | 127.85324 | 1 | 1 | 1 | 1 | 1 | 1 | 4 | 5 | 5 | 3 | 3 | 3 | 3 | 3 | 3 | 3 | 3 | 3 | 3 | 3 | 3 | 3 | 3 | 3 | 3 | 3 |
| Okinawa | Kerama | 26.20314 | 127.26717 | 3 | 3 | 3 | 3 | 3 | 3 | 4 | 3 | 3 | 3 | 3 | 4 | 5 | 4 | 4 | 4 | 3 | 1 | 1 | 1 | 1 | 2 | 2 | | | |
| Okinawa | Kume | 26.33156 | 126.82681 | 4 | | 1 | | | | | | | | | | | 1 | 4 | 3 | 1 | 1 | 1 | 1 | 1 | 1 | 1 | 1 | 1 | |
| Miyako | Miyako | 24.99089 | 125.28617 | | | | | | | | | | | | | | | 4 | 4 | 4 | 5 | 4 | 4 | 4 | 5 | 4 | 4 | 3 | |
| Yaeyama | Yaeyama (Hatoma, Ishigaki, Sekisei) | 24.31366 | 124.02195 | 2 | 2 | 2 | 2 | 2 | 1 | 1 | 1 | 1 | 1 | 1 | 1 | 1 | 3 | 3 | 3 | 3 | 3 | 4 | 4 | 5 | 5 | 5 | 4 | 4 | 3 |
| Yaeyama | Hateruma | 24.07135 | 123.77215 | | | | | | 1 | 1 | 1 | 1 | 1 | 1 | 1 | 1 | 1 | 1 | | | 3 | 3 | 4 | | | | | | |

**Fig. 9.6** The oldest picture of *Acanthaster planci* sensu lato, taken at Yuikojima, Amami Oshima, in 1956, courtesy of Dr. Shohei Shirai

### 9.2.2 The 1960s: No *A. planci* sensu lato Population Outbreaks but Possible Northward Migration (>29° N)

In the 1960s, people first became aware of the value and importance of coral reefs as a source of leisure activities for tourists. In parallel, anthropogenic impact on coral reefs dramatically increased due to coastal development (Oki 2014). Agriculture in the catchment areas changed from pineapple to sugarcane fields, and severe terrestrial sediment discharge emerged. In the middle of the high-economic growth period of the 1960s, the coral of the Nansei Islands was in recovery, because no intensive *A. planci* sensu lato population outbreaks occurred during this period (Ministry of the Environment and Japanese Coral Reef Society 2004; Oki 2014).

While no intensive *A. planci* sensu lato population outbreaks took place in Japan during the 1960s, possible northward migration occurred in northern temperate areas. For example, no official record of *A. planci* sensu lato existed in areas further north of Amami Oshima before the first *A. planci* sensu lato was found at Kodakara Island (29.21°N) and Nakanoshima on the Tokara Islands in 1953 (Tokioka 1953), although the inquiry survey revealed a small outbreak around Kashiwa Island (33.77°N), southwest Shikoku, in 1946. *A. planci* sensu lato was first officially recorded at Okinoshima (32.70°N), the southeastern most part of Shikoku Island, by Mr. Shohei Shirai in 1959 (Uchinomi 1962). In the same year, a small juvenile *A. planci* sensu lato was also found at Shirahama (33.65°N) in the south of Wakayama by Mr. Torao Yamamoto (Uchinomi 1962). Subsequently, however, neither juvenile nor adult *A. planci* sensu lato were observed in Wakayama until 1970 (Nature Conservation Bureau, Environment Agency 1974). Similarly, the first *A. planci* sensu lato was found in Cape Sata (30.98°N), the southernmost part of Kyushu Island, in 1962 (Kurata 1982). Around the same time, the possible northward migrations of other coral reef organisms, such as *Protoreaster nodosus Ferdina ocellata*, *Phyllacanthus dubis*, *Salmacis bicolor*, *Toxopneustes elegans*, *Hoterocentrotus mamillatus*, *Metalia spartacus*, *Phyrella fragilis* (Uchinomi 1962), and *Culcita novaeguineae* (Nikaido 1963), were also observed in the south of Shikoku and Honshu.

According to the Japan Meteorological Agency website (*Web site 1), which summarizes annual average surface water temperature data from around Japan from 1914 to 2014, the temperature around Kyushu and Shikoku has increased by 1.2 °C over that 100-year period. The increase in water temperature is more remarkable during the winter (almost 1.5 °C around Shikoku and Tokai). Figure 9.7a, b shows these winter water temperature trends at a regional scale over this 100-year period in comparison to the first observations and population outbreaks of *A. planci* sensu lato in temperate areas. Note that the temperature is averaged over a relatively large marine area (Fig. 9.7c), and absolute water temperature values are not directly correlated with *A. planci* sensu lato survival. Overall the increase in water temperature seems to coincide with the first observations of *A. planci* sensu lato and higher population outbreak frequency in temperate areas in Japan. However, the local temperature greatly depends on the path of the Kuroshio Current in a given year. Many coral reef organisms around the Ryukyu Islands that have relatively long larval durations (2–7 weeks, Yamguchi 1973), such as *A. planci* sensu lato and *C. novaeguineae*, spawn during summer when the water temperature exceeds 28 °C (e.g., Yasuda et al. 2009, 2010). After the settlement, at least 2 years are required for *A. planci* sensu lato to mature into adults (Yamaguchi 1974); therefore, juveniles must survive the cold winter. Yamaguchi (1987) speculated that even though the actual lethal temperature for *A. planci* sensu lato juveniles is not known, it should be roughly 14–15 °C. Therefore, winter temperatures may regulate the survival of nonacult *Acanthaster* populations at the northern limits of its distribution (Yamaguchi 1987). The recent increase in water temperature may have facilitated *A. planci* sensu lato migration northward, together with that of other marine animals, including corals (Yamano et al. 2011).

### 9.2.3 The 1970s and 1980s: The First "Confirmed" Successive *A. planci* sensu lato Population Outbreaks in Japan

*A. planci* sensu lato population outbreaks in Japan became more intensive and extensive in the 1970s. Although the Japanese government spent over 600 million yen to remove *A. planci* sensu lato during this period (Yamaguchi 1986), killing over ten million starfish (see the summary of control

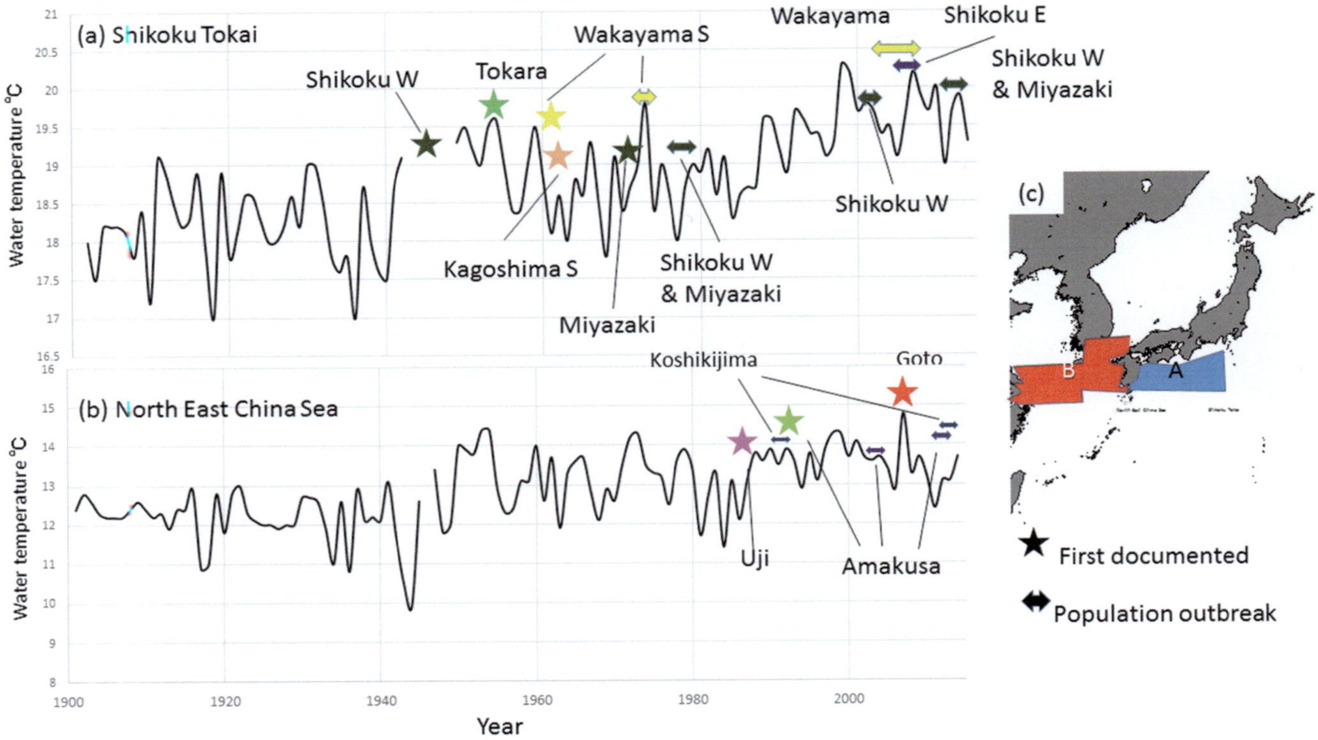


**Fig. 9.7** Average water temperature during the winter months since 1904 in the (**a**) Shikoku Tokai region and (**b**) North East China Sea region (see (**c**)). The first observations of *Acanthaster planci* sensu lato occurrence (*stars*) and outbreaks (*double-headed arrows*) are shown. Temperature data are taken from the Japan Meteorological Agency (http://www.data.jma.go.jp/kaiyou/data/shindan/a_1/japan_warm/japan_warm.html)

efforts from the 1970s to 2004 in Fig. 9.8), the control effort failed, with all but a few corals in Okinawa disappearing before 1985 (Yamaguchi 1986). Chronic and successive population outbreaks were observed along the Nansei Islands and temperate areas (further north than Osumi Islands, >30.40°N). Population outbreaks in the southern parts of Kyushu (31.47°N), Shikoku (32.76°N), Wakayama (33.47°N), and even Miyake Island (34.04°N) were observed, representing the northernmost habitats for *A. planci* sensu lato.

Possible primary population outbreaks (possible origins of the secondary outbreaks) were observed after 1969 at the following locations: Onna village on western Okinawa Island (Yamaguchi 1986), Ie Island (27.44°N, 10 km west of Okinawa Island, Takemoto 2005), Kume Island (Environment Agency 1974), and Hatoma Island (Environment Agency 1974; Okinawa Prefectural Tourism and Development Public Corporation Foundation 1976). Population density peaked at Ie Island in 1972, with approximately 120,000 starfish being removed (Takemoto 2005). In 1973, roughly 1,800,000 starfish were killed along the western coast of Okinawa Island (Okinawa Prefectural Tourism Development Bureau 1976). On Kume Island, 97,500 (within 10 days) and 80,360 (within 6 days) starfish were killed in 1972 and 1973, respectively (Nature Conservation Bureau, Environment Agency 1974), and the population outbreak lasted at least until 1975 (Marine Parks Center of Japan 1987). Subsequent population outbreaks occurred in Yoron Island (23 km north of Okinawa Island), with 309,000 starfish being killed in 1973 (Natural Environmental Bureau, Ministry of the Environment 2003), and Iheya Island (27.04°N) approximately 30 km northwest of Okinawa Island, with 12,000 starfish being killed in 1973 and 260,000 starfish being removed in 1975 and 1976 (Takemoto 2005). From 1975 to early 1998, small but persistent population outbreaks were observed at Tokunoshima (27.87°N, 90 km north of Yoron 27.05°N), with about 2,000 starfish being removed each year since 1976 (Natural Environmental Bureau, Ministry of the Environment 2003, Council of Management for Coral Reef Conservation in Amami Islands 2014, Takemoto 2005). In Amami Oshima, *A. planci* sensu lato numbers peaked in 1980, with 91,602 starfish being removed (Natural Environmental Bureau, Ministry of the Environment 2003; Oki 2014; Council of Management for Coral Reef Conservation in Amami Islands 2014).

Yamaguchi (1987) pointed out that the population outbreaks in Okinawa might have also intensified in Amami Oshima during the early 1980s, when winter water temperatures were higher in the northern Ryukyu Islands.

During the 1970s, population outbreaks in temperate waters were observed for the first time. Around Ashizuri-Uwakai (33.10°N, see Fig. 9.5), in the southern part of

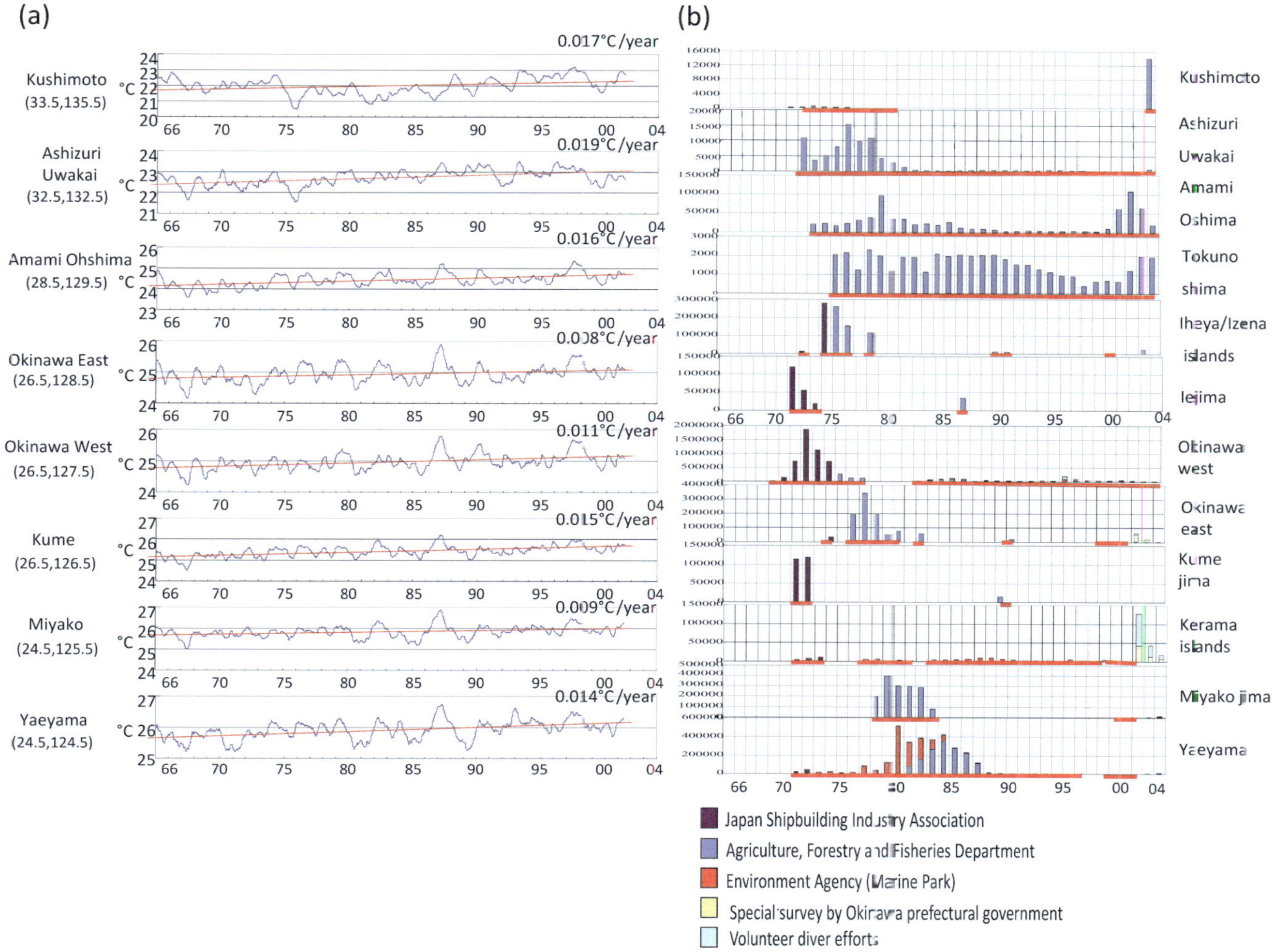


**Fig. 9.8** One-year moving average of water temperature from the Japan Meteorological Agency (**a**) and *Acanthaster planci* sensu lato control efforts in Japan (**b**) from 1970 to 2004. Years of reported control efforts are shown as red underbars (Modified from Takemoto 2005)

Shikoku, more than 10,000 starfish were killed in 1973 (Natural Environmental Bureau, Ministry of the Environment 2003). At this time, local people found ten times more starfish in the area of southern Kochi (32.41°N) than in the northern Ehime area (33.10°N). The Kuroshio Current directly approaches the southern Kochi Prefecture, while only a small branch of the current intrudes into the northern Ehime area. Similarly, a small-scale population outbreak was observed at Kushimoto (33.48°N, south Wakayama, Honshu) for the first time in 1973, with 370 starfish being removed in 1974 (Natural Environmental Bureau, Ministry of the Environment 2003). Interestingly, corals were not much damaged by the small outbreak in 1974 because most of the starfish appeared at the southern part of Cape Shionomisaki (33.25°N) and Cape Sumisaki (33.26°N, western side of Shionomisaki Cape) where small numbers of tabular corals were present at the time (Nature Conservation Bureau, Environment Agency 1974). Some corals were present in neighboring areas, including the tabular *Acropora* spp., which is a preferred food resource of *A. planci* sensu lato; however, *A. planci* sensu lato individuals were not found initially. Because *A. planci* was never found on the eastern side of the Kii Peninsula (or even the eastern side of Cape Shionomisaki), which faces the opposite side of the Kuroshio Current and is colder, *A. planci* sensu lato distribution in Kushimoto is primarily determined by where the Kuroshio Current brings larvae, followed by settlement (Nature Conservation Bureau, Environment Agency 1974). The *A. planci* sensu lato removed at Kushimoto in 1973 ranged from juveniles (5 cm) to adults of over 3 years of age (39 cm), implying that successive recruitment may have occurred (either self-seeding in Kushimoto and/or continuous larval supply from the south) (Nature Conservation Bureau, Environment Agency 1974).

In 1973, several individuals of *A. planci* sensu lato were collected for the first time at both Miyazaki and south Oita,

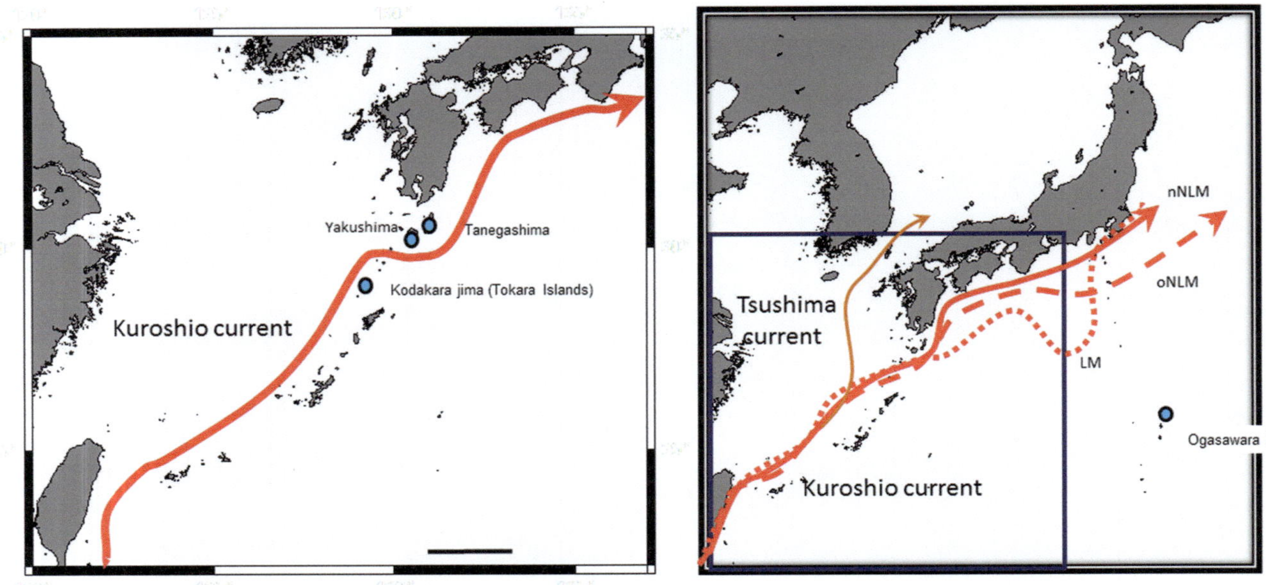


**Fig. 9.9** Typical Kuroshio Current paths (*right*). *LM* large meander path far from the temperate shore, except for Miyake Island, *oNLM* offshore non-large meander path far from the shore, *nNLM* nearshore non-large meander path (Modified from Miyazawa 2004; Yasuda et al. 2009)

eastern Kyushu (Ministry of the Environment and Japanese Coral Reef Society 2004). In early 1978, the first population outbreaks of possibly 1- to 2-year-old *A. planci* sensu lato (9–10 cm in length) were also observed further north, at Miyake Island (Moyer and Tanaka 1978). Moyer and Tanaka (1978) suggested that the variety in size of *A. planci* sensu lato found in 1978 implies that larvae transported from the south reproduced in Miyake Island, increasing their numbers by self-seeding. Yamaguchi (1986) suggested that these population outbreaks were caused by larval transportation via the warm tropical Kuroshio Current after it changed its path to pass along the coasts of Shikoku and Miyake Island. Yamaguchi (1987) also noted that the first *A. planci* sensu lato was recorded at Miyake Island (34.05°N) in 1977, coinciding with a period of warmer years associated with the large Kuroshio Current's meandering approach toward Miyake Island (see Fig. 9.9).

The spread of *A. planci* sensu lato outbreaks was detected along both northward and southward population migratory paths. For example, population outbreaks were observed at Ikema Island (24.94°N) and Yabiji (24.94°N) in 1973 (Okinawa Prefectural Tourism Development Bureau 1976) and then from 1979 to 1984, with almost 400,000 starfish being killed at Miyako (24.75°N) in 1981 (Yamaguchi 1984; Takemoto 2005). The Yaeyama area, located further south of Okinawa Island, had an outbreak in 1981, with approximately 500,000 starfish being removed (Kamezaki et al. 1987). However, the density of *A. planci* sensu lato remained relatively high in this region until 1989 (Yaeyama Environmental Network by Yaeyama Fishermen's Cooperative *Web site 2). Hateruma Island (24.07°N), located to the southwest of Sekisei Lagoon, was also infested in 1984 (Ministry of the Environment and Japanese Coral Reef Society 2004). This outbreak could have been a secondary outbreak following the primary outbreak at Sekisei Lagoon (24.32°N).

Notably, almost all corals along Okinawa Island, Kume Island, and other adjacent small islands disappeared during these outbreaks (Yamaguchi 1984; Nature Conservation Division, Department of Cultural and Environmental Affairs, Okinawa Prefectural Government 2006; Yamaguchi 1986). Yamaguchi (1984) suggested the delineation of regular monitoring sites for the early detection of *A. planci* sensu lato population outbreaks to save important reefs, which is critical for tourism, biodiversity, and research.

### 9.2.4 The 1990s: Few *A. planci* sensu lato Outbreaks but Further Northward Migration

Few intensive population outbreaks newly began in the 1990s (Table 9.3, Nomura 2004; Oki 2014; Takemoto 2005; Natural Conservation Division Okinawa Prefectural Government 2006; Ministry of the Environment and Japanese Coral Reef Society 2004). Instead, chronic outbreaks with origins from the 1970s were observed on Yaeyama, Okinawa Island, Chibishi in Kerama (west of Okinawa Island), Yoron, and Tokunoshima. Notably, coral reefs in Japan were severely damaged by coral bleaching in

1998 (Ministry of the Environment and Japanese Coral Reef Society 2004).

According to an issue of a local newspaper, the Okinawa Times, published on November 7, 1996, approximately 81,000 starfish were killed at Onna Village (the western part of Okinawa Island) in 1996, and high densities have been continually detected around the village until the present (Nakamura et al. 2014, Yasuda personal observation). A huge starfish population was observed at Atta, on the eastern part of Okinawa Island (an average of 367.5 starfish per 10-min swim), and at Chibishi in Kerama in 1997 (Nature Conservation Division, Department of Cultural and Environmental Affairs, Okinawa Prefectural Government 2006; Shimoike 2000). In the Yaeyama area, 1000–2000 starfish per year were continuously removed from 1990 to 1994 (Yaeyama Environmental Network by Yaeyama Fishermen's Cooperative, *Web site 2, Takemoto 2005), but these numbers were much lower than those documented in the 1980s (e.g., 270,000 in 1982, Yaeyama Fishermen's Cooperative *Web site 2).

The Tsushima Current is a smaller branch of the Kuroshio Current that flows along the western side of Kyushu (Fig. 9.9). This current might have brought *A. planci* sensu lato larvae to this temperate area in the 1990s. An increased *A. planci* sensu lato population around the Uji Island (31.21°N), located at the west of southern Kyushu, was first observed in 1987 (Nature Conservation Bureau, Environment Agency and Marine Parks Center of Japan 1994). A population outbreak was also detected at Koshiki Island (31.85°N), located at slightly further north of Uji Island, in 1990 (Nature Conservation Bureau, Environment Agency and Marine Parks Center of Japan 1994). In 1994, *A. planci* sensu lato was first observed in Amamkusa (32.19°N), located at further north of Koshiki Island (Ministry of the Environment, Japanese Coral Reef Society 2004). In 2007, *A. planci* sensu lato was first observed by a local diver at Fukue Island in the Goto Islands (32.75°N, Local diver's website *Web site 3). Although annual data on the path of the Tsushima Current and its associated currents along the west side of Kyushu are not available, it is possible that *A. planci* sensu lato larvae dispersed along these currents, facilitated by the increase in water temperature (Fig. 9.7). A comparison of the fish species at Wakasa Bay, on the western side of Kyushu along the Tsushima Current, between the early 1970s and middle 2000s, suggested that southern fish species such as the rockfish, *Sebastiscus marmoratus*, have significantly increased, while those indigenous to northern waters have decreased, highlighting the effect of global warming around this region (Masuda 2007). Kobayashi et al. (2006) also reported that subtropical species such as slipper lobster *Scyllarides haanii*, brown-tipped sea star *Thromidia catalai*, and blue-lipped sea krait *Laticauda laticaudata* have been recently observed in the area due to global warming.

### 9.2.5 The 2000s: Further Intensive, Persistent *A. planci* sensu lato Population Outbreaks Throughout Southern Japan

In the 2000s, *A. planci* sensu lato population outbreaks in the northern temperate area of Japan seemed more chronic and occurred at comparable or greater intensity than in the 1970s (Table 9.3). Consequently, it was difficult to differentiate between primary and secondary outbreaks. Furthermore, *A. planci* sensu lato population outbreaks were observed for the first time in several new areas.

#### 9.2.5.1 Areas Between Kume Island and Okinawa Island (Around 26°N)

Population outbreaks occurred in the Kerama Islands from 2001 until 2006, peaking in 2002 (Taniguchi 2012). Over 120,000 starfish were killed around the Kerama Islands, with volunteers removing almost two-thirds of the individuals in 2002 (Takemoto 2005, Fig. 9.8). A persistent population outbreak was also observed at Onna village in western Okinawa. Over 20,000 starfish were removed in 2001 and 2002, but the outbreak gradually subsided from 2003 to 2008 (Nakamura et al. 2014).

A population outbreak occurred around Kume Island in 2003 and peaked in 2004, after the population outbreaks in western Okinawa and the Yaeyama region, including the Sekisei Lagoon (Ministry of the Environment and Japanese Coral Reef Society 2004; Natural Conservation Division Okinawa Prefecture 2011; Nature Conservation Division, Department of Cultural and Environmental Affairs, Okinawa Prefectural Government 2006).

#### 9.2.5.2 From Okinawa Island to Amami Oshima

In 2000, a starfish population in the southern part of Amami Oshima (Setouchi-cho and Uken-mura) began to increase, peaking in 2002, when over 100,000 starfish were removed and corals disappeared (Oki 2014; Council of Management for Coral Reef Conservation in Amami Islands [H16-H25, H represents Heisei, Japanese Era name, 2004–2013] 2014). The population outbreak in Amami Oshima lasted until 2007 (Oki 2014; Council of Management for Coral Reef Conservation in Amami Islands [H16-H25] 2014), with the *A. planci* sensu lato population almost disappearing by 2015. Fortunately, corals are now recovering in this area (Fig. 9.10). A starfish population in Tokunoshima (situated between Okinawa Island and Amami Oshima) also began to increase in 2002, doubling in 2003 (Oki 2014; Council of Management for Coral Reef Conservation in Amami Islands [H16-H25] 2014).

#### 9.2.5.3 Yaeyama Region

In the Yaeyama region, which is in the upstream area of the Kuroshio Current, signs of a population increase were

**Fig. 9.10** Changes in coral cover observed at Saneku Point, Amami, through a devastating population outbreak of *Acanthaster planci* sensu lato from 2002 to 2007 and its recovery in 2015 (The 2002 and 2007 images are courtesy of Katsuki Oki)

observed in 2002, and nearly 3000 individuals of *A. planci* sensu lato were removed from Yaeyama in 2003 (Yaeyama Environmental Network by Yaeyama Fishermen's Cooperative *Web site 2). The population size suddenly increased in 2008 and remained at the high density until 2012, when 287,421 individuals were removed (Yaeyama Environmental Network by Yaeyama Fishermen's Cooperative *Web site 2). The population seemed to decrease from 2013 to 2015.

A population outbreak was observed around Miyako Island in 2004, and starfish remained at relatively high density until 2012 (Biodiversity Center of Japan, Natural Environmental Bureau Ministry of the Environment 2015). Intensive population outbreaks were observed in 2004, 2007, and 2011 around Miyako Island, with a persistent population outbreak observed from 2005 to 2008 at Yabiji, 5 km north of Miyako Island.

#### 9.2.5.4 Temperate Areas

A population outbreak in Kinko Bay (south of Kyushu) began in 2004, and 300–500 starfish were removed (Minaminihon Shinbun Newspaper 2004). A moderate population outbreak was observed from 2007 to 2008, and an intensive population outbreak began in 2011 (Biodiversity Center of Japan, Natural Environmental Bureau, Ministry of the Environment 2015). A population outbreak started in Kushima (south of Miyazaki) at the end of 2011 and peaked in 2013. Subsequently, possible secondary outbreaks were observed approximately 25 km north at Tsuki Island and Nichinan in 2014 and 2015, possibly via a branch current of the Kuroshio flowing northward from Kushima to Nichinan. In summer 2016, three divers still found 45 individuals (mostly >30 cm) per day around Nichinan, where almost no hard corals were left uneaten (Yasuda personal observation).

In Koshiki Island, a population outbreak suddenly started in 2010, with 13 fishermen collecting more than 1000 starfish in a day (a blog by a local person in Koshiki Island, *Web site 4) in 2011. However, according to a local fisherman, Mr. Gensui Shimono, the number of *A. planci* sensu lato dramatically decreased from 2012 onward, with just 15 individuals being found by eight divers per year in 2014. Along the western side of Kyushu, a population outbreak was observed for the first time around Amakusa from 2002 to 2003. This outbreak occurred 8 years after the first *A. planci* sensu lato was found in the Amakusa area (Yomiuri Shinbun Oct. 3, 2010). The population density between 2002 and 2003 was not very high (only hundreds of starfish were removed), and the individuals of starfish were of similar size to each other, suggesting a single mass recruitment from the south (Natural Environmental Bureau, Ministry of the Environment 2003). However, the second population outbreak in Amakusa, from 2009 to 2010, was approximately ten times larger than the previous one (over 2000 starfish were removed). At the time, different age classes of starfish were detected (including juveniles), implying a degree of local self-seeding and/or successive migration from the south (Yomiuri Shinbun Oct. 3, 2010).

Population outbreaks more intense than those observed in the 1970s (almost 50 times as many starfish removed) began around the southern part of Kushimoto (Wakayama) in 2004 and continued until 2008 (Biodiversity Center of Japan, Natural Environmental Bureau, Ministry of the Environment 2015). However, signs of potential outbreaks were detected starting in 1999 (Nomura 2004). Like the outbreaks in the 1970s, the 2004 outbreak started on the western side of Cape Shionomisaki, which is very close to the path of the Kuroshio Current. However, because of higher water temperatures (Fig. 9.8), a local researcher speculated that the large population in 2004 might have been promoted by the self-seeding of the local population. Specifically, the population was the offspring of a smaller population in 2002 that had already increased in size since 1999 (Natural Environmental Bureau, Ministry of the Environment 2005). The populations observed in 1999 and 2002 were likely formed by larval clouds transported from southern population outbreaks.

#### 9.2.5.5 Daito Islands

The first known population outbreak at Minami Daito Island (located approximately 360 km east of Okinawa Island) was recorded in 2007, and the outbreak had ended by 2011 (Biodiversity Center of Japan, Natural Environmental Bureau, Ministry of Environment 2015). A local who has dived around Minami Daito Island for the last 30 years, Mr. Hisao Kohama, claimed he had not observed or heard *A. planci* sensu lato occurring around Minami Daito Island before the outbreak in the 2000s but said that *A. planci* sensu lato may have occurred at very low density in the past. However, a few years after major successive population outbreaks around the Okinawa Islands (although he could not remember the actual year), the number of *A. planci* sensu lato increased near the Minami Daito Island (especially around its northern part) to the extent that they were piling up on top of each other for a 3-year period (probably from 2004 to 2007). Although this island is located 360 km from Okinawa Island and is far from the main Kuroshio path, a GPS-equipped buoy released from western Okinawa Island reached Minami Daito Island within a month (Nakamura et al. 2015), implying that larval dispersal from western Okinawa Island to Minami Daito Island is possible.

### 9.2.6 Possible Source Origins of Secondary Population Outbreaks in Japan

Successive outbreaks (category >4 in Table 9.2) were not observed for from 1992 to 1994 in Japan (Table 9.3). Therefore, at least two prolonged successive population outbreaks have occurred in Japan: one from 1969 to 1991, beginning in western Okinawa and Kume Islands, and another from 1995 to 2015 (possibly ongoing at 2016), beginning in western Okinawa. The patterns of these two successive population outbreaks are similar: both started in the region around western Okinawa and progressed to populations in the south (Yaeyama) and north (Amami and temperate areas). The timing of the population outbreaks in Ashizuri-Uwakai and Kushimoto was also similar in both successive outbreaks. Nevertheless, it is not easy to differentiate between primary and secondary outbreaks in Japan because prolonged successive outbreaks also imply self-seeding, which obscures the primary or secondary nature of the outbreak. Thus, I have summarized the possible source populations for each new population outbreak (≥2 in Table 9.3) based on the timing of documented outbreaks in Japan, on the assumption that new outbreaks were caused by larval recruits from other outbreak populations (Table 9.4). Because a period of almost 2 years is required for *A. planci* sensu lato to reach maturity if corals are abundant, larval dispersal from a source population would need at least 2 years before a secondary outbreak. While the counter-Kuroshio Current occurs along the Ryukyu Islands, currents running from temperate to tropical areas have not been documented. Therefore, I have listed all *A. planci* sensu lato populations that have been documented at upstream of the Kuroshio and counter–Kuroshio Currents for each possible secondary outbreak. In addition, I have counted how many times the population of a particular Japanese site (>3) became the possible source origin for another outbreak (>2) that occurred 2 years after the original outbreak (Table 9.4). The areas with the highest likelihood of providing source populations were west Okinawa (44 times) and Yaeyama (40 times) (Table 9.4). If I selected the population geographically closest to the area of a new population outbreak as the most likely source origin (i.e., those shaded in Table 9.4), the highest frequency was detected for west Okinawa (22 times), followed by Amami Oshima (11 times) (Table 9.4). Therefore, one plausible source origin of secondary outbreaks in Japan is west Okinawa. The Amami Oshima population also plays an important role as a "hub population" that may connect population outbreaks from tropical to temperate areas, and it is likely critical for successive population outbreaks in temperate areas. However, more sophisticated methods (such as biophysical modeling) are required to estimate larval dispersal distances from west Okinawa to other places.

### 9.2.7 Places Where Population Outbreaks Have Not Been Observed in Japan

After examining the records of population outbreaks throughout Japan, I identified at least two regions (the Osumi Islands and Ogasawara) where *A. planci* sensu lato population outbreaks have not been observed (or at least recognized), although the starfish has been confirmed to be distributed at these sites (Fig. 9.5). These locations appeared to have been colonized by *A. planci* sensu lato or experienced an increase in population size after WWII (Kurata 1984; Tokioka 1953; Uchinomi 1962).

The Ogasawara Islands are isolated volcanic islands in the Pacific that formed approximately 50 million years ago and are situated roughly 1000 km south of Tokyo, Honshu. No records of *A. planci* sensu lato exist from the Ogasawara Islands before WWII, and it was first officially recorded in 1968 (Kurata 1984). A maximum of 23 individuals were found by local divers in 1979, after the large meandering Kuroshio path (LM in Fig. 9.9) approached the islands (Kurata 1984). In 1994, Tachikawa found a few individuals (Ministry of the Environment and Japanese Coral Reef Society 2004). Subsequently, *A. planci* sensu lato was rarely observed until 2011, when a single individual was observed at Chichi Island in Ogasawara (Biodiversity Center of Japan,

**Table 9.4** Possible sources of secondary outbreaks and its frequency in Japan

| Possible Source | | Seconday Sink | Outbreak Year |
|---|---|---|---|
| Yoron | → | Okinawa W | 1942 |
| Iheya/Izena | → | Amami | 1955 |
| Yaeyama | → | Amami | 1955 |
| Amami | → | Okinawa W | 1957 |
| Amami | → | Miyako | 1957 |
| Okinawa W | → | Yoron | 1972 |
| Kume | → | Yoron | |
| Yaeyama | → | Yoron | |
| Okinawa W | → | Kerama | 1972 |
| Kume | → | Kerama | |
| Yaeyama | → | Kerama | |
| Yaeyama | → | Miyako | 1973 |
| Okinawa W | → | Miyako | |
| Kume | → | Miyako | |
| Okinawa W | → | Iheya/Izena | 1973 |
| Yaeyama | → | Iheya/Izena | |
| Kume | → | Iheya/Izena | |
| Okinawa W | → | Yoron | 1973 |
| Kume | → | Yoron | |
| Yaeyama | → | Yoron | |
| Okinawa W | → | Amami | 1973 |
| Yaeyama | → | Amami | |
| Kume | → | Amami | |
| Okinawa W | → | Miyazaki | 1973 |
| Yaeyama | → | Miyazaki | |
| Kume | → | Miyazaki | |
| Okinawa W | → | Ashizuri-Uwakai | 1973 |
| Yaeyama | → | Ashizuri-Uwakai | |
| Kume | → | Ashizuri-Uwakai | |
| Okinawa W | → | Kushimoto | 1973 |
| Yaeyama | → | Kushimoto | |
| Kume | → | Kushimoto | |
| Okinawa W | → | Yoron | 1973 |
| Yaeyama | → | Yoron | |
| Kume | → | Yoron | |
| Okinawa W | → | Iheya/Izena | 1973 |
| Yaeyama | → | Iheya/Izena | |
| Kume | → | Iheya/Izena | |
| Amami | → | Okinawa E | 1975 |
| Yoron | → | Okinawa E | |
| Iheya/Izena | → | Okinawa E | |
| Okinawa W | → | Okinawa E | |
| Kerama | → | Okinawa E | |
| Kume | → | Okinawa E | |
| Miyako | → | Okinawa E | |
| Yaeyama | → | Okinawa E | |
| Hateruma | → | Okinawa E | |
| Amami | → | Miyazaki | 1976 |
| Yoron | → | Miyazaki | |
| Iheya/Izena | → | Miyazaki | |
| Okinawa W | → | Miyazaki | |
| Kerama | → | Miyazaki | |
| Kume | → | Miyazaki | |
| Yaeyama | → | Miyazaki | |
| Amami | → | Kagoshima S | 1976 |
| Yoron | → | Kagoshima S | |
| Iheya/Izena | → | Kagoshima S | |
| Okinawa W | → | Kagoshima S | |
| Kerama | → | Kagoshima S | |
| Kume | → | Kagoshima S | |
| Yaeyama | → | Kagoshima S | |

| Possible Source | | Seconday Sink | Outbreak Year |
|---|---|---|---|
| Ashizuri-Uwakai | → | Miyake | 1978 |
| Amami | → | Miyake | |
| Tokunoshima | → | Miyake | |
| Yoron | → | Miyake | |
| Iheya/Izena | → | Miyake | |
| Okinawa W | → | Miyake | |
| Kerama | → | Miyake | |
| Yaeyama | → | Miyake | |
| Yaeyama | → | Miyako | 1979 |
| Okinawa W | → | Miyako | |
| Okinawa E | → | Miyako | |
| Tokunoshima | → | Miyako | |
| Yoron | → | Miyako | |
| Iheya/Izena | → | Miyako | |
| Tokunoshima | → | Okinoerabu | 1982 |
| Amami | → | Okinoerabu | |
| Yoron | → | Okinoerabu | |
| Okinawa E | → | Okinoerabu | |
| Kerama | → | Okinoerabu | |
| Miyako | → | Okinoerabu | |
| Yaeyama | → | Okinoerabu | |
| Miyako | → | Okinawa W | 1983 |
| Yaeyama | → | Okinawa W | |
| Kerama | → | Okinawa W | |
| Okinawa E | → | Okinawa W | |
| Yoron | → | Okinawa W | |
| Tokunoshima | → | Okinawa W | |
| Yaeyama | → | Hateruma | 1984 |
| Miyako | → | Hateruma | |
| Kerama | → | Hateruma | |
| Tokunoshima | → | Uji | 1987 |
| Amami | → | Uji | |
| Kerama | → | Uji | |
| Okinawa W | → | Uji | |
| Yoron | → | Uji | |
| Yaeyama | → | Uji | |
| Hateruma | → | Uji | |
| Amami | → | Koshiki | 1990 |
| Tokunoshima | → | Koshiki | |
| Kerama | → | Koshiki | |
| Yaeyama | → | Koshiki | |
| Kerama | → | Iheya/Izena | 1990 |
| Amami | → | Iheya/Izena | |
| Tokunoshima | → | Iheya/Izena | |
| Yaeyama | → | Iheya/Izena | |
| Kerama | → | Kume | 1990 |
| Amami | → | Kume | |
| Tokunoshima | → | Kume | |
| Yaeyama | → | Kume | |
| Kerama | → | Okinawa E | 1991 |
| Yaeyama | → | Okinawa E | |
| Amami | → | Okinawa E | |
| Tokunoshima | → | Okinawa E | |
| Kerama | → | Okinawa W | 1996 |
| Tokunoshima | → | Okinawa W | |
| Amami | → | Okinawa W | |
| Kerama | → | Okinawa E | 1997 |
| Amami | → | Okinawa E | |
| Amami | → | Ashizuri-Uwakai | 2000 |
| Okinawa W | → | Ashizuri-Uwakai | |
| Kerama | → | Ashizuri-Uwakai | |

| Possible Source | | Seconday Sink | Outbreak Year |
|---|---|---|---|
| Okinawa W | → | Iheya/Izena | 2002 |
| Amami | → | Iheya/Izena | |
| Amami | → | Kushimoto | 2002 |
| Okinawa W | → | Kushimoto | |
| Okinawa W | → | Okinawa E | 2002 |
| Amami | → | Okinawa E | |
| Amami | → | Amakusa | 2002 |
| Okinawa W | → | Amakusa | |
| Okinawa W | → | Yaeyama | 2003 |
| Kerama | → | Yaeyama | |
| Amami | → | Yaeyama | |
| Amami | → | Kagoshima S | 2003 |
| Okinawa W | → | Kagoshima S | |
| Kerama | → | Kagoshima S | |
| Okinawa W | → | Yaeyama | 2003 |
| Kerama | → | Yaeyama | |
| Amami | → | Yaeyama | |
| Okinawa E | → | Daoito-jima | 2004 |
| Okinawa W | → | Daoito-jima | |
| Kerama | → | Daoito-jima | |
| Amami | → | Daoito-jima | |
| Tokunoshima | → | Daoito-jima | |
| Izena/Iheya | → | Daoito-jima | |
| Okinawa E | → | Kume | 2004 |
| Okinawa W | → | Kume | |
| Kerama | → | Kume | |
| Amami | → | Kume | |
| Tokunoshima | → | Kume | |
| Izena/Iheya | → | Kume | |
| Okinawa W | → | Miyako | 2004 |
| Kerama | → | Miyako | |
| Okinawa E | → | Miyako | |
| Iheya/Izena | → | Miyako | |
| Ashizuri-Uwakai | → | Kushimoto | 2004 |
| Amami | → | Kushimoto | |
| Tokunoshima | → | Kushimoto | |
| Iheya/Izena | → | Kushimoto | |
| Okinawa E | → | Kushimoto | |
| Okinawa W | → | Kushimoto | |
| Kerama | → | Kushimoto | |
| Amami | → | Okinoerabu | 2005 |
| Tokunoshima | → | Okinoerabu | |
| Okinawa W | → | Okinoerabu | |
| Okinawa E | → | Okinoerabu | |
| Iheya/Izena | → | Okinoerabu | |
| Kerama | → | Okinoerabu | |
| Yaeyama | → | Okinoerabu | |
| Yaeyama | → | Muroto | 2006 |
| Miyako | → | Muroto | |
| Kume | → | Muroto | |
| Okinawa W | → | Muroto | |
| Okinawa E | → | Muroto | |
| Kerama | → | Muroto | |
| Tokunoshima | → | Muroto | |
| Ashizuri-Uwakai | → | Muroto | |
| Amami | → | Kagoshima S | 2007 |
| Okinawa W | → | Kagoshima S | |
| Okinawa E | → | Kagoshima S | |
| Kerama | → | Kagoshima S | |
| Kume | → | Kagoshima S | |
| Miyako | → | Kagoshima S | |
| Yaeyama | → | Kagoshima S | |

| Possible Source | | Seconday Sink | Outbreak Year |
|---|---|---|---|
| Yaeyama | → | Hateruma | 2008 |
| Miyako | → | Hateruma | |
| Kerama | → | Hateruma | |
| Okinawa W | → | Hateruma | |
| Kushimoto | → | Miyake | 2008 |
| Ashizuri-Uwakai | → | Miyake | |
| Amami | → | Miyake | |
| Tokunoshima | → | Miyake | |
| Okinawa W | → | Miyake | |
| Kerama | → | Miyake | |
| Miyako | → | Miyake | |
| Yaeyama | → | Miyake | |
| Amami | → | Amakusa | 2008 |
| Tokunoshima | → | Amakusa | |
| Okinawa W | → | Amakusa | |
| Kerama | → | Amakusa | |
| Miyako | → | Amakusa | |
| Yaeyama | → | Amakusa | |
| Okinawa W | → | Yoron | 2009 |
| Miyako | → | Yoron | |
| Yaeyama | → | Yoron | |
| Hateruma | → | Yoron | |
| Kagoshima S | → | Koshiki | 2010 |
| Tokunoshima | → | Koshiki | |
| Okinawa W | → | Koshiki | |
| Miyako | → | Koshiki | |
| Yaeyama | → | Koshiki | |
| Hateruma | → | Koshiki | |
| Tokunoshima | → | Kagoshima S | 2010 |
| Okinawa W | → | Kagoshima S | |
| Miyako | → | Kagoshima S | |
| Yaeyama | → | Kagoshima S | |
| Hateruma | → | Kagoshima S | |
| Ashizuri-Uwakai | → | Kushimoto | 2011 |
| Yoron | → | Kushimoto | |
| Okinawa W | → | Kushimoto | |
| Miyako | → | Kushimoto | |
| Yaeyama | → | Kushimoto | |
| Amakusa | → | Goto | 2011 |
| Okinawa W | → | Goto | |
| Miyako | → | Goto | |
| Yaeyama | → | Goto | |
| Hateruma | → | Goto | |
| Ashizuri-Uwakai | → | Kushimoto | 2011 |
| Okinawa W | → | Kushimoto | |
| Miyako | → | Kushimoto | |
| Yaeyama | → | Kushimoto | |
| Ashizuri-Uwakai | → | Miyazaki | 2011 |
| Okinawa W | → | Miyazaki | |
| Miyako | → | Miyazaki | |
| Yaeyama | → | Miyazaki | |
| Hateruma | → | Miyazaki | |
| Okinawa W | → | Kerama | 2011 |
| Hateruma | → | Kerama | |
| Miyako | → | Kerama | |
| Yaeyama | → | Kerama | |
| Okinawa W | → | Yoron | 2011 |
| Miyako | → | Yoron | |
| Yaeyama | → | Yoron | |
| Hateruma | → | Yoron | |

| Possible source for secondary outbreak | Frequency |
|---|---|
| Okinawa W | 44 |
| Yaeyama | 40 |
| Amami | 29 |
| Kerama | 28 |
| Izena/Iheya | 22 |
| Tokunoshima | 19 |
| Miyako | 18 |
| Kume | 16 |
| Okinawa E | 10 |
| Yoron | 10 |
| Hateruma | 9 |
| Ashizuri^Uwakai | 7 |
| Amakusa | 1 |
| Kagoshima S | 1 |
| Kushimoto | 1 |

| Geographically closest source populations | Frequency |
|---|---|
| Okinawa W | 22 |
| Amami | 11 |
| Kerama | 8 |
| Ashizuri^Uwakai | 6 |
| Yaeyama | 4 |
| Tokunoshima | 3 |
| Izena/Iheya | 2 |
| Okinawa E | 2 |
| Yoron | 1 |
| Amakusa | 1 |
| Kagoshima S | 1 |
| Kushimoto | 1 |

All possible source locations where population outbreak took place two years before the initiation of other location were listed unless it is physically impossible. Geographically nearest populations are highlighted in yellow

Natural Environmental Bureau Ministry of Environment 2015). The Ogasawara Islands are geographically isolated from the Ryukyu Islands and are not usually in the path of the main Kuroshio Current; however, the regional and local hydrodynamic patterns are largely unknown (The Ministry of the Environment and Japanese Coral Reef Society 2004). In the case of broadcast spawner coral species, Nakajima et al. (2012) found significant population differentiation between Okinawan and Ogasawara populations, indicating limited larval dispersal between the two regions. Population outbreaks have been reported in the North Mariana Islands, approximately 1000 km south of the Ogasawara Islands. Although no study has directly compared populations in Ogasawara and other Pacific Islands, previous studies using microsatellites (Yasuda et al. 2009) and mitochondrial DNA (Houk et al. 2007; Vogler et al. 2013) have indicated genetic isolation between Pacific Islands, suggesting limited larval dispersal between them. Therefore, larval dispersal from the Mariana Islands to the Ogasawara Islands is likely also limited. Kayanne et al. (2012) reported that the number of coral species is much smaller in the Ogasawara Islands than in the Mariana and Okinawan Islands, implying that the larval supply of coral species from the Mariana Islands to the Ogasawara Islands is limited.

The water temperature range of these islands is optimal (19–28 °C) for both larval and adult *A. planci* sensu lato (Ministry of the Environment and Japanese Coral Reef Society 2004). A few rivers steadily discharge into the sea from these islands; thus, the nutrient level is similar to that of other subtropical areas in Japan, such as north Okinawa Island as a whole (The Ministry of the Environment and Japanese Coral Reef Society 2004), although the nutrient concentration from the land is higher in the winter than in the summer due to limited precipitation during the summer season in the Ogasawara Islands (Nohara et al. 2009). The lower nutrient concentration due to limited precipitation during the spawning period of *A. planci* sensu lato may help prevent outbreaks in Ogasawara Islands. Another difference between the Ogasawara Islands and other coral reef areas in Japan that have experienced *A. planci* sensu lato population outbreaks is the absence of secondary population outbreaks originating from Okinawa and other areas, such as Mariana Islands. Therefore, in addition to *A. planci* sensu lato occurring at low density on the islands, secondary larval recruitment is much less than in other regions, preventing population outbreaks.

In contrast, Tanegashima and Yaku Island (Osumi Islands) are volcanic islands situated between Kyushu and the Amami Islands along the Kuroshio Current. The main stream of the Kuroshio Current flows from the East China Sea to the Pacific through these islands. The hard coral cover and number of coral species in the Osumi Islands are slightly lower than in the Ryukyu Islands, but there is higher soft coral coverage (WWF Japan 2008). Yaku Island is famous for its world heritage old cedar forest, which is supported by the highest annual precipitation in Japan, and is visited by 300,000–700,000 tourists annually, with at least 13 diving shops. An *A. planci* sensu lato was officially recorded on Yaku Island in 2003 (Nature Conservation Division, Okinawa Prefectural Government 2003). A local diver for the diving boat service Katsushinmaru, Mr. Masaru Takeishi, stated that just two individuals of *A. planci* sensu lato were observed over the last 8 years: one in 2010 and another in 2011 at Isso, in the north of Yaku Island. While basic information is more limited in Tanegashima, population outbreaks have definitely not occurred between 2003 and 2015 (Biodiversity Center of Japan, Natural Environmental Bureau, and Ministry of Environment 2015). A local diver for the Sea-Mail diving service in Tanegashima, Mr. Tetsuro Hayashi, only found one to two *A. planci* sensu lato individuals per year when the water temperature was high, although he has dived at several coral-rich places around Tanegashima almost daily since 1997.

Population outbreaks in Kyushu, Shikoku, Honshu, and even Miyake Island have been partly attributed to larval dispersal by the Kuroshio Current. The strong flow of the Kuroshio Current is believed to directly pass these islands, while the strong ocean currents around the Osumi Islands lead to the presence of few benthic species, including corals, around these islands (Hirata 1967). Juvenile *A. planci* sensu lato has a high mortality rate in the field (Keesing and Halford 1992). The growth rate of juvenile *A. planci* sensu lato dramatically increases when its food is changed from coral algae to corals (Birkeland and Lucas 1990). Therefore, without enough corals for juvenile *A. planci* sensu lato to grow quickly, mortality will be high, preventing population outbreaks. Alternatively, despite some larval transport, the survival rates of *A. planci* sensu lato juveniles that settle around these islands may be low. For instance, Mr. Tetsuro Hayashi in Tanegashima once found many small *A. planci* sensu lato juveniles (<1 cm length, probably settled that year) beneath the table corals at Nakase in 2012; however, no subsequent adult population outbreak occurred. The coral reef communities in Tanegashima are in relatively good condition; thus, predators of juvenile or adult *A. planci* sensu lato may be abundant, preventing population outbreaks. In an inquiry survey in the Okinawa Islands, a fisherman on Kume Island mentioned that, in years when the trumpet shell is abundant, the number of *A. planci* sensu lato is reported to be small (Marine Parks Center of Japan 1987). Similarly, Shirai (1956) noted that a local specialist in the collection of great green turbans (*Turbo marmoratus*) on Amami Oshima mentioned whenever he found a trumpet shell, *A. planci* sensu lato were always nearby. He also often found *A. planci* sensu lato inside the stomachs of trumpet shells (8–9 of 10 trumpet shells). These stories show that local people in Okinawa and

Amami Oshima thought the trumpet shell to be an important predator for suppressing the population density of *A. planci* sensu lato. However, it is also possible that trumpet shells vigorously ate *A. planci* sensu lato because the population density of the starfish was very high at that time.

Whatever the case, questions remain about why population outbreaks are rarely observed on other islands. Thus, the mechanisms/processes leading to population outbreaks need to be identified, and comparative analyses of these non-outbreak regions may provide answers.

## 9.3 Conclusions

The answers for the questions are the following:

1. Is *A. planci* sensu lato migrating toward the north like other coral species?

   I considered it is highly possible. While there is no record of *A. planci* sensu lato in temperate area between 100 and 50 years ago, now it is quite common and conspicuous. Given the increasing water temperature, it is also possible that such poleward migration is at least partly related to global warming. Further careful observations including fluctuation of the Kuroshio Current are needed.
2. Has the intensity of population outbreaks increased in recent years?

   Yes for temperate areas but not obvious for tropical ones. It was expected that intensity of population outbreaks in temperate areas such as Kushimoto, Ashizuri-Uwakai, and western side of Kyushu regions has increased based on Table 9.3. Reason for this is unclear though the sizes of population seem to be increasing even during non-outbreak periods which would be also related to the abundance of corals and less mortality of *A. planci* sensu lato during winter.
3. Is the western Okinawa population really the source of other population outbreaks in Japan?

   Considering the historical patterns (Table 9.4), yes. It is still unclear, however, how much distance large numbers of larvae can normally spread between islands causing secondary outbreaks. Population genetic methods that clarify relatively long-term migration patterns showed genetic homogeneity along *A. planci* sensu lato populations in Japan though the methods for the moment cannot distinguish secondary outbreak phenomena from a small number of migration that often occurred for evolutionally time scale.

   As a whole, it is highly possible that the impact of coral predation by *A. planci* sensu lato especially in temperate area in Japan has been increasing for several decades. Further studies regarding the balance between frequency of population outbreak of *A. planci* sensu lato and increase of corals in temperate area are critical for considering coastal conservation strategy. As the Kuroshio Current plays an important role for the transport of larvae, future examination between past ocean current pathways and the timing and spread of secondary outbreak using numerical simulation would improve our understanding of the patterns of secondary outbreaks. Furthermore, developing some methods for quantifying local self-recruitment would be important to prevent local chronicle outbreak. Direct estimation of larval identification and density in the field during spawning periods have become possible recently (Yasuda et al. 2015; Suzuki et al. 2016; Uthicke et al. 2015). Further improvement of such technique would help to gain more quantitative information about self-recruitment and make early detection of population outbreak possible in the future.

**Acknowledgments** I am grateful to Dr. Ken Okaji for reviewing and providing many significant detailed comments about this manuscript. I also thank Mr. Eiji Yamakawa and Mr. Megumu Tsuchikawa for sampling *A. planci* sensu lato on Okinawa and Ogasawara Islands. I am also grateful to everyone who provided information about *A. planci* population outbreaks. In particular, I thank Mr. Katsuki Oki and Mr. Hisao Kohama from the Minami Daitojima Fishery Corporation, Mr. Masaru Takeishi from Yakushima, Mr. Gensui Shimono from Koshiki-jima Fishery Corporation, Mr. Tetsuro Hayashi from Sea-Mail in Tanegashima, Ms. Emiko Ando from Kagoshima Environmental Research and Service, Dr. Shu Nakachi from the Biological Institute on Kuroshio, and Ms. Toshiko Nishibe. Sampling for genetic analysis was conducted within the framework of a comprehensive management program of the crown-of-thorns starfish outbreaks in Okinawa Prefecture funded by the Nature Conservation Division, Department of Cultural and Environmental Affairs, Okinawa Prefectural Government. This research was also supported by the Environment Research and Technology Development Fund (4RF-1501) of the Ministry of the Environment, Japan, and the Program to Disseminate Tenure Tracking System from the Ministry of Education, Culture, Sports, Science and Technology (MEXT).

## References

Asahi Shimbun Newspaper (2004) Sep. 12 (in Japanese)

Benzie J (1999) Genetic structure of coral reef organisms: ghosts of dispersal past. Am Zool 39:131–145

Biodiversity Center of Japan, Natural Environmental Bureau, Ministry of Environment (2015) Compiled Report 2008–2012 Monitoring-site 1000 (in Japanese)

Biodiversity Center of Japan, Natural Environmental Bureau, Ministry of the Environment (2013) Monitoring sites 1000 (in Japanese)

Biodiversity Center of Japan, Natural Environmental Bureau, Ministry of the Environment (2014) Monitoring sites 1000 (in Japanese)

Birkeland C (1982) Terrestrial runoff as a cause of outbreaks of *Acanthaster planci* (Echinodermata: Asteroidea). Mar Biol 69:175–185

Birkeland C, Lucas J (1990) *Acanthaster planci*. Major management problem of coral reefs. CRC Press, Boston

Council of Management for Coral Reef Conservation in Amami Islands (16-H25) (2014) Compiled report of coral reef monitoring and control effort for crown of thorn starfish (in Japanese)
Endean R (1974) *Acanthaster planci* on the Great Barrier Reef. Proc 2nd Int Coral Reef Symp 1:563–576
Haszprunar G, Spies M (2014) An integrative approach to the taxonomy of the crown-of-thorns starfish species group (Asteroidea: Acanthaster): a review of names and comparison to recent molecular data. Zootaxa 25:271–284
Hirata K (1967) Distribution of raised reef limestones and present reef-building activity in Takara-jima and Kodakara-jima, Nansei Islands. Jpn Sci Reps Kagoshima Univ 16:75–107. (In Japanese)
Houk P, Bograd S, van Woesik R (2007) The transition zone chlorophyll front can trigger Acanthaster planci outbreaks in the Pacific Ocean: historical confirmation. J Oceanogr 63:149–154
Jangoux M, Aziz A (1984) Les asterides (Echinodermes) du centre-ouest de l'ocean Indien (Seychelles, Maldives et iles Mineures). Bull Mus Nat d'Hist Nat Paris Ser 6(A)(4):857–884
Kamezaki N, Nomura K, Ui S (1987) Population dynamics of stony corals and *Acanthaster planci* in Sekisei Lagoon, Okinawa (1983–1986). Mar Parks J 74:12–17
Katoh M, Hashimoto K (2003) Genetic similarity of outbreak populations of crown-of-thorns starfish (*Acanthaster planci*) that were 15 years apart in Okinawa, Japan. Coral Reefs 22:178–180
Kayanne H, Hongo C, Okaji K, Ide Y, Hayashibara T, Yamamoto H, Mikami N, Onodera K, Ootsubo T, Takano H, Tonegawa M, Maruyama S (2012) Low species diversity of hermatypic corals on an isolated reef, Okinotorishima, in the northwestern Pacific. Galaxea, J Coral Reef Stud 14:73–95
Keesing J, Halford A (1992) Importance of postsettlement processes for the population dynamics of *Acanthaster planci* (L.). Marine Freshw Res 43:635–651
Kobayashi T, Hori S, Doi H, Kawano M (2006) Noteworthy phenomena on the marine organisms in the coastal region of Yamaguchi Prefecture, southwestern Japan Sea. Bull Yamaguchi Pref Fish Res Ctr 4:19–56
Kurata (1984) *Acanthaster planci* in Ogasawara Islands. Mar Parks J 61:7–9. (in Japanese)
Lucas J (1973) Reproductive and larval biology of *Acanthaster planci* (L.) in Great Barrier Reef waters. Micronesica 9:197–203
Lucas J, Jones M (1976) Hybrid crown-of-thorns starfish (*Acanthaster planci* X *A. brevispinus*) reared to maturity in the laboratory. Nature 263:409–412
Marine Parks Center of Japan (1987) Report for study of revealing factors of population dynamics of organisms in marine ecosystem and its management method (S 61). (in Japanese)
Masuda R (2007) Seasonal and interannual variation of subtidal fish assemblages in Wakasa Bay with reference to the warming trend in the sea of Japan. Environ Biol Fish 82:387–399
Ministry of the Environment and Japanese Coral Reef Society (2004) Coral reefs of Japan. Japan Wildlife Research Center, Tokyo
Ministry of the Environment, Natural Environmental Bureau (2005) Investigation for management policy report (Survey on marine organisms living in Kushimoto Marine Park Region). (In Japanese)
Mitsukuri K (1903) Sample collecting travel report of Amami Oshima and Okinawa. Zool Mag 15:176. (in Japanese)
Miyazaki Nichinichi Shinbun (2012) (in Japanese)
Moran P (1986) The *Acanthaster* phenomenon. Oceanogr Mar Biol Ann Rev 24:379–480
Moran PJ, De'ath G (1992) Estimates of the abundance of the crown-of-thrones starfish *Acanthaster planci* in outbreaking and non-outbreaking populations on reefs within the Great Barrier Reef. Mar Biol 113:509–515
Moyer JT, Tanaka T (1978) Crown-of-thorns starfish invades Miyake-jima. Memorical Biol Station Miyake-jima 44:17. (in Japanese)
Nakachi S (2007) Discovery of *Acanthaster brevispinus* at Okinoshima Sukumo. Current 27(7) 6. (in Japanese)
Nakajima Y, Nishikawa A, Iguchi A, Sakai K (2012) Regional genetic differentiation among northern high-latitude island populations of a broadcast-spawning coral. Coral Reefs 31:1125–1133
Nakamura M, Okaji K, Higa Y, Yamakawa E, Mitarai S (2014) Spatial and temporal population dynamics of the crown-of-thorns starfish, *Acanthaster planci*, over a 24-year period along the central west coast of Okinawa Island, Japan. Mar Biol 161 2521–2530
Nakamura M, Kumagai NH, Sakai K, Okaji K, Ogasawara K, Mitarai S (2015) Spatial variability in recruitment of acroporid corals and predatory starfish along the Onna coast, Okinawa, Japan. Mar Ecol Prog Ser 540:1–12
Natural Conservation Division, Department of Cultural and Environmental Affairs Okinawa Prefectural Government (2011) Report for renovation project for coral reef resources (H22) (in Japanese)
Natural Environmental Bureau, Ministry of the Environment (2003) The report of conservation project for marine parks (H14) (in Japanese)
Natural Environmental Bureau, Ministry of the Environment (2005) Investigation on management policy. A report for investigation of marine organism distribution in Kushimoto Marine Park (in Japanese)
Nature Conservation Bureau, Environment Agency (1973) Research on shallow marine ecosystem. Research on the mechanisms and management for the population outbreak of crown-of-thorns starfish (in Japanese)
Nature Conservation Bureau, Environment Agency (1974) The sequel to the research on shallow water ecology. Mechanism of crown-of-thorns starfish outbreak and its countermeasures (in Japanese)
Nature Conservation Bureau, Environment Agency and Marine Parks Center of Japan (1994) The Report of the Marine Biotic Environment Survey in the 4th National Survey on the Natural Environment Vol. 3 Coral Reefs (in Japanese)
Nature Conservation Division, Department of Cultural and Environmental Affairs, Okinawa Prefectural Government (2006) Guideline for crown-of-thorns starfish. (in Japanese)
Nature Conservation Division, Okinawa Prefectural Government (2003) Report on implementation of reef check, fiscal 2003 (in Japanese)
Nikaido M (1963) The tropical starfish, *Culcita novaeguineae* Müller et Troschel has been found in the sea near Okinoshima, Kochi Prefecture, Japan. Zool Mag 72:163–164
Nishida M, Lucas JS (1983) Genetic differences between geographic populations of the crown-of-thorns starfish throughout the Pacific region. Mar Biol 98:359–368
Nohara S, Satake K, Takase T, Kurokawa M (2009) Characteristics of nutrient of island and coastal waters in Izu and Ogasawara Islands. Jpn Jpn J Limnol 70:225–238. (in Japanese)
Nomura K (2004) Coral communities in Kushimoto (13). Mar Pavilion 33:1. (in Japanese)
Nomura K, Kimura T, Kawagoe H (2001) The introduction of the spot check method as a simple monitoring method of large coral reef area, and that practical example in the Sekisei Lagoon, Southern Ryukyus. Mar Parks J 131:5–12. (in Japanese)
Oki K (2014) Changes of coral reefs in Amami Islands-Historical consideration of population outbreak of crown-of-thorns starfish (in Japanese)
Okinawa Prefectural Tourism Development Bureau (1976) Influence of *Acanthaster planci* on coral reef ecosystem, concerning heavy infestations of *Acanthaster planci* (in Japanese)
Pratchett MS, Caballes CF, Rivera-Posada JA, Sweatman HPA (2014) Limits to understanding and managing outbreaks of crown-of-thorns starfish (*Acanthaster* spp.) In: Hughes RM, Hughes DJ,

Smith IP (eds) Oceanogr. Mar Biol Annu Rev: An Annual Review, vol 52. CRC Press, Boca Raton, pp 133–200

Saba M, Irimura S (2002) Hitode Guidbook. Hankyu Communications. (in Japanese)

Shimoike K (2000) Changes in coral coverage and occurrence of the crown-of-thorns starfish around Akajima Island from 1989 to 1999. 1:19–21 (in Japanese)

Shirai S (1956) Letter from Amami, No2. Collect Breed 18:301–307. (in Japanese)

Suzuki G, Yasuda N, Ikehara K, Fukuoka K, Kameda T, Kai S, Nagai S, Watanabe A, Nakamura T, Kitazawa S, Bernardo L, Natori T, Kojima M, Nadaoka K (2016) Detection of a high-density Brachiolaria-stage larval population of crown-of-Thorns Sea star (Acanthaster planci) in Sekisei Lagoon (Okinawa, Japan). Diversity 8:9

Tada M (1982) Transition of sea of Ashizuri over 20 years. Mar Parks J 56:3–5. (in Japanese)

Takahashi N (1986) Tolerance and reactions of *Acanthaster planci* against lowered temperatures. Undergraduate thesis, Dept. of Marine Sciences, Univ. of the Ryukyus. 17pp. (in Japanese)

Takemoto N (2005) Analysis of the present data of *Acanthaster planci* and larval sampling in the field aiming for revealing the mechanisms of population outbreak of crown-of-thorns starfish. Mini-dissertation (in Japanese)

Taniguchi H (2012) Five years after the outbreak of crown-of-thorns starfish: the present state and future prospects of coral reefs around Akajima Island. Midoriishi 23:20–24. (in Japanese)

Timmers MA, Andrews KR, Bird CE, deMaintenton MJ, Brainard RE, Toonen RJ (2011) Widespread dispersal of the crown-of-thorns sea star, *Acanthaster planci*, across the Hawaiian Archipelago and Johnston Atoll. J Mar Biol 2011:10

Timmers M, Bird C, Skillings D, Smouse P, Toonen R (2012) There's no place like home: Crown-of-Thorns Outbreaks in the Central Pacific are regionally derived and independent events. PLoS ONE 7:e31159

Tokioka T (1953) Invertebrate fauna of the intertidal zone of the Tokara Islands *V. Echinodermata*. Seto Marine Biol Lab 3:144–148

Uchiumi H (1962) Recent evidence for the northward extension of the range of some tropical echinoderms in Japanese waters. Zool Mag 71:102–108. (in Japanese)

Uthicke S, Doyle J, Duggan S, Yasuda N, McKinnon A (2015) Outbreak of coral-eating Crown-of-Thorns creates continuous larvae cloud over 320km of the Great Barrier Reef. Scientific Report. 5, Article number: 16885

Vogler C, Benzie J, Lessios H, Barber P, Worheide G (2008) A threat to coral reefs multiplied? Four species of crown-of-thorns starfish. Biol Lett 4:696–699

Vogler C, Benzie J, Barber PH, Erdmann MV, Ambariyanto, Sheppard C, Tenggardjaja K, Gérard K, Wörheide G (2012) Phylogeography of the crown-of-thorns starfish in the Indian Ocean. PLoS One 7:e43499

Vogler C, Benzie JAH, Tenggardjaja K, Ambariyanto, Barber PH, Wörheide G (2013) Phylogeography of the crown-of-thorns starfish: genetic structure within the Pacific species. Coral Reefs 32:515–525

Web site 1: the Japan Meteorological Agency (n.d.) http://www.data.jma.go.jp/kaiyou/data/shindan/a_1/japan_warm/japan_warm.html

Web site 2: Yaeyama Environmental Network by Yaeyama Fishermen's cooperative (n.d.) http://www.churaumi.net/onihitode/onihitode2.html

Web site 3: Local diver in Fukue jima. (n.d.) http://fukuejima.la.coocan.jp/sea-kyoku-hi/oni-hitode.html

Web site 4: A blog by local diver in Koshikijima (n.d.) http://blog.livedoor.jp/super_sea_story_pro-koshiki/archives/52367441.html

WWF Japan (2008) Nansei Islands biological diversity field research report. http://www.wwf.or.jp/activities/upfiles/Nansei_Is_BDreport_opt.pdf (In Japanese)

Yaeyama Mainichi Shinbun (2010) May 27

Yamaguchi M (1973) Early life histories of coral reef asteroids, with special reference to *Acanthaster planci* (L.) In: Jones OA, Endean R (eds) Biology and geology of coral reefs. Academic, New York, pp 369–387

Yamaguchi M (1974) Growth of juvenile *Acanthaster planci* (L.) in the laboratory. Pac Sci 28:123–138

Yamaguchi M (1981) Crown-of-thorns starfish and coral reefs. Anim Nat 11:28–35. (in Japanese)

Yamaguchi (1984) Population outbreak of *Acanthaster planci* in Okinawa and its counterplan. 1–10 (Brief notes in Japanese)

Yamaguchi (1986) *Acanthaster planci* infestations of reef and coral assemblages in Japan: a retrospective analysis of control efforts. Coral Reefs 5:23–30

Yamaguchi (1987) Occurrences and persistency of *Acanthaster planci* pseudo-population in relation to oceanographic conditions along the Pacific coast of Japan. Galaxea 6:277–288

Yamano H, Sugihara K, Nomura K (2011) Rapid poleward range expansion of tropical reef corals in response to rising sea surface temperatures. Geophys Res Lett 38:L04601. https://doi.org/10.1029/2010GL046474

Yamazato K (1969) *Acanthaster planci*, the coral predator. Konnichi Ryukyu 13:7. (in Japanese)

Yamazato K, Kiyan T (1973) Reproduction of *Acanthaster planci* in Okinawa. Micronesica 9:185–195

Yasuda N, Nagai S, Hamaguchi M, Lian CL, Nadaoka K (2006) Development of microsatellite markers for the crown-of-thorns starfish *Acanthaster planci*. Mol Ecol Notes 6:141–143

Yasuda N, Nagai S, Hamaguchi M, Okaji K, Gérard K, Nadaoka K (2009) Gene flow of *Acanthaster planci* (L.) in relation to ocean currents revealed by microsatellite analysis. Mol Ecol 18:1574–1590

Yasuda N, Ogasawara K, Kajiwara K, Ueno M, Oki K, Taniguchi H, Shinichiro K, Okaji K, Nadaoka K (2010) Latitudinal differentiation of crown-of-thorns starfish (*Acanthaster planci*) in the reproduction patterns through the Ryukyu Island archipelago. Plank Benth Res 5:156–164

Yasuda N, Nagai S, Taquet C, Nadaoka K, Suharsono, Susato HA (2011) Reef connectivity of *Acanthaster* sp. in the Coral Triangle Region. In: Polymorephisms JSoD (ed) DNA Polymorphisms. Toyo Shoten, p134–138

Yasuda N, Taquet C, Nagai S, Yoshida T, Adjeroud M (2014) Genetic connectivity of the coral-eating sea star *Acanthaster planci* during the severe outbreak of 2006–2009 in the Society Islands, French Polynesia. Mar Ecol 36:668–678

Yokochi H, Ogura M (1987) Spawning period and discovery of juvenile Acanthaster planci (L.) (Echinodermata: Asteroidea) at northwestern Iriomote-Jima, Ryukyu Islands. B Mar Sci 41:611–616

Yomiuri Newspaper (2007) Oct. 14

# Studies on Mesophotic Coral Ecosystems in Japan

# 10

Frederic Sinniger and Saki Harii

**Abstract**

In the last years, mesophotic coral ecosystems (MCEs) received increased attention from the research community. MCEs exist at depths between 30–40 m and until 100 m or deeper. While MCEs were already reported from Japan nearly 50 years ago, mesophotic research in Japan did not progress as in other countries like the USA or Australia. Nonetheless, over the years, several interesting studies were conducted on the MCEs found in southern Japan on various fields from geology to biology, and the momentum of Japanese mesophotic research increased in the recent years. In this chapter, we will summarize the knowledge on MCEs in Japan in different domains such as their distribution, their biodiversity, their occurrence in the fossil record, recent findings on the reproduction of mesophotic corals, their potential ecological role in a global change context, threats menacing these ecosystems, and future directions of MCE research in Japan.

**Keywords**

Deep coral reefs • Distribution • Biodiversity • Fossil corals • Ecology

## 10.1 Introduction to Mesophotic Coral Ecosystems

The vast majority of coral reef research has focused on corals found in shallow waters. However, coral ecosystems extend much deeper, where light intensity represents only a few percent of that measured in the first meters under the surface. In this "twilight zone", light-dependent corals can form rich and complex ecosystems. These ecosystems found from 30–40 m down to over 100 m (165 m in some very transparent waters of the central Pacific) are usually referred to as mesophotic coral ecosystems or MCEs (Hinderstein et al. 2010). In the recent literature, the term MCE is also sometimes used to refer to deep environments not necessarily dominated by zooxanthellate corals but by sponges or algae. Although the strict definition of MCE is still under discussion (and in our opinion the terms mesophotic algal ecosystems and mesophotic sponges ecosystems should be introduced for situations where these organisms are the main habitat builders), the shared characteristic of these different ecosystems is the presence of light-dependent sessile organisms despite a largely reduced light availability compared to the overlaying shallow waters. Therefore, it is not surprising that "shallow" mesophotic communities have recently been found in areas with particularly low light penetration (Muir and Wallace 2015). Likewise, in addition to eurybathic species (species tolerating a wide range of depths) some species can be observed both in the mesophotic zone and in shaded parts of shallow reefs. Despite many gaps remaining in our understanding of mesophotic ecology, parameters other than light likely define MCEs. For example, cave ecosystems, although light limited and potentially sharing some species with MCEs, should be treated independently as they depend on different hydrodynamic, geomorphological, and biological processes. Nevertheless, overlapping species between shallow and mesophotic depths have become of great interest for researchers following the hypothesis that

F. Sinniger (✉) • S. Harii
Sesoko Station, Tropical Biosphere Research Center, University of the Ryukyus, Motobu, Okinawa, Japan
e-mail: fredsinniger@hotmail.com


A. Iguchi, C. Hongo (eds.), *Coral Reef Studies of Japan*, Coral Reefs of the World 13,
https://doi.org/10.1007/978-981-10-6473-9_10

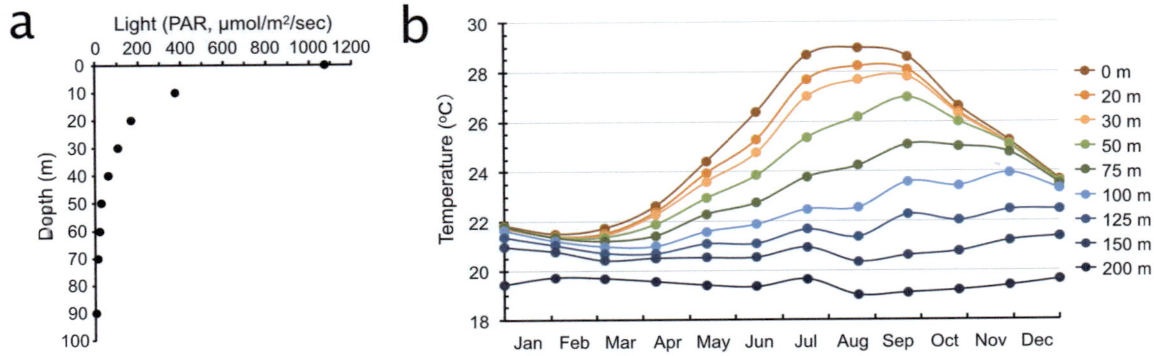


**Fig. 10.1** Light and temperature from the surface to mesophotic depths. (**a**) Light measured on a sunny day offshore of Sesoko Island in March. (**b**) Monthly average temperatures over multiple years west of Okinawa Island between 26–27 °N and 127–128 °E (Source: Japan Oceanographic Data Center, http://www.jodc.go.jp/service.htm)

"deep reefs could serve as refuge for shallow coral species under thermal stress" Glynn (1996). In addition to the excitement of discovering unknown communities and ecosystems, this hypothesis contributed significantly to the development of mesophotic research by motivating researchers to investigate on the relations between shallow and mesophotic corals and alerting decision makers and funding agencies on the potential importance of MCEs in a global context of climate change. Paradoxically, despite some of the first mesophotic coral ecosystems having been discovered in Japan (Yamazato 1972; Yamazato et al. 1967), Japanese MCEs remain relatively poorly explored (in comparison with the Great Barrier Reef, Hawaii, or the Caribbean Sea). However, this situation is progressively changing with reports of Japanese mesophotic corals increasing in recent years.

### 10.1.1 Environmental Conditions

In the Ryukyu archipelago, light penetration varies greatly depending on the sites and the seasons. Except in the relatively large islands of Amami, Okinawa, Ishigaki and Iriomote, the small size and limited watershed of most islands of the archipelago limits the potential for river sedimentary inputs. However, the extensively developed agriculture often accompanied by poor measure to prevent soil erosion was shown to affect shallow reefs (Omija 2004; Hongo and Yamano 2013; Yamano et al. 2015) and likely impacts coastal MCEs as well even though such impact remains to be investigated. At one of the most studied site near Sesoko Island offshore of the main island of Okinawa, the light (measured as Photosynthetic Active Radiation, PAR) reaching a diverse upper mesophotic community at 40 m represents 7.7 % of the light reaching the surface in average (Prasetia et al. 2016). Further offshore, only 5.5 % of the surface light was measured[1] near a deeper mesophotic community at 40 m and 0.3 % at 90 m. At this location, zooxanthellate scleractinian corals are still found at a depth of 70 m, reached by only 0.8 % of the surface irradiance (i.e., 8.6 µmol photon $m^{-2} \bullet s^{-2}$) (Fig. 10.1a).

Seawater temperature fluctuates seasonally down to 150 m depth. In winter, the water column is relatively homogenized with average temperatures[2] between 21.4 and 20.4 °C from the surface to 150 m depth. However in summer, while temperature below 150 m remains relatively constant around 19 °C, average temperatures in the first 100 m water depth range from 29 to 22.5 °C (Fig. 10.1b). Probably, the most peculiar environmental phenomenon in the region is the temperature inversion occasionally observed in winter, when mesophotic water is warmer than the surface water. Despite such events being limited in time and sporadic, they are recurrent enough for the monthly average temperatures in both December and January over several years showing slightly higher values at 50 and 75 m depth compared to 20 and 30 m (between 0.02 and 0.1 °C). Such events were also reported in punctual measurements recently made in the area (Prasetia et al. 2016). While leading to an unstable situation, this pattern of warmer deeper water can be easily understood when considering the geographical setting of the region. The Ryukyu archipelago receives warm water from Kuroshio, and this inflow of tropical water allows the high diversity of coral species to develop at this rather high latitude (24–29° latitude) and even at higher latitudes along the coasts of Kyushu, Shikoku, and Honshu islands in "mainland" Japan. In winter, however, cold North wind blowing over the archipelago leads to extreme heat flux between the warm water (around 18–20 °C) and the atmosphere (down to 8 °C or less). Heat fluxes were measured in winter as low as 231 cal $\bullet$ $cm^{-2} \bullet day^{-1}$ in warm periods up to 1,584 cal $\bullet$ $cm^{-2} \bullet day^{-1}$ in cold (North wind)

[1] Data measured in March 2016.

[2] Monthly average temperatures over multiple years west of Okinawa Island between 26–27 °N and 127–128 °E (source: Japan Oceanographic Data Center, http://www.jodc.go.jp/service.htm).

periods (Agee and Howley 1977). In this context, MCEs can be relatively isolated from the rapid cooling of the surface layer of the ocean. Whether this situation is unique to specific locations and topographic configuration, as well as the influence of the deeper waters of the Kuroshio on MCEs remain to be investigated.

## 10.2 Distribution of Mesophotic Coral Ecosystems in Japan

Currently, in Japan, MCEs have been reported only from the Ryukyu archipelago (Fig. 10.2). In addition to the lack of dedicated studies on mesophotic communities at higher latitudes in Japan, light and temperature conditions may limit the maximum depth for zooxanthellate coral communities to develop (Muir et al. 2015). Indeed, even though isolated coral colonies (unless indicated otherwise, here and in the following sections, the term "coral" refers to zooxanthellate corals) can be found deeper, the bottom of the reef slope has been shown to be shifting from 80 m depth in Hateruma Island, the most southern island in Japan, to only 30 m depth at Yaku Island (Yakushima) on the northern edge of the Ryukyu archipelago (Hori 1990) suggesting a shallowing trend from South to North.

MCEs can be found in three main types of bottom topography: (a) nearshore vertical walls, (b) nearshore deep terraces and gentle slopes, and (c) topographic heights such as banks and pinnacles (Fig. 10.3). The walls limit light availability while providing hard substrates suitable for corals and other photosynthetic benthic cnidarians to prosper. Along these drop-offs (Fig. 10.3a), sciaphilic communities, such as gorgonians and antipatharians, more frequent in deeper environments offshore can develop at relatively shallower depths (Fig. 10.4a). Sponges and other benthic organisms follow the same patterns, for example, the foraminifer *Cycloclypeus carpenteri*, most frequently found between 60 and 70 m depth off Motobu Peninsula (Hohenegger 1994; Hohenegger and Yordanova 2001), is relatively abundant already at 45 m at the bottom of vertical drop-offs along the main part of Okinawa Island (Fig. 10.4b).

Deep terraces, banks, and pinnacles (possibly including submerged fossil reefs) represent the other types of bottom topography suitable for MCE development (Fig. 10.3b, c). While most of these environments remain to be explored, they cover a vast area along the Ryukyu archipelago and

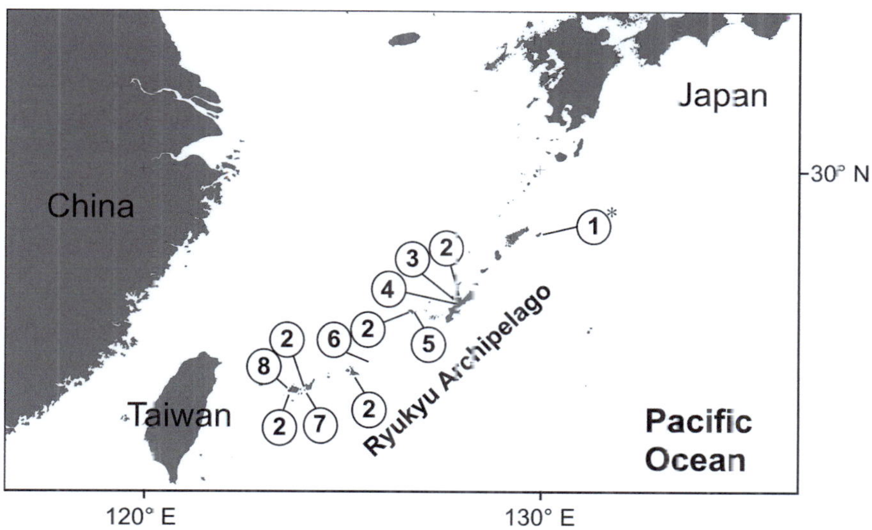


**Fig. 10.2** Map of the area. The numbers correspond to the different reports of MCEs. (*1*) Kikai Island, Humblet et al. (2014) (*fossil*); (*2*) Okinawa, Kume, Miyako, Ishigaki Islands, Yamazato et al. (1967); (*3*) Off Sesoko Island, Sinniger et al. (2013); (*4*) Off Okinawa Island, Ohara et al. 2013; (*5*) Kume Island, Kimura et al. (2011); (*6*) Off Miyako Island, Arai et al. (2016); (*7*) Sekisei lagoon, Humblet et al. (2009a); (*8*) Amitori Bay, Iriomote Island, Nishihira and Veron (1995), Sinniger (2016)

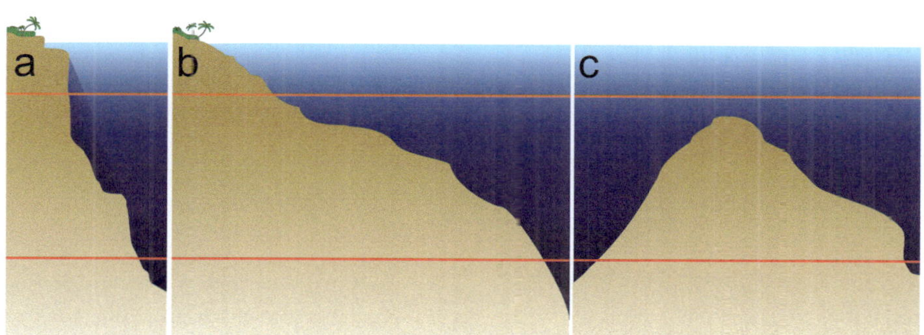


**Fig. 10.3** Different topographies of MCEs in Japan. (**a**) Nearshore vertical walls, (**b**) nearshore deep terraces and gentle slopes, and (**c**) topographic heights. The *orange* and *red lines* represent 30 m and 100–150 m depth respectively

**Fig. 10.4** (**a**) Gorgonians, sponges, algae, and a coral colony on a nearshore vertical wall on the east coast of Okinawa Island, 45 m; (**b**) *Cycloclypeus carpenteri* at the bottom of a vertical wall, east coast of Okinawa Island, 50 m; (**c**) *Seriatopora*, other corals and coralline algae on a gentle slope offshore of Sesoko Island, 40 m; (**d**) *Leptoseris*, sponges, algae and an antipatharian on the top of a "Sone" offshore of Okinawa Island, 75 m

potentially provide a much larger habitat than current shallow coral reef habitats. Often located further out of the shade from the islands, the higher light availability compared to this found on the vertical walls of the islands may explain the diversity of scleractinian corals found in some of these MCEs (Fig. 10.4c, d). The main distinction between the nearshore features (gentle slopes and deep terraces, Fig. 10.3b) compared to topographic heights (Fig. 10.3c) is the source of corals inhabiting these MCEs. In nearshore MCEs, in addition to larval recruitment, fragments of coral colonies can tumble down the slope following strong storms or other destructive actions on shallow reefs and eventually adapt to the new environment. On the other hand, topographic heights rely exclusively on larval recruitment, a constraint that likely affects the biodiversity encountered in these MCEs.

Preliminary observations of the diversity both at species and community levels suggest that the Ryukyu archipelago may figure among the high mesophotic coral biodiversity regions in the world. While drop-offs promote the development of deep species in shallower areas, some shallow species can be found in the deep MCEs located offshore where light availability is not reduced by terrestrial sedimentary input or shade from the islands. Despite data missing for most of the Ryukyu archipelago, already a few of such MCEs have been reported.

### 10.2.1 Central Ryukyu Islands

#### 10.2.1.1 Okinawa Island

In 1967, during the submersible survey from the government of the Ryukyus, corals were observed during two dives along

**50 Years Ago: Pioneer Observations of MCEs in the Ryukyu Archipelago**

The first observation of Japanese MCEs was made by the Government of the Ryukyus (1967). Between September and November 1967, a manned submersible "Yomiuri Gou" survey was organized by the Government of the Ryukyu Islands (now a part of Japan) to assess marine resources in the deeper coastal areas around the Ryukyu archipelago. This survey explored five islands (and a total of 21 sites) in the archipelago at depths from 16 to 290 m. The main objective was to investigate fisheries resources (e.g., fishes, crustaceans, precious corals); therefore, numerous sites investigated did not cover suitable environments for mesophotic corals (i.e., too deep or sandy substrates). Nevertheless, mesophotic corals were observed at nine sites during the ascent phase of several dives (Yamazato et al. 1967). These pioneering observations are among the earliest in situ sightings of mesophotic corals in the world and overall three reef zones were defined. In zone 1 (30–50 m), corals were dominant with little amount of coralline algae found between them. Coral coverage on reefs reduced to 20–50 % in zone 2 (50–70 m) while algal presence increased. In this zone, the bottom topography is often composed of topographic features resembling shallow spurs and grooves perpendicular to the coastline. On the shelf of Miyako Island, a similar type of feature strongly influenced by tidal currents was described by Tsuji (1993) and referred to as "ridges and furrows." In this text, we will use the term "deep spurs and grooves" to refer to these structures occurring at mesophotic depths and reported by Yamazato et al. (1967). Corals are distributed mainly on the top of the ridges, while coralline algae cover the sides, and the bottom of the grooves is covered by either sand or gravels (that can be encrusted by various benthic fauna or algae). Finally in zone 3 (70–100 m), corals were rare with less than 20 % coverage and lower species richness while large foraminifers were commonly encountered. This zone shows similarities to the deep sea observed on the shelf margin (below the mesophotic zone) with the occurrence of gorgonians and antipatharians (black corals). However, these three zones are not always present or equally developed at the different sites. The deepest coral was reported from a depth of 102 m and constitutes the deepest observation of a zooxanthellate scleractinian in Japan so far (Yamazato 1972); unfortunately no details were provided on this sighting. However, the deepest corals identified in the report are some *Pachyseris* and *Leptoseris* reported from 98 to 100 m depth off Ishigaki Island. This maximum depth distribution of corals is shallower than other North Pacific Ocean islands such as Hawaii, the Mariana Islands, or Johnston Atoll but comparable to the deepest corals found in American Samoa and deeper than most records of Atlantic mesophotic corals (summarized in Baker et al. 2016). Yamazato et al. (1967) also discussed more specifically the bathymetric distribution of hermatypic corals. In his discussion, they compared maximum depth of coral distribution between the Ryukyu archipelago and observations from Bikini Atoll (Wells 1954). Despite the limited data available on the subject, they suggest that the difference in the depth of the deepest corals (100 m in Japan vs 144 m in Bikini Atoll) could be due to bathymetric distribution of temperatures and limited light availability at higher latitudes.

the Kouri Sone (a "Sone" is a Japanese fisherman term to define bathymetric height similar to a shoal or a knoll in English). This location, originally referred to as "Kei reef," was unique in comparison of other coral sites explored as macroalgae represented the dominant benthic organisms (Yamazato et al. 1967). Between 20 and 30 m, *Sargassum* sp. covered most of the substrate and was progressively replaced by *Spatoglossum stipitatum* (referred to as *Zonaria stipitata*) between 30 and 50 m. Plate *Acropora*, *Pocillopora*, and *Dipsastraea*/*Favites*-like corals (referred as "faviid-like") along with some massive corals were found in the shallowest part of the surveyed area (around 20 m depth). At 40 m, only *Acropora*-like and *Dipsastraea speciosa*-like corals were reported, and at 50 m *Leptoseris*-like corals were observed. Coral occurrence increased until 50 m depths, reaching coverage of about 10 % corresponding to a ratio of 1:1 between corals and macroalgae. Between 50 and 70 m coverage of both coral and macroalgae decreased and the deepest corals were observed around 70 m depth. Sponges and hydrozoans were sampled on rocks at 70 m, and below 90 m, octocorals (*Anthogorgia*, *Callogorgia*, *Dendronephthya*, *Siphonogorgia*), the black coral *Cirrhipathes spiralis*, and stylasterid corals were observed. At this location, no deep spurs and grooves were observed.

During the same submersible survey, a second site offshore of the west coast of the central part of Okinawa Island showed very high densities of corals (branching *Acropora* sp., *Millepora* sp., and *Turbinaria* sp. or *Montipora*-like plate corals) from 40 to 55 m depth with 100 % coral coverage (Yamazato et al. 1967). Between 55 and 65 m, the coverage decreased with smaller coral colonies (10–20 cm diameter). Below coverage continued to decrease with rare colonies of *Leptoseris* at 87 m being the deepest corals reported at this site. Among other organisms, *Siphonogorgia* and other gorgonians occurred below 65 m. From 90 to 110 m, the topography was rather flat and sandy with rocky outcrops covered in sponges, gorgonians, and other octocorals such as *Telesto* and *Ellisella*, attracting swarms of small fishes (*Apogon*)

More recently, a diverse mesophotic coral community was reported from 40 to 50 m depth between Sesoko Island and Motobu Peninsula (Sinniger et al. 2013). At 40 m, this community dominated by *Pachyseris speciosa*, *Seriatopora hystrix*, and the rare *Acropora tenella* is developing in a nearly flat area characterized by low-elevation rocky outcrops (the main substrate for most corals, including *Pachyseris*) separated by patches of sand or rubbles (the main habitat for unattached *Acropora tenella* colonies) (Fig. 10.4c). *Seriatopora hystrix* can be found on both rubbles and hard substrates. The seaward margin of this area gently slopes down to 50–55 m where a progressive transition to sand occurs, with a different species assemblage found in this sandy rubble area. In addition to widely distributed corals such as *Seriatopora* or *Pachyseris*, species more specifically encountered in this environment include *Euphyllia paradivisa* (Eyal et al. 2016), *Madracis kirbyi*, as well as several *Alveopora* and *Leptoseris* species. On the shallower side of this MCE, the topography progressively slopes up and connects with the typical Okinawan shallow-reef communities.

In addition, in the same region, Hohenegger et al. (1999) compared large light-dependent symbiotic foraminifer communities between two transects at nearby sites with distinct topographies and environmental conditions. The northern transect lies north of Sesoko Island at depths ranging from 3 to 100 m. The proportion of living corals at 40 m was reported at slightly over 40 % before decreasing to nearly 0 % at 50 m. The southern transect on the other hand is located west of Sesoko Island along the slope of a well-formed reef crest at depths of 5–100 m. Coral cover along this transect was significantly lower at all depths except near the surface, and no living corals were found at 40 m or below. Following the observation that the water was less transparent (higher light attenuation coefficient) at the northern transect despite the highest coral coverage, Hohenegger et al. (1999) suggested that the high abundance of corals at mesophotic depths is related to higher nutrient inputs. While the role of nutrients remains to be investigated in Japanese mesophotic corals, different hydrological conditions and high rates of sedimentation also likely affect the abundance of corals in the south transect. The topography of the sites may also influence the coral community structure. While water was slightly more turbid in the northern transect, this transect receives light throughout the day as it is never shaded by Sesoko Island. The slope is also less steep, especially between 30 and 40 m. The southern transect being located on the west side of Sesoko Island this area is more shaded during the morning hours (as the measurements of light intensity in Hohenegger et al. (1999) were made at noon, this effect is impossible to detect based on their data). The southern transect is also more exposed to strong storms and typhoons which may destroy corals both through direct intense water motion and through the fall of debris from the shallow reef down the slope. In this context, a steeper slope will negatively affect the presence of corals. Based on the data presented in Hohenegger et al. (1999), the high proportion of sand at all depths, but especially below 30 m, provides the simplest explanation for the lower abundance of corals. However, despite providing the simplest explanation, substrate characteristics may not always be the most determinant parameter affecting coral distribution. Indeed, hard substrate does not necessarily involve high coral cover and that factors other than substrate type must play a role in controlling coral distribution.

On the west coast of Okinawa Island, another upper mesophotic community was reported (Ohara et al. 2013). This MCE, named Ryugu Reef, consists in the lower part of a larger coral community extending from shallower depths (White et al. 2013). Between 30 and 42 m, *Pachyseris* largely dominates the community, while below 42 m the sandy substrate is not suitable for extensive development of corals.

#### 10.2.1.2 Kume Island

During the 1967 submersible survey, further west of the main Okinawa Island, a site on the southern side of Kume Island was investigated at depths ranging between 30 and 290 m (Yamazato et al. 1967). At mesophotic depths, the seafloor can be divided into three geomorphological zones: an irregular steep forereef slope (30–50 m), deep spurs and grooves (50–70 m), and a gentle slope consisting of sandy gravels (70–100 m). The mesophotic coral community in the upper zone was dominated by table *Acropora* along with *Millepora* sp. and coralline algae. The rough rocky surface was mostly covered by corals (about 90 % coverage) at 35 m and plate corals reached up to 70–80 cm diameter. Coral cover and colonies diameter decreased with increasing depth. No abrupt shift in coral diversity was observed with the transition to the spurs and groove zone, and the decreasing trend in coral abundance and size with increasing depth continues. Several *Dipsastraea speciosa* and one *Mycedium elephantotus* (referred to as *M. tenuicostatum*) were sampled at 68 m where the coral cover was estimated to be about 20 %. Other corals, possibly *Acropora* and *Porites*, were observed at this depth along with hydrozoans, sponges, and macroalgae such as *Zonaria*. Below this second zone, on the gentle slope down to 100 m, some patches of rubbles composed of 20–30 cm diameter limestone gravel host some corals, especially "*Dipsastraea*-like" colonies along with gorgonians, hydrozoans, sponges, and macroalgae such as *Padina* and *Peyssonnelia*. Large *Cycloclypeus* foraminifers were observed around 90 m depth. The bottom of this zone, at 102 m depth, consists of rubbles with *Peyssonnelia* and coralline algae, and it is at this depth that the deepest coral was observed. Although no details are available on this observation, it still constitutes the record depth for a zooxanthellate scleractinian in Japan nowadays. The high transparency of the water (50 m horizontal visibility was measured at 130 m depth) at this site may explain this deep finding.

Near the site explored by Yamazato et al. (1967), in the southeastern part of Kume Island, a dense coral community dominated by branching *Acropora horrida* was discovered in 2010 (Kimura et al. 2011). This upper mesophotic community extends from 15 m down to 40 m and covers a small mound rising out of a sandy slope, with a live coral coverage up to 90 % between 25 and 30 m (Fujita et al. 2012). In addition to the dominant *A. horrida*, several other species such as *A. latistella* or *Favites* were observed (Kimura et al. 2011).

### 10.2.2 Sakishima Islands (South Ryukyu Islands)

#### 10.2.2.1 Miyako Island

In the 1967 submersible survey, two of the three explored sites covered mesophotic depths (Yamazato et al. 1967). At the first site, off Tomori, the mesophotic slope appears to exhibit a similar zonation as observed in Kume Island, with a steep reef slope and rough hard substrate above 50 m giving way to a deep spurs and grooves topography between

50 and 70 m and finishing by a gentle sedimentary slope covered by sand and gravels until 100 m. However, as the study started at the depth of 50 m, detailed comparison of corals in the upper mesophotic zone cannot be made. Between 50 and 70 m, corals cover about 30 % of the substrate with table *Acropora*, *Dipsastraea speciosa*-like corals, *Leptoseris*, and *Millepora*. The remaining substrate is colonized by black corals (*Cirrhipathes*), large gorgonians, sponges, and coralline algae. At 60 m *Leptoseris scabra* was observed, and at 70 m algae such as *Halimeda* or *Codium minus* (referred to as *Codium mamillosum* var. *minus*) and coralline algae are mainly reported. Below this zone, no corals were observed on the gravels and small rocks, yet some black corals (*Cirrhipathes*) and octocorals such as fan-shaped gorgonians or *Ellisella* sea-whip were found among macroalgae (e.g., *Aneurianna*, *Halimeda* sp., *Ulva* sp., *Padina* sp.).

In the second site, off Kurima Island, beside Miyako Island, the coral slope extends until 110 m contrary to the previous site off Tomori where a gentle sandy slope occurred between 100 and 120 m (Yamazato et al. 1967). While the survey extended down to a depth of 280 m, corals occurred only in the upper part of the zone surveyed. As observed in Kume Island and off Tomori, the bottom of the reef slope is forming deep spurs and grooves until 70 m before giving way to a sand and gravel slope. Only a very narrow band of corals was observed as the lower limit of coral distribution was established at 70 m. At this depth, scattered colonies of several corals were observed and some specimens collected, such as *Dipsastraea speciosa*, *Goniastrea* sp., *Porites* (referred to as *Porites* (*Synaraea*) *hawaiiensis*), and *Pachyseris speciosa*. Below the sand and gravel, only small gorgonians, sponges, and other benthic organisms along with coralline algae were found.

More recently, a large submerged reef structure was reported near Miyako Island (Arai et al. 2016). This topographic structure extends up to 1 km in length and about 500 m in width at depths between 56 and 80 m. Between 56 and 66 m, the geological structure exhibits clear spurs and grooves similar to those observable in the shallow reefs around the Ryukyu archipelago. The benthic covers appeared to be dominated by algae, soft corals, and sponges. While no scleractinian corals were reported, a colony of potentially encrusting coral can be seen in one of the illustrations although the resolution of the image is not sufficient for confident identification. Rhodoliths occur on the deeper terraces neighboring this submerged reef at depths from 100 to 130 m.

Despite the apparent low coral cover, the complex topography of this area combined to its isolation from neighboring islands make it an interesting target for further studies of the mesophotic communities inhabiting submerged reef structures.

#### 10.2.2.2 Ishigaki Island

In Ishigaki Island, two sites were explored by the submersible survey of the government of the Ryukyus (Yamazato et al. 1967). One submersible dive, just out of Sekisei Lagoon barely overlapped with the mesophotic zone as the dive begun at a depth of 213 m and ended at 98 m. Yet colonies of *Pachyseris speciosa* and *Leptoseris* together with large foraminifers and sponges were found down to 100 m on rubbles in a sand/gravel flat area. Shallower depths were surveyed in the other site as the dive at the entrance of Nagura Bay, off the Cape Osaki, started at a depth of 90 m and finished at around 40 m, with still some observations on the slope until 32 m on the way up. Compared the other sites explored, the visibility was considerably less, with only 12 m horizontal visibility at 78 m. Nonetheless, dense coral coverage was also observed in the upper part of the survey. Between 32 and 43 m, the reef slope was essentially composed of round table *Acropora* and discoidal *Cyphastrea* sp. colonies. *A. longicyathus* and *C. serailia* were collected at 41 m. At 40 m, the coral coverage was estimated to be around 80 %, and living corals and coralline algae were observed down to 55 m. Below this depth, the bottom presents a different topography with rocky outcrops separated by sandy patches on a gentle slope extending to 90 m. In this environment, sponges, octocorals (*Dendronephthya*, *Siphonogorgia*, *Ellisella*, and *Echinogorgia*), antipatharians (*Cirrhipathes spiralis*, *Antipathes* sp.), bryozoans, and hydrozoans dominated the fauna. Around 85 m, large *Melithaea* colonies were observed.

Two more recent surveys have been conducted around Ishigaki Island. Below 30 m depth north of Ishigaki, off Yonehara, low scleractinian coverage (at most 10 %) was reported (Iryu et al. 1995). At this site, the coral assemblage was dominated by foliaceous species such as *Pachyseris speciosa* or *Leptoseris scabra*. Later, in 2004, a ROV survey of along the southern margin of the Sekisei Lagoon southwest of Ishigaki Island revealed higher scleractinian coverage at mesophotic depths (Humblet et al. 2009a). Table *Acropora* spp. dominated the scleractinian assemblage comprising also encrusting, plate-like and foliaceous corals between 40 and 55 m. In contrast with the previous study further north, the scleractinian coverage in this area reached up to 35 % at 50 m. Below 55 m, coral coverage drastically decreased and the assemblage was dominated by encrusting and laminar growth forms, including *Pachyseris speciosa*. Other benthic organisms included sponges, octocorals and, below 60 m, the large foraminifer *Cycloclypeus carpenteri*. Below this depth and down to the limit of the study (100 m), scleractinian corals became rare with only a few isolated *Leptoseris* colonies observed between 80 and 85 m. On the other hand, other benthic invertebrates such as sponges and octocorals became abundant.

#### 10.2.2.3 Iriomote Island

The southernmost site explored was near Kanokawa, south of Iriomote Island (Yamazato et al. 1967). This dive explored a portion of the slope between 58 and 90 m. Deep spurs and grooves shaped the bottom down to 68 m. Several disc-shaped corals were found on the top and sides of the spurs, with *Porites*, *Dipsastraea speciosa*, and table *Acropora* identified from visual observations only. Gorgonians and antipatharians were also present. At the bottom of the grooves, rounded rubbles 20–30 cm in diameter were covered with macroalgae and coralline algae. Between 68 and 80 m, only shallow grooves remained with sand or gravel as main substrate. Large foraminifers became abundant at 72 m although their number was reduced on sandy bottom. *Pachyseris speciosa* was the deepest coral found at 75 m (Yamazato 1972; Yamazato et al. 1967). Macroalgae such as *Padina* or *Halimeda* were found on the rubbles down to 85 m, along with gorgonians and other invertebrates.

After a gap of nearly 30 years in Okinawan mesophotic coral research, another dense community of mesophotic corals was found by Dr. H. Yokochi in Amitori Bay around 50 m depth (Nishihira and Veron 1995). This community is dominated by different *Leptoseris* species showing a distinct depth zonation in their distribution (Yokochi unpublished Sinniger 2016). A dense thicket of *Leptoseris amitoriensis* occupies the deepest part of this community. As suggested by its name, this species was first found in Amitori Bay and described based on specimens from the extensive stands found at a depth of around 50 m. Deeper than about 55 m, the bottom of Amitori Bay consists of a gentle slope composed of fine sediments down to the maximum depth of the bay. Apart from *L. amitoriensis*, another *Leptoseris* species, *L. gardineri*, coexists in the deepest zone of this ecosystem. Above the community of foliose *Leptoseris*, the minute and crispy *L. papyracea* forms a dense population that covers a high proportion of the substrate. Further up, another high coverage population of *Seriatopora* connects upper mesophotic and shallow reef. Despite a rather low diversity in comparison with the high coral coverage, the density of corals and the zonation displayed by the different species make this MCE one of the most spectacular in Japan.

## 10.3 Biodiversity

The main difficulties to study MCE biodiversity are the same as for most other studies on remote environments: access and sampling issues. Historically sampling at mesophotic depths has been made by trawling or dredging. However, these methods are very destructive and not ideal for sampling on hard and irregular substrates due to the risks of losing or destroying the sampling gear. Development of scuba diving allowed more targeted and selective sampling although scuba diving at mesophotic depths requires specific training and equipment (e.g., mixed-gas diving). When examining scleractinian coral diversity, another significant problem arises from the fact that numerous zooxanthellate corals are known to change their morphology in order to adapt to different environmental conditions. Considering that most taxonomic identifications of scleractinian corals are based on morphological characters, such variability is a serious impediment to species identification of coral samples collected in the low light environment of MCEs. Therefore, species identification of mesophotic corals relies heavily on the combination of microstructure observations, molecular biology, and comparisons with shallow samples. Within the Ryukyu archipelago, morphological characteristics of the different corals encountered on MCEs seem relatively conserved within species, and the development of an accurate reference database may be of crucial help for identifying mesophotic corals in the future and should be prioritized. However, until such reference database can be established, considerable taxonomic work must be engaged to solve taxonomic uncertainties one by one.

### 10.3.1 *Leptoseris*

*Leptoseris* is the emblematic mesophotic coral genus. The type locality of one species, *L. amitoriensis*, is located in the Ryukyu archipelago. In Japan, *L. amitoriensis* (Fig. 10.5a) was described from Amitori Bay in Iriomote Island and officially reported only from the Yaeyama Islands, in the south of Okinawa Prefecture. However, future surveys in enclosed bays in other islands in the prefecture may reveal a broader distribution of this species. This fragile species forms foliose colonies with more or less divided fronds. While the fronds never get as narrow as in *L. gardineri*, another species found in Amitori Bay (Nishihira and Veron 1995), the largest fronds of older colonies can be confused with *L. foliosa*, and the specific status of *L. amitoriensis* in regard to other *Leptoseris* species is now undergoing a reexamination using both morphological and molecular approaches. Such approach has been successfully applied to clarify relations between agaricids in Hawaii (Luck et al. 2013); however, preliminary results suggest that the situation in the Ryukyu archipelago is more complex, with numerous genotypes and intermediate morphotypes. The whole *Leptoseris* genus deserves a revision, and work on the original type specimens of *L. amitoriensis* and freshly collected samples is still ongoing.

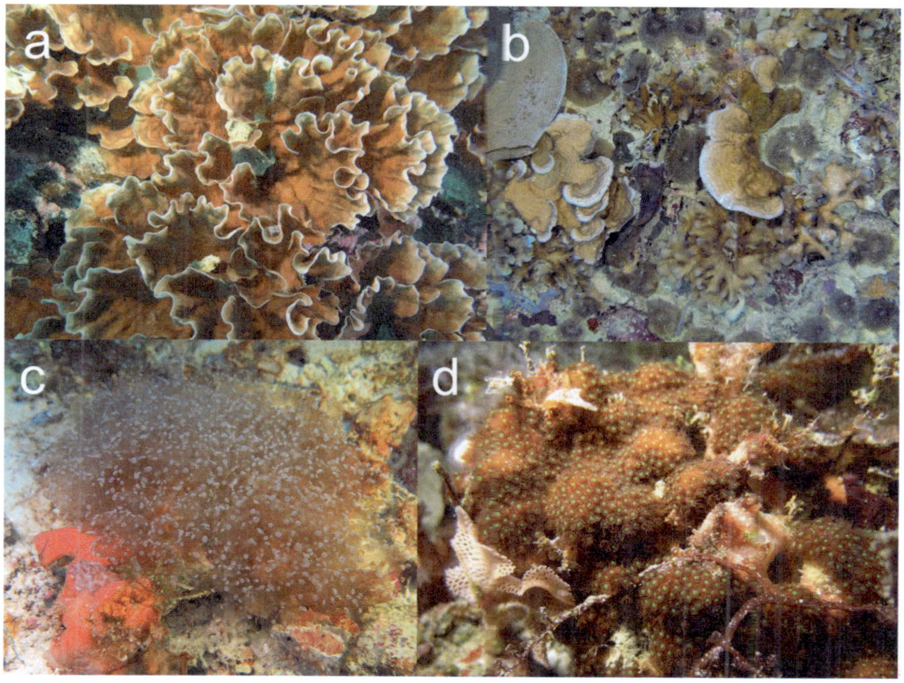


**Fig. 10.5** (**a**) *Leptoseris amitoriensis*, Ishigaki Island, 40 m; (**b**) Two different morphotypes of *Pachyseris* among *Leptoseris* sp. and unidentified corallimorpharians, "typical" *P. speciosa* on the *top left above* another morphotype that may be similar to *P. foliosa*, east coast of Okinawa Island, 56 m; (**c**) *Euphyllia paradivisa*, offshore Sesoko Island, 55 m depth; (**d**) *Madracis kirbyi*, offshore Sesoko Island, 55 m

### 10.3.2 *Pachyseris*

*Pachyseris* corals are commonly encountered in reef slopes, deeper areas of the reef as well as on shallow reefs in shaded places. Only a few species from this genus are currently described and four out of the five existing species were reported from Japan (Nishihira and Veron 1995; Ohara et al. 2013). *Pachyseris foliosa* was reported for the first time in Japan based on samples collected off Okinawa Island (Ohara et al. 2013). However, when comparing figures from Ohara et al. (2013) with the original description of Veron (1990) and *P. foliosa* (all from Papua New Guinea) in Veron (2000), the similarities are inconclusive. Recently, among numerous *Pachyseris* morphotypes present in the region, we found some that could correspond to the same species reported in Ohara et al. (2013) (Fig. 10.5b). While different from the holotype of *P. foliosa*, they are similar to the additional image provided in the original description (see Fig. 77 in Veron 1990). This exemplifies the need for deeper studies to resolve relationships between *Pachyseris* species and morphotypes.

In addition, with increasing attention paid to mesophotic corals, the number of species seeing their distribution range expanding and new species discovery will increase without doubts. One additional example of such expansion is the recent finding of *Euphyllia paradivisa* in Okinawa Island (Fig. 10.5c). This species was previously known only from a limited part of the coral triangle and now is known from MCEs in both Okinawa and the Red Sea (Eyal et al. 2016). Likewise, *Madracis kirbyi*, a presumably rare species, not recorded from Japan, was recently found at mesophotic depths offshore of Motobu Peninsula (Fig. 10.5c). These findings confirm the need for further biodiversity surveys of the mesophotic coral fauna in Okinawa. Beside the discovery of new species and the reassessment/revision of the geographic range of known species, such surveys might also provide useful information on the ecology and abundance of various known coral species. For example, *Lobophyllia* (formerly *Australomussa*) *rowleyensis* is a supposedly rare coral according to Nishihira and Veron (1995). However, in both Okinawa and Kume Islands, this species is regularly encountered around 40 m depth. This suggests that instead of being a rare species, this species simply preferentially inhabits environments that were rarely investigated.

## 10.4 Fossil Mesophotic Corals

Quaternary fossil reef deposits are extensively distributed in the Central and South Ryukyu Islands, and the age of coral reef initiation in this region is estimated to be 1.45–1.65 Ma (Iryu et al. 2006). The rate of tectonic subsidence/uplift, the amplitude and frequency of sea level changes and the timing of coral reef development are key factors controlling the nature and distribution of fossil reef deposits in the Ryukyu Islands (Iryu et al. 2006; Yamamoto et al. 2006). The resulting successions of Pleistocene strata (Ryukyu Group) contain a diverse and abundant coral fauna associated with a wide range of reef environments, including MCEs (Humblet and Iryu 2014; Humblet et al. 2009b).

**Kumejima Expedition 2009**

Beside scleractinian corals, MCEs host a large diversity of benthic organisms, and the difficulties in accessing MCEs resulted in a similar lack of knowledge on the taxonomy and distribution of other benthic organisms at mesophotic depths, perhaps compounded by the lesser interest in so-called lower taxa (e.g., polychaetes, bryozoans, ascidians, etc.). All over the world, MCE diversity research focus mainly on corals, sponges, fishes, and algae. In the Ryukyu archipelago, one international expedition was organized in 2009 by Dr. T. Naruse from the University of the Ryukyus to explore the biodiversity of Kume Island, with a primary focus on other groups than those cited above.

Crustacean specialists largely dominated the scientific team, although several other taxonomic groups were also included in this survey. Mesophotic depths were not specifically targeted; however, trawlings and dredgings were performed over the whole mesophotic depth range (Naruse 2012). Because of the destructive nature of the sampling gear and the targeted organisms, mainly coral rubbles slopes were sampled using trawling or dredging. Diving explorations focused on shallower depths with only a few rare incursions to the upper mesophotic zone and a few samples collected down to 55 m. Although numerous samples remain to be studied, over 50 new species were discovered (Naruse et al. 2012), and numerous new distribution records were published for crustaceans, echinoderms, or fishes associated with coral rubbles environments at mesophotic depths around the island (e.g., Chan and Hayashi 2012; Chen et al. 2012; Osawa 2012; Ota 2012). Probably as a result of the sampling gear used, no mesophotic scleractinian corals were collected during the expedition. However, several new plumulariid octocorals were found and are in the process of being formally described (Nonaka and Sinniger, unpublished). In addition to their value in improving the understanding of marine biodiversity, these findings highlight the lack of knowledge on the diversity of organisms inhabiting mesophotic depths.

Fossil mesophotic coral assemblages are identified based on the analogy with the taxonomic composition of modern mesophotic communities, as well as their association with other key indicators of deep coral reef habitats such as deep coralline algal species (often forming rhodoliths below 50 m in the Ryukyu Islands) or the large foraminifer *Cycloclypeus carpenteri*, commonly found below 65 m (Iryu et al. 1995).

Pleistocene coral assemblages that can be interpreted as fossil MCEs have been reported from Kikai Island in the northern Central Ryukyus, Okinawa Island in the southern Central Ryukyus and Irabu Island in the South Ryukyus (Humblet and Iryu 2014; Humblet et al. 2009b). Keeping in mind the limitations of the data currently available, we can identify some general features of fossil deep coral assemblages. The fossil MCEs described in these three locations share a common characteristic of being mainly composed of thin laminar morphologies, i.e., encrusting, platy, or foliaceous. *Porites* and/or *Montipora* (not always distinguishable in the fossil record) appear to be important components of these assemblages. *Turbinaria* also seems to be widespread. There are however notable differences among deep coral assemblages described in the three islands. For example, *Echinophyllia* was found only in fossil MCEs on Irabu and Kikai Island. *Galaxea astreata* and *Pachyseris speciosa* are significant species in southern Okinawa and Kikai Island, respectively. *Stylocoeniella* (potentially confused with a *Montipora* species) and *Leptoseris* are common on Irabu Island. The diversity of merulinids also varies from one site to another. Surprisingly no mesophotic *Acropora* species has yet been reported from the fossil record. Differences in community structure of MCEs described in the fossil record likely reflect the spatial or depth-related variations in taxonomic compositions displayed by modern mesophotic communities.

## 10.5 Sexual Reproduction of Mesophotic Corals in Okinawa

Scleractinian corals release their gametes (eggs and sperm) or shed their planula larvae into the water column (see Chap. 7). After their planula larvae disperse and settle onto the substrata, they metamorphose into polyps and start their sessile life. While there are numerous studies on sexual reproduction of shallow corals, little is known on the reproduction of mesophotic corals (but see Holstein et al. 2015, 2016; Prasetia et al. 2016, 2017a, b; Eyal-Shaham et al. 2016). Sexual reproduction plays an important role in maintaining and extending coral distribution. In the context of the Deep Reef Refugia Hypothesis (DRRH, see Sect. 10.6), the main possibility for deep corals to recolonize degraded shallow reefs is through larval dispersal (Bongaerts et al. 2010). It is therefore crucial to study and understand the sexual reproduction of mesophotic corals including fecundity, reproductive seasonality, timing of spawning and larval characteristics (such as behavior and settlement competency periods), and vertical connectivity between shallow and mesophotic reefs.

In Indo-Pacific region and the Ryukyu archipelago in particular, *Acropora tenella* was the first mesophotic spawning coral species to be studied for its reproductive biology

**Fig. 10.6** (**a**) Egg bundles of *Acropora tenella* (in the background) spawn in aquarium in June 2016, (**b**) planula of *Seriatopora hystrix* released from a 40 m depth colony, the white bar represents 400 μm, (**c**) 11 months old juvenile colony of *S. hystrix* deployed in a shallow reef (about 3 m depth) on Sesoko Island, intervals between marks on the scale represent 0.5 mm (Photographs **b**, **c** courtesy of R. Prasetia)

(Prasetia et al. 2016, 2017a). It is also the first investigation on sexual reproduction in the genus *Acropora* at mesophotic depths in the world. This species is a hermaphrodite with oogenesis and spermatogenesis occurring for 11–12 and 5–6 months, respectively. The gametogenesis is relatively longer compared to shallow *Acropora* species, probably because of lower seawater temperature in spring and summer and lower light intensity compared to shallow water (Prasetia et al. 2016). In addition, the fecundity (number of eggs per polyps) and the gonad index (number of eggs per square centimeter) are among the lowest reported for an acroporid coral. The longer gametogenesis and lower reproductive yields may reflect the limited energy available at the depths where this species is found (around 40 m). Except for the longer gametogenesis, the reproductive behavior of *A. tenella* is not significantly different form other shallow *Acropora* species. The rapid maturation of gametes at the end of spring seems to be correlated with a rapid increase in seawater temperature. The spawning is expected to occur at night in both June and July, like shallow acroporid corals in the same area, (around Sesoko Island, northern Okinawa). Although there is no record on in situ spawning, aquarium observations showed that released eggs are positively buoyant (Prasetia et al. 2017a) (Fig. 10.6a). This observation raises questions on how gametes fertilize at the surface of the ocean and how larvae return to the deeper part of the reef. Larval dispersal ranges of corals are controlled by their settlement competency period, position in the water column, and currents that actually deliver the larvae (Harii and Kayanne 2003). In shallow reefs, eggs of many corals including *Acropora* are positively buoyant due to their high lipid contents (Arai et al. 1993; Harii et al. 2007) and flushed by wind-driven current (Willis and Oliver 1990; also see Chap. 7). The potential challenges of fertilization 40 m above the spawning origin combined to the reported low fecundity and low gonad index may explain the rarity of this species and its absence from most mesophotic sites investigated in the region.

*Seriatopora hystrix* is another common mesophotic species that is the focus of several studies in Okinawa in regard to the DRRH (see Sect. 10.6). This species is widely known as a brooder in shallow reefs (Atoda 1951) and is a good model to examine shallow reef recovery from mesophotic populations by larval dispersal. *S. hystrix* from 40 m offshore Sesoko Island has been observed to release planula larvae at least during the summer months (Prasetia et al. 2017b, Fig. 10.6b). Preliminary results suggest that larvae can survive at least up to several months in shallow reef conditions (Fig. 10.6c), although light intensity appears to be a strongly limiting parameter for direct transplant from upper mesophotic to shallow reefs.

These first studies on mesophotic coral reproduction open the path to further research on the larval behavior, dispersal, and genetic connectivity in order to understand better the true potential of MCEs to contribute to shallow reef recovery and also to understand the interactions between different MCEs and there evolution through time.

## 10.6 Deep Reef Refugia Hypothesis

One of the major drivers of the fast-growing upper mesophotic research effort in the last years comes from the Deep Reef Refugia Hypothesis (DRRH). The hypothesis initially stated that several environments, including deep reefs, could serve as refuge for shallow species threatened by ocean warming (Glynn 1996). The deeper waters being less affected by seasonal temporary extreme warming events, shallow coral species may be able to survive catastrophic warming events in these remote habitats. The hypothesis was further refined by stating that to serve as refugia, MCEs should be also serving as a source of recolonization for shallow reefs (Bongaerts et al. 2010). Recent studies (Serrano et al. 2016; Bongaerts et al. 2017; Holstein et al. 2015, 2016) suggest that the potential for mesophotic corals to reseed shallow reefs differs between species and locations, and investigations on the physiological adaptation and depth specialization, symbiotic association and vertical connectivity are contributing to estimate the potential and the challenges of a mesophotic source of larvae for shallow reefs.

The Ryukyu archipelago and especially the main Island of Okinawa and neighboring locations such as Sesoko Island were strongly affected by massive coral bleaching events occurring in 1998 and 2001 (Loya et al. 2001). These events lead to a long-term reorganization of the coral species assemblages in the shallow reefs (van Woesik et al. 2011). One of the species concerned is *S. hystrix*. In Okinawa, *S. hystrix* is normally relatively common in shallow reefs. However, being a species highly sensitive to bleaching, it is strongly threatened by global warming. As a result, *S. hystrix* disappeared from the shallow reefs around Sesoko Island following the massive bleaching events of 1998 and 2001 (van Woesik et al. 2011). The frequent occurrences of *S. hystrix* at mesophotic depths in the archipelago, especially near Sesoko Island (Sinniger et al. 2013, 2017) (Fig. 10.4c), provide an excellent model to test the ecological role of MCEs in Okinawa. Although fine scale population genetics to assess connectivity between shallow and deep *S. hystrix* is still ongoing, first observations of various morphotypes and genotypes do not support the hypothesis of a clonal deep population and the observation of larval release in deep colonies suggest an active sexual reproduction of deep colonies (Sinniger et al. 2017, also Sect. 10.5). Therefore, the first part of the DRRH can be validated as this species actually survived in deeper parts of the reef. However, as *S. hystrix* was not yet found again in the shallow reefs around Sesoko Island, it remains to be seen if and how deep refuge populations can recolonize depleted shallow reefs. Such research will define whether Okinawan deep reefs are more likely to act as refugia or as traps from which species cannot get out.

## 10.7 Threats on Mesophotic Coral Ecosystems in Japan

In addition of providing refuge to shallow corals during thermal stress, MCEs also have been suggested to be less affected by powerful storms and destructive wave action. However, the effect of typhoons on deeper reef communities may be significant in the Pacific, for example, through mechanical destruction of coral communities by massive coral colonies and boulder tumbling down reef slopes during typhoons (Harmelin-Vivien 1994). More recently typhoon-induced damages have been observed in the Great Barrier Reef in Australia (Bongaerts et al. 2013). In Okinawa, extensive damages have been reported for a mesophotic *Acropora horrida* community offshore of Kume Island as well as at the upper edge (31 m) of a mesophotic *Pachyseris* community on the west coast of Okinawa Island.

In Kume Island, the typhoons #2 (Songda) and #9 (Muifa) tracked near/passed over a dense *A. horrida* community (see Sect. 10.2) on the 28th of May and between the 4th and 6th of August 2011, respectively. A survey conducted 1 month after this event showed extensive damages resulting in a loss of coral cover (from over 90 % to 20–30 %) and the invasion of dead corals by filamentous cyanobacteria (Fujita et al. 2012). However, Fujita et al. (2012) also found surviving fragments already reattached to the substrate or fused with other coral branches. This finding combined to the absence of bleaching and burial by sediments suggests a strong potential for rapid recovery of this community.

On the west coast of Okinawa Island, typhoon #16 (Jelawat) passed hit Ryugu Reef (see Sect. 10.2) on the 29th of September 2012. White et al. (2013) found that the "deep" (32 m) community of foliose *Pachyseris* corals suffered more damages than the shallower (25 m) and more diverse coral community. While the dominance of *Pachyseris* at the deeper site may have hindered observable community composition shifts, at the shallower site branching *Acropora* coverage was reduced and the proportion of fungiids increased. At both deep and shallow site, coral rubbles covered extensive area of the reef. However, as was the case in Kume Island 1 year earlier, a few months after the typhoon passage, signs of recovery could be observed in the *Pachyseris* community.

These two studies in Okinawa demonstrate that, while MCEs can be heavily impacted by typhoons, they show a strong resilience and recovery potential. Whether this conclusion holds MCEs in general or reflects the relative remoteness of these MCEs from anthropogenic influences remains to be tested. But these two studies already provide a good baseline to evaluate the impact of typhoons on mesophotic corals and their subsequent recovery.

Other threats are likely endangering MCEs in Okinawa, such as the impact of *Acanthaster* (crown of thorns sea star, see Chap. 9), sedimentation, nutrient enrichment, or shoreline modifications, and climate change. However, until now no study investigated these threats on mesophotic corals in Japan.

Considering the paucity of knowledge on how human activities (e.g., coastal development, pollution) affect MCEs and the rapid development of the islands of the archipelago (including landfills for various uses), a proactive conservation approach should be applied until adequate knowledge on the distribution, biodiversity, ecology, and connectivity of MCEs is acquired. Failing to do so may lead to the dramatic destruction of these ecosystems due to our ignorance as to how they respond to environmental changes or even where they are located.

## 10.8 Future Perspectives for Japanese Mesophotic Research

As highlighted in the recent UNEP report on MCEs (Baker et al. 2016), the priority for mesophotic research is to locate and identify MCEs to allow understanding the conditions required by such rich ecosystems to develop and survive. This is especially important in Japan in regard to the area suitable for mesophotic habitats. The high diversity of MCEs encountered until now in the Ryukyu archipelago would suggest the presence of a high biodiversity of mesophotic organisms in these ecosystems. In this respect, large multidisciplinary biodiversity surveys, across a broad depth range similar to the Kumejima Expedition 2009, or focusing on mesophotic depths, are highly needed to identify the biodiversity of these remote ecosystems. Identifying the organisms inhabiting these ecosystems is an absolute pre-requisite to any further study on the biology or ecology of mesophotic corals and associated organisms.

In regard to the Deep Reef Refugia Hypothesis, while it was shown that some species could be relatively protected from massive bleaching occurring in shallow reefs, the role of these corals can play in the recovery of shallow reefs is not yet known. Research on the physiological adaptation of deep corals to shallow reef conditions, horizontal, and vertical connectivity are highly needed in order to understand the real potentials of MCEs to contribute to shallow reef recovery. Moreover, Japanese mesophotic corals being likely connected to other MCEs in the West Pacific, international collaborations are essential in order to obtain the most complete and accurate knowledge on these mysterious ecosystems.

**Acknowledgments** We thank Dr. M. Humblet (Nagoya University) for his precious contribution to this chapter and Mr. Rian Prasetia (University of the Ryukyus) for the information and photographs on mesophotic coral reproduction. This chapter was written within the frame of the KAKENHI project No. 16H02490.

## References

Agee E, Howley R (1977) Latent and sensible heat flux calculations at the air-sea interface during AMTEX 74. J Appl Meteorol 16(4):443–447

Arai I, Kato M, Heyward A, Ikeda Y, Iizuka T, Maruyama T (1993) Lipid composition of positively buoyant eggs of reef building corals. Coral Reefs 12(2):71–75. https://doi.org/10.1007/BF00302104

Arai K, Matsuda H, Sasaki K, Machiyama H, Yamaguchi T, Inoue T, Sato T, Takayanagi H, Iryu Y (2016) A newly discovered submerged reef on the Miyako-Sone platform, Ryukyu Island Arc, Northwestern Pacific. Mar Geol 373:49–54. https://doi.org/10.1016/j.margeo.2016.01.007

Atoda K (1951) The larva and postlarval development of the reef-building corals. V. *Seriatopora hystrix* Dana. Sci Rep Tohoku Univ Ser 4:33–39

Baker E, Puglise K, Harris P (2016) Mesophotic coral ecosystems—a lifeboat for coral reefs. The United Nations Environment Programme and GRID-Arendal, Nairobi/Arendal, p 98

Bongaerts P, Ridgway T, Sampayo EM, Hoegh-Guldberg O (2010) Assessing the 'deep reef refugia' hypothesis: focus on Caribbean reefs. Coral Reefs 29(2):309–327. https://doi.org/10.1007/S00338-009-0581-X

Bongaerts P, Muir P, Englebert N, Bridge T, Hoegh-Guldberg O (2013) Cyclone damage at mesophotic depths on Myrmidon Reef (GBR). Coral Reefs 32(4):935–935. https://doi.org/10.1007/s00338-013-1052-y

Bongaerts P, Riginos C, Brunner R, Englebert N, Smith SR, Hoegh-Guldberg O (2017) Deep reefs are not universal refuges: reseeding potential varies among coral species. Sci Adv 3(2):e1602373. https://doi.org/10.1126/sciadv.1602373

Chan BK, Hayashi R (2012) Epibiotic barnacles (Crustacea: Cirripedia: Thoracica) collected by the Kumejima 2009 Expedition, with descriptions of two new species. Zootaxa 3367(1):21–48

Chen I-S, Suzuki T, Shao K-T (2012) A new deepwater goby of the genus *Discordipinna* Hoese & Fourmanoir, 1978 (Teleostei: Gobiidae) from Kumejima of the Ryukyus, Japan. Zootaxa 3367(1):274–280

Eyal G, Eyal-Shaham L, Cohen I, Tamir R, Ben-Zvi O, Sinniger F, Loya Y (2016) *Euphyllia paradivisa*, a successful mesophotic coral in the northern Gulf of Eilat/Aqaba, Red Sea. Coral Reefs 35(1):91–102. https://doi.org/10.1007/s00338-015-1372-1

Eyal-Shaham L, Eyal G, Tamir R, Loya Y (2016) Reproduction, abundance and survivorship of two *Alveopora* spp. in the mesophotic reefs of Eilat, Red Sea. Sci Rep 6. https://doi.org/10.1038/srep20964

Fujita Y, Kimura T, Shioiri A, Naruse T (2012) Typhoon damage of large-scaled coral communities dominated by *Acropora horrida* (Dana, 1846) (Scleractinia, Acroporidae) in the mesophotic zone off Kumejima Island, the Ryukyu Islands, Japan. Biol Mag Okinawa 50:61–66

Glynn PW (1996) Coral reef bleaching: facts, hypotheses and implications. Glob Chang Biol 2(6):495–509. https://doi.org/10.1111/j.1365-2486.1996.tb00063.x

Harii S, Kayanne H (2003) Larval dispersal, recruitment, and adult distribution of the brooding stony octocoral *Heliopora coerulea* on Ishigaki Island, southwest Japan. Coral Reef 22:188–196. https://doi.org/10.1007/s00338-003-0302-9

Harii S, Nadaoka K, Yamamoto M, Iwao K (2007) Temporal changes in settlement, lipid content and lipid composition of larvae of the spawning hermatypic coral *Acropora tenuis*. Mar Ecol Prog Ser 346:89–96. https://doi.org/10.3354/Meps07114

Harmelin-Vivien ML (1994) The effects of storms and cyclones on coral reefs: a review. J Coast Res:211–231

Hinderstein LM, Marr JCA, Martinez FA, Dowgiallo MJ, Puglise KA, Pyle RL, Zawada DG, Appeldoorn R (2010) Theme section on "Mesophotic Coral Ecosystems: Characterization, Ecology, and Management". Coral Reefs 29(2):247–251. https://doi.org/10.1007/S00338-010-0614-5

Hohenegger J (1994) Distribution of living larger foraminifera NW of Sesoko-Jima, Okinawa, Japan. Mar Ecol 15(3–4):291–334. https://doi.org/10.1111/j.1439-0485.1994.tb00059.x

Hohenegger J, Yordanova E (2001) Displacement of larger foraminifera at the western slope of Motobu Peninsula (Okinawa, Japan). PALAIOS 16(1):53–72

Hohenegger J, Yordanova E, Nakano Y, Tatzreiter F (1999) Habitats of larger foraminifera on the upper reef slope of Sesoko Island, Okinawa, Japan. Mar Micropaleontol 36(2):109–168. https://doi.org/10.1016/S0377-8398(98)00030-9

Holstein DM, Smith TB, Gyory J, Paris CB (2015) Fertile fathoms: deep reproductive refugia for threatened shallow corals. Sci Rep 5. https://doi.org/10.1038/srep12407

Holstein DM, Paris CB, Vaz AC, Smith TB (2016) Modeling vertical coral connectivity and mesophotic refugia. Coral Reefs 35(1):23–37. https://doi.org/10.1007/s00338-015-1339-2

Hongo C, Yamano H (2013) Species-specific responses of corals to bleaching events on anthropogenically turbid reefs on Okinawa Island, Japan, over a 15-year period (1995–2009). PLoS One 8(4):e60952. https://doi.org/10.1371/journal.pone.0060952

Hori N (1990) Coral reefs in Japan: the peripheral zone of the world coral reefs. In: Tropics in Japan: natural history of Coral Islands. Kokin-Shoin Ltd., Tokyo, pp 3–22

Humblet M, Iryu Y (2014) Pleistocene coral assemblages on Irabu-jima, South Ryukyu Islands, Japan. Paleontol Res 18(4):224–244. https://doi.org/10.2517/2014PR020

Humblet M, Furushima Y, Iryu Y, Tokuyama H (2009a) Deep zooxanthellate corals in the Ryukyu Islands: insight from modern and fossil reefs. In: 27th IAS meeting of sedimentologists, Alghero, Italy, Medimond international proceedings, p 174

Humblet M, Iryu Y, Nakamori T (2009b) Variations in Pleistocene coral assemblages in space and time in southern and northern Central Ryukyu Islands, Japan. Mar Geol 259(1):1–20. https://doi.org/10.1016/j.margeo.2008.12.002

Iryu Y, Nakamori T, Matsuda S, Abe O (1995) Distribution of marine organisms and its geological significance in the modern reef complex of the Ryukyu Islands. Sediment Geol 99(3):243–258. https://doi.org/10.1016/0037-0738(95)00047-C

Iryu Y, Matsuda H, Machiyama H, Piller WE, Quinn TM, Mutti M (2006) Introductory perspective on the COREF Project. Island Arc 15(4):393–406. https://doi.org/10.1111/j.1440-1738.2006.00537.x

Kimura T, Shimoike K, Suzuki G, Nakayoshi I, Shioiri A, Tabata A, Tabata Y, Fujita Y, Zayasu Y, Yamano H, Namizaki N, Yokoi K, Ogasawara K, Yasumura S (2011) Large scale communities of *Acropora horrida* in the mesophotic zone off Kume Island, Okinawa. J Jpn Coral Reef Soc 13(1):43–45

Loya Y, Sakai K, Yamazato K, Nakano Y, Sambali H, van Woesik R (2001) Coral bleaching: the winners and the losers. Ecol Lett 4(2):122–131. https://doi.org/10.1046/j.1461-0248.2001.00203.x

Luck DG, Forsman ZH, Toonen RJ, Leicht SJ, Kahng SE (2013) Polyphyly and hidden species among Hawai'i's dominant mesophotic coral genera, *Leptoseris* and *Pavona* (Scleractinia: Agariciidae). PeerJ 1:e132. https://doi.org/10.7717/peerj.132

Muir PR, Wallace CC (2015) A rare 'deep-water'coral assemblage in a shallow lagoon in Micronesia. Mar Biodivers 1–2. https://doi.org/10.1007/s12526-015-0419-4

Muir PR, Wallace CC, Done T, Aguirre JD (2015) Limited scope for latitudinal extension of reef corals. Science 348(6239):1135–1138. https://doi.org/10.1126/science.aaa1605

Naruse T (2012) The KUMEMEJIMA 2009 Expedition: station data and collecting sites. Zootaxa 3367(1):8–20

Naruse T, Chan T-Y, Tan HH, Ahyong ST, Reimer JD (2012) Scientific results of the Kumejima marine biodiversity expedition—Kumejima 2009. Zootaxa 3367:1–280

Nishihira M, Veron J (1995) Hermatypic corals of Japan. Kaiyusha, Tokyo

Ohara T, Fujii T, Kawamura I, Mizuyama M, Montenegro J, Shikiba H, White KN, Reimer JD (2013) First record of a mesophotic *Pachyseris foliosa* reef from Japan. Mar Biodivers 43(2):71–72. https://doi.org/10.1007/S12526-012-0137-0

Omija T (2004) Terrestrial inflow of soils and nutrients. Coral Reefs Jpn 47:64–68

Osawa M (2012) *Raymunida* Macpherson & Machordom, 2000 (Crustacea: Decapoda: Anomura: Munididae) from the KUMEJIMA 2009 Expedition in the Ryukyu Islands, Japan. Zootaxa 3367(1):134–144

Ota Y (2012) Gnathiidae from Kumejima Island in the Ryukyu Archipelago, southwestern Japan, with description of three new species (Crustacea: Isopoda). Zootaxa 3367(1):79–94

Prasetia R, Sinniger F, Harii S (2016) Gametogenesis and fecundity of *Acropora tenella* (Brook 1892) in a mesophotic coral ecosystem in Okinawa, Japan. Coral Reefs 35:53–62. https://doi.org/10.1007/s00338-015-1348-1

Prasetia R, Sinniger F, Harii S (2017a) First record of spawning in the mesophotic *Acropora tenella* in Okinawa, Japan. Galaxea J Coral Reef Stud 19:1–2

Prasetia R, Sinniger F, Hashizume F, Harii S (2017b) Reproductive biology of the deep brooding coral Seriatopora hystrix: implications for shallow reef recovery. Plos One. https://doi.org/10.1371/journal.pone.0177034

Serrano XM, Baums IB, Smith TB, Jones RJ, Shearer TL, Baker AC (2016) Long distance dispersal and vertical gene flow in the Caribbean brooding coral *Porites astreoides*. Sci Rep 6:21619. https://doi.org/10.1038/srep21619

Sinniger F (2016) Ryukyu Archipelago, Japan. In: Baker EK, Puglise KA, Harris PT (eds) Mesophotic coral ecosystems – a lifeboat for coral reefs? The United Nations Environment Programme and GRID-Arendal, Nairobi/Arendal, pp 50–62

Sinniger F, Morita M, Harii S (2013) "Locally extinct" coral species *Seriatopora hystrix* found at upper mesophotic depths in Okinawa. Coral Reefs 32(1):153–153. https://doi.org/10.1007/S00338-012-0973-1

Sinniger F, Prasetia R, Yorifuji M, Bongaerts P, Harii S (2017) Seriatopora diversity preserved in upper mesophotic coral ecosystems in Southern Japan. Front Mar Sci 4:155. https://doi.org/10.3389/fmars.2017.0015

The Government of the Ryukyus (1967) Report on the fisheries resources investigation. The Government of the Ryukyus, Naha

Tsuji Y (1993) Tide influenced high energy environments and rhodolith-associated carbonate deposition on the outer shelf and slope off the Miyako Islands, southern Ryukyu Island Arc, Japan. Mar Geol 113(3–4):255–271. https://doi.org/10.1016/0025-3227(93)90021-M

van Woesik R, Sakai K, Ganase A, Loya Y (2011) Revisiting the winners and the losers a decade after coral bleaching. Mar Ecol Prog Ser 434:67–76. https://doi.org/10.3354/meps09203

Veron J (1990) New scleractinia from Japan and other Indo-West Pacific countries. Galaxea 9(2)

Veron JE (2000) Corals of the world, vol 1–3. Australian Institute of Marine Science, Townsville, pp 404–405

Wells JW (1954) Recent corals of the Marshall Islands: an ecologic and taxonomic analysis of living reef-and non-reef-building corals at Bikini and other Marshall Islands atolls, vol 260, Geological Survey Professional Paper, vol I. US Government Printing Office, Washington, DC

White KN, Ohara T, Fujii T, Kawamura I, Mizuyama M, Montenegro J, Shikiba H, Naruse T, McClelland T, Denis V (2013) Typhoon damage on a shallow mesophotic reef in Okinawa, Japan. PeerJ 1:e151. https://doi.org/10.7717/peerj.151

Willis BL, Oliver JK (1990) Direct tracking of coral larvae: implications for dispersal studies of planktonic larvae in topographically complex environments. Ophelia 32(1–2):145–162. https://doi.org/10.1080/00785236.1990.10422029

Yamamoto K, Iryu Y, Sato T, Chiyonobu S, Sagae K, Abe E (2006) Responses of coral reefs to increased amplitude of sea-level changes at the Mid-Pleistocene Climate Transition. Palaeogeogr Palaeoclimatol Palaeoecol 241(1):160–175. https://doi.org/10.1016/j.palaeo.2006.06.014

Yamano H, Satake K, Inoue T, Kadoya T, Hayashi S, Kinjo K, Nakajima D, Oguma H, Ishiguro S, Okagawa A (2015) An integrated approach to tropical and subtropical island conservation. J Ecol Environ 38(2):271–279. https://doi.org/10.5141/ecoenv.2015.028

Yamazato K (1972) Bathymetric distribution of corals in the Ryukyu Islands. In: Proceedings of the symposium on corals and coral reefs. Marine Biological Association of India, Cochin, pp 121–133

Yamazato K, Kamura S, Shinsho N (1967) Distribution of benthic organisms along the Ryukyu coast. In: Report on the fisheries resources investigation. The Government of the Ryukyus, Naha, p 299

# The Hydrodynamic Impacts of Tropical Cyclones on Coral Reefs of Japan: Key Points and Future Perspectives

11

Chuki Hongo

**Abstract**

Tropical cyclones are extreme geodynamic events that cause serious damage to coral reefs and coral communities, including mechanical destruction of reef builders and changes in reef topography. This chapter reviews hydrodynamic research into the effects of tropical cyclones on coral reefs and coral communities of Japan. Although much research has focused on the effects of changing wind speeds, hydrodynamic parameters such as wave height, wave period, and wave velocity also have a direct influence on reef evolution. More data related to wave activity must be collected if we are to develop a better understanding of the hydrodynamic impacts of tropical cyclones on coral reefs. In Japan, an observation network that records wave height and wave period has been maintained since the 1970s because the region is especially prone to tropical cyclones. In addition, the possible hydrodynamic impacts of tropical cyclones on coral reefs are taken into account assuming that these storms will become more intense in the future.

**Keywords**

Tropical cyclones • Coral reef • Coral • Hydrodynamic research • Wave height • Wave period • Japan

## 11.1 Introduction

Tropical cyclones (TCs) are large-scale natural phenomena characterized by extreme winds and torrential rainfall. The high winds generate extreme wave height, wave length, and water velocity. At the scale of the coral community, the hydrodynamic action causes direct mechanical disturbance of corals (e.g., breakage of branching corals, overturning of massive corals, dislodgement of tabular corals; Fabricius et al. 2008). De'ath et al. (2012) concluded that TCs were the principal cause of coral loss on the Great Barrier Reef between 1985 and 2012. At the reef scale, TCs often cause changes in reef morphology. For example, coral rubble and reef rocks can be transported from a reef slope to a reef crest (Goto et al. 2011; Fig. 11.1). In Funafuti Atoll, the formation of storm ridge with a height of 3.5 m occurred after Cyclone Bebe in 1972 (Maragos et al. 1973). The torrential rainfall associated with TCs leads to an increase in sediment volumes and nutrients carried by rivers into the ocean, which in turn affects coral reefs. The increase in turbidity and sedimentation levels reduces irradiance, and the change in water quality causes a decrease in coral cover (Fabricius 2005). However, TCs can also cause a reduction in sea surface temperature via vertical mixing driven by strong winds (Kuroyama et al. 2000), and these cooling events can help to reduce the risk of coral bleaching (Manzello et al. 2007).

An increase in TC intensity in the late twenty-first century is likely to have severe impacts on coral reefs. Although it remains uncertain whether past changes in TC activity exceed the variability expected from natural causes, the globally averaged TC intensity is expected to increase by between 2 % and 11 % by 2100 (Knutson et al. 2010).

C. Hongo (✉)
Department of Chemistry, Biology, and Marine Science, University of the Ryukyus, Nishihara, Okinawa, Japan
e-mail: g123001@sci.u-ryukyu.ac.jp


A. Iguchi, C. Hongo (eds.), *Coral Reef Studies of Japan*, Coral Reefs of the World 13,
https://doi.org/10.1007/978-981-10-6473-9_11

**Fig. 11.1** (**a**) Large reef boulders (long axis >1 m) deposited on the reef crest at Amami Island in the Ryukyu Islands. (**b**) Two fresh reef boulders (long axis >0.4 m) deposited on the reef crest at Kudaka Island in the Ryukyu Islands. Typhoon 1203 (Mawar) approached within about 200 km of the island on June 5, 2012 (The photograph was taken on June 6, 2012)

According to high-resolution downscaling experiments using a cloud-resolving model, the number of super typhoons each year will increase from three at present to 12 by the late twenty-first century (Tsuboki et al. 2015). Correlatively, maximum wind speeds will increase from 74 m/s at present to 88 m/s by the late twenty-first century in the northwest Pacific as a consequence of global warming (Tsuboki et al. 2015). This forecast implies that corals will be severely damaged by hydrodynamic disturbance. Moreover, average 1 h rainfall rates associated with TCs will also increase in the future climate, from 8.9 mm/h at present to 11.1 mm/h by the late twenty-first century (Tsuboki et al. 2015). Such torrential rainfall will probably cause an increase in the supply of fluvial sediments and nutrients to coral reefs.

Ongoing global and local environmental changes are likely to accelerate the decline of coral reefs under attack from more intense TCs. Some studies have shown that ocean acidification may lead to reduced coral skeletal density (Cooper et al. 2008). For example, coral skeletal density declined from 1.32 g/cm$^3$ in 1998 to 1.25 g/cm$^3$ in 2003, equivalent to 0.36 %/year, based on X-ray sclerochronological analysis of massive *Porites* from the northern Great Barrier Reef over a 16-year period (1988–2003; Cooper et al. 2008). Moreover, ocean acidification is expected to cause a decrease in cementation in reef frameworks and an increase in bioerosion (Manzello et al. 2008). In the eastern tropical Pacific, which is characterized by a relatively low pH, low calcium carbonate saturation state (Ω), and high $pCO_2$ that affects upwelling and the mixing of $CO_2$-enriched deep waters with surface layers, the proportion of cement is positively related to Ω (~5 % cement for Ω of 2.49 at Galápagos and ~13% of cement for Ω of 3.50 at Gulf of Chiriqui) but inversely related to bioerosion rate (25.4 kg $CaCO_3$/m$^2$/year at Galápagos, 13.6 kg $CaCO_3$/m$^2$/year at Gulf of Chiriqui; Manzello et al. 2008). Madin et al. (2012) used a mechanistic framework based on the responses of individual-level demographic rates to changes in the physical and chemical environments to show that tabular corals (*Acropora hyacinthus*) are prone to large and rapid declines in coral cover, as well as population collapse. This relates primarily to their increased vulnerability to TC-induced dislodgment as a consequence of ocean acidification. Furthermore, increasing amounts of nutrients and sediments supplied by river discharge cause a decrease in coral skeletal density (Dunn et al. 2012; Sowa et al. 2014). Sowa et al. (2014) used an analysis of the X-ray sclerochronology of massive *Porites* from Shiraho reef on Ishigaki Island, Okinawa, to demonstrate that coral skeletal density decreased from approximately 1.8–1.3 g/cm$^3$ with decreasing distance from reef crest to coast, as a result of river discharge containing red soil and nutrients. Both environmental and coral skeletal changes probably increase the risk of dislodgement and breakage of corals by TCs.

This chapter reviews the key points of research into the impacts of TCs on coral reefs and coral communities around Japan and also discusses the possible future effects on coral communities of the ongoing attack from TCs.

## 11.2 Tropical Cyclones and Their Impacts on Coral Reefs Around Japanese Reef Province

The terms "hurricane, cyclone, and typhoon" describe the same phenomenon, the names of which change according to its geographic location (World Meteorological Organization, https://www.wmo.int/pages/mediacentre/factsheet/tropical-cyclones.html; accessed July 16, 2015). In the western North Atlantic, central and eastern North Pacific, Caribbean Sea, and Gulf of Mexico, these storms are known as hurricanes, whereas in the western North Pacific they are known as typhoons, and in the Bay of Bengal and Arabian Sea they are

**Table 11.1** Classification scale for tropical cyclones

| International category | Saffir-Simpson scale | 1-min average wind speed (knot) |
|---|---|---|
| Tropical depression | – | <33 |
| Tropical storm | – | 34–47 |
| Severe tropical storm | – | 48–63 |
| Typhoon/hurricane/cyclone/severe tropical cyclone | 1 | 64–82 |
| | 2 | 83–95 |
| | 3 | 96–112 |
| | 4 | 113–136 |
| | 5 | ≧137 |

called cyclones. In the western South Pacific and southeast Indian Ocean, they are called severe tropical cyclones, but in the southwest Indian Ocean just tropical cyclones. Depending on their maximum sustained wind speed, tropical cyclones are classified into five categories according to Table 11.1. Tropical cyclones are often graded using the Saffir-Simpson Hurricane Wind Scale (Simpson and Riehl 1981), which is based on the 1-min sustained wind speed (Table 11.1). This scale estimates potential property damage. In the western North Pacific, the term "super typhoon" is used for tropical cyclones with sustained winds exceeding 130 knots. In Japan, the term "typhoon" is used for tropical cyclones with 10-min sustained wind speeds exceeding 34 knots.

There are several major factors controlling the development of TCs, namely, (1) the presence of a pre-existing low surface pressure, (2) a warm and moist tropical atmosphere that is conducive to thunderstorm development, (3) ocean surface temperatures greater than about 26 °C so that sufficient moisture and heat can be supplied into the low pressure area to sustain thunderstorm development, and (4) a sufficient distance from the equator so that air will tend to spiral inward at low levels toward the lower pressure and outward at upper levels away from the high pressure (Pielke and Pielke 1997). TCs usually dissipate over land or colder oceans. TCs are a normal part of marine and coastal environments in the latitudinal bands 7–25° north and south of the equator, i.e., about 70 % of the tropical seas in which coral reefs occur (Done 2011).

Many coral reefs and coral communities around Japan are susceptible to the effects of TCs because Japan is especially prone to TCs (Oouchi et al. 2006; Balaguru et al. 2012). In Japan, an annual average of 11.5 TCs approach within 300 km of an observation station, and of these, an annual average of 2.8 TCs made landfall across Japan over the period of 1951–2014 (Japan Meteorological Agency, http://www.data.jma.go.jp/fcd/yoho/typhoon/statistics/index.html; accessed July 16, 2015). In particular, the Ryukyu Islands and the Pacific coasts of Japan at Kyushu, Shikoku, and Honshu experience TCs every year. Coral reefs and coral communities are also present in these areas because coral larvae are transported by the Kuroshio Current and the Tsushima Warm Current (Kayanne et al. 2004; Yamano et al. 2014). According to the global best track data for TCs over the period of 1982–2012, the location of TC maximum intensity has followed a poleward migration with rates of 53 km/10 years in the Northern hemisphere (Kossin et al. 2014). This implies that coral reefs and coral communities in the high-latitude zones, such as Japan, will be affected by TCs in the near future.

## 11.3 Key Points of Hydrodynamic Studies on Japanese Coral Reefs

Many studies of the direct impact of TCs on coral reefs have focused on wind speeds (e.g., Fabricius et al. 2008). An increase in the wind speed of TCs causes an increase in the damage sustained by coral communities (e.g., Done 1992). In fact, corals and coral reefs are directly affected by ocean waves generated by winds. Therefore, we need to focus on the role of higher water energy gradients on corals and coral reefs. In Japan, the Japan Meteorological Agency (JMA) maintained an observation system for ocean waves that recorded significant wave height (*Hs*) and significant wave period (*Ts*) data using an ultrasonic-wave-based wave gauge at 11 stations over the period of 1974–2012. In addition, these data were obtained using observation buoys from 2000 to the present day and by a radar-based wave meter at six stations on land from 2012 to the present day. Moreover, analyzed three wave parameters (*Hs*, *Ts*, and wave direction) and two wind parameters (speed and direction) based on the observed data are provided by the JMA twice a day at 26 sites. The wave observation network (the Nationwide Ocean Wave Information Network for Ports and HArbourS: NOWPHAS) has been maintained by the Ministry of Land, Infrastructure, Transport and Tourism (MLIT) since the 1970s. *Hs* and *Ts* data, as well as the wave spectrum, were obtained using an ultrasonic-wave-based wave gauge and a mooring-GPS buoy at 76 stations in 2013 (Kawaguchi et al. 2015). Sixteen GPS buoys, which measure the vertical motion of the mooring buoys caused by changes in sea surface elevation every second using the RTK-GPS technology, are included at the stations on the coast of the Pacific (Kawaguchi et al. 2015). Among these 76 stations, *Hs* and *Ts* were obtained every 20 min at 70 stations and every 2 h at 6 stations (Kawaguchi et al. 2015) Consequently, the NOWPHAS wave observation system provides *Hs* and *Ts* for 5 sites on coral reefs in Ryukyu Islands and for 21 sites on coral reef communities in Japan (Fig. 11.2).

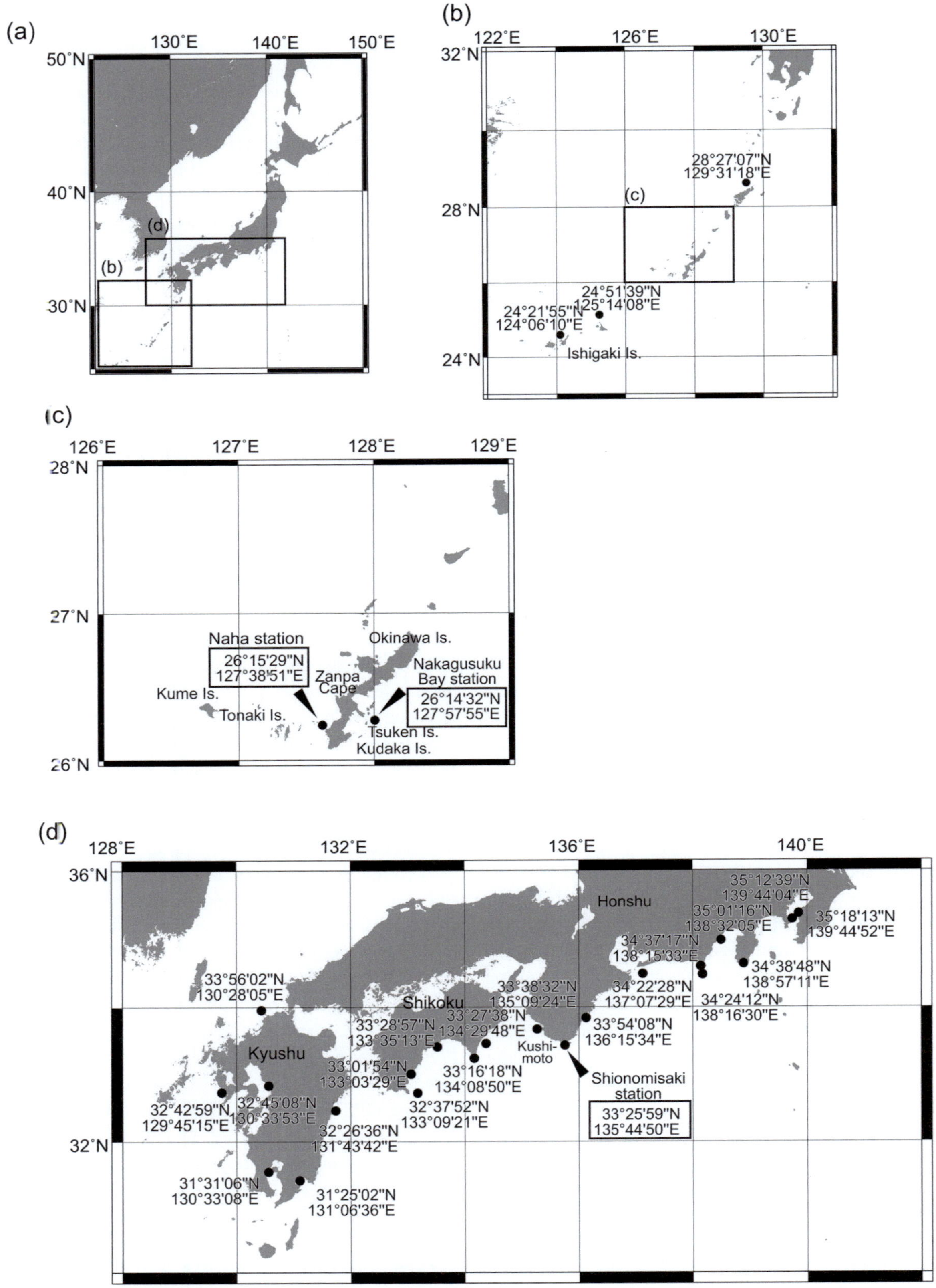


**Fig. 11.2** The NOWPHAS observational sites for significant wave height (*Hs*) and significant wave period (*Ts*) for 5 sites on coral reefs in Ryukyu Islands and 21 sites on coral reef communities in Japan. Kawaguchi et al. (2015) show 76 observational sites in Japan

The hydrodynamic regime in and around a reef system is complex because various driving forces exist that generate or modify the wave height, wave period, wave velocity, wave-induced acceleration, and currents (Hearn 2011). A shallow water depth on fringing reefs, with their reef flats attached directly to the island shorelines, significantly affects the hydrodynamic characteristics (Tamura et al. 2007). Additionally, making field observations of hydrodynamic variables during TCs is difficult and dangerous work; consequently, our knowledge of the hydrodynamic regimes associated with fringing reefs remains limited. However, in Japan, abundant data related to the hydrodynamic effects of TCs has been collected using instruments such as wave gauges and wave velocity meters (e.g., Egashira et al. 1985; Nakaza et al. 1994; Nadaoka et al. 2001; Yamamoto et al. 2006; Tamura et al. 2007) and water tank experiments (e.g., Takayama et al. 1977; Nakaza et al. 1990, 1994; Takahashi et al. 2011). For example, Yamamoto et al. (2006) observed wave characteristics around Shiraho reef on Ishigaki Island before and after Typhoon 0509 (Matsa). Their data showed that wave directions in the shallow lagoon were controlled by the three-dimensional reef topography (i.e., reef crest and channel) throughout the observation period. Nakaza and Hino (1988) observed an abrupt increase in water level on the shallow reef during Typhoon 8613 (Vera), and they referred to this phenomenon as the "bore-like surf beat" or "Tsunami-like surf beat." This phenomenon has also been described using water tank experiments by Nakaza et al. (1990, 1994). Moreover, there is a close relationship between wave height in the outer ocean and water level in the shallow lagoon; for example, if the wave heights in the outer ocean range from about 3 m to about 10 m, the water level in the shallow lagoon will increase by about 0.5 m to about 2 m (Nakaza et al. 1994).

Wave simulations are one of the key to understand the hydrodynamic regime on coral reefs during TCs. In Japan, for example, Tamura et al. (2007) indicated that a new three-dimensional simulation model developed using SDS-Q3D showed good agreement with the observed data from Shiraho reef on Ishigaki Island. For example, the simulation indicated that the large channel penetrating deep into the reef attracts the tide-averaged mean flow, even from distant areas of the reef. The model was improved as follows: (1) the SDS-Q3D model was reformulated by modifying the vertical integration in the Galerkin procedure to allow for tidal variation and wave setup, (2) a high-resolution bathymetry map obtained from the Ikonos satellite imagery was used, (3) a moving boundary scheme was applied to properly express the exposure of the reef crest at low tide, and (4) an energy balance equation with a wave-breaking model was introduced to evaluate the transformation of directional irregular waves in and near the reef under the combined effects of wave shoaling, refraction, and breaking. In addition, the CADMAS-SURF (Super Roller Flume for Computer Aided Design of Maritime Structure) simulation model is a specialized numerical wave tank model that was released for open use in 2001 (Coastal Development Institute of Technology 2001). The model can be used to assess the threshold of destruction for structures (e.g., sea walls) and contributes to coastal management because it calculates wave velocity, wave acceleration, and wave form (enabling the user to analyze wave height, reflected waves, and wave overtopping). Detailed reef structures (e.g., a reef crest and patch reefs) are also set in the model (Fig. 11.3). Moreover, the software package that runs the model is easily installed on a personal computer. The model has been successfully applied to coral reefs (e.g., Hongo et al. 2012; Watanabe et al. 2016). For example, upper reef slopes and reef crests play a role in the reduction of wave energy during TCs (Fig. 11.4). The breaking points are controlled by the relationship between the wave height and water depth. As a wave breaks on the upper reef slope to reef crest, the rapid decrease in wave velocity in the shallow lagoon is commensurate with the energy loss associated with the wave breaking. In fact, on Ibaruma reef (Ishigaki Island), the CADMAS-SURF model forecast that wave velocity changed from 2.7–3.4 m/s at the reef crest to 0 m/s at the bottom of the shallow lagoon during a simulated TC ($Hs$ = 10 m and $Ts$ = 15 s; Hongo et al. 2012). Recently, the CADMAS-SURF/3D simulation model was released by Coastal Development Institute of Technology (Coastal Development Institute of Technology 2010). The model will be applied to coral reefs.

## 11.4 The Hydrodynamic Impact of TCs on Coral Reefs in Japan: Modern and Paleorecords

TCs can have both depositional and erosional effects on coral reefs (Stoddart 1971; Maragos et al. 1973; Scoffin 1993). From approximately 10,000 to 6000 years ago, coral reefs in the Ryukyu Islands were in catch-up mode as the sea level rose rapidly. Initial reef facies consisted of in situ massive, branching, tabular, and corymbose corals (Kayanne et al. 2004; Kan 2011), and these reefs developed gradually by expanding in response to rising sea levels. In Ryukyu Islands, from 6000 to 4000 years ago, the upward coral growth stopped as the sea level stabilized. The reefs then expanded laterally to form reef crests and corals were transported landward to the lagoons. For example, on Kudaka Island, one of the Ryukyu Islands, reef boulders (long axis >1 m) were transported from reef crest to shallow lagoon at a rate of 8.78 $m^3$/year by TCs (Onda 1999). Onda (1999) estimated that boulders deposited by TCs contributed to vertical accumulation at a rate of at least 6.48 mm/year in the shallow lagoon. The global average rate of vertical

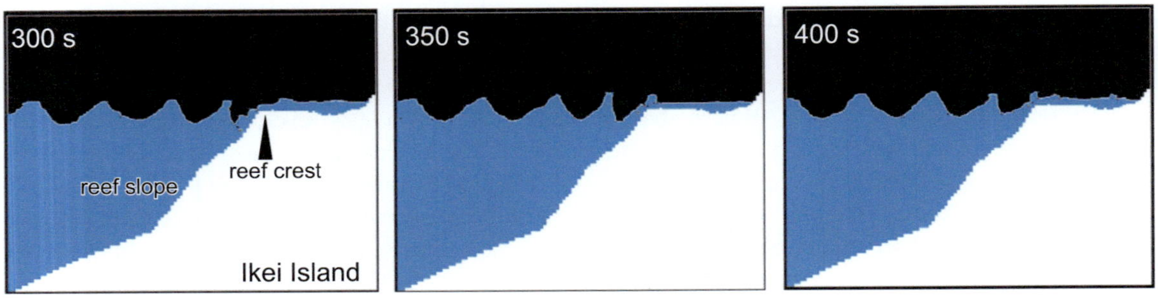


**Fig. 11.3** Snapshots of water level at Ikei Island calculated using the CADMAS-SURF software. This figure showed that a model TC (incident wave height = 7.1 m and wave period = 13 s) strikes the reef

**Fig. 11.4** Schematic diagram of a fringing reef on the Ryukyu Islands. As a wave breaks on the upper reef slope to reef crest, the decrease in wave height in the shallow lagoon behind the reef is commensurate with the energy loss associated with the wave breaking

accumulation of reef flats under high-energy conditions is 10–13 mm/year, and this is presumably controlled by TCs (Montaggioni and Braithwaite 2009). In addition, on Tonaki Island, one of the Ryukyu Islands, paleo-grooves were filled with transported tabular *Acropora* due to the action of TCs and the lagoons gradually became shallow and flat (Kan et al. 1997). Hatenohama cays on Kume Island, Okinawa, are an unvegetated cay composed mainly of sand and gravel (Hasegawa 1990). Typhoon 8613 (Vera) approached Kume Island on August 26,1986, and had a minimum central pressure of 925 hPa and maximum wind speed of 46 m/s (Digital Typhoon, http://agora.ex.nii.ac.jp/digital-typhoon/; accessed September 13, 2015). A thickness of approximately 2 m of sediment was eroded from the cays by the TC waves (Hasegawa 1990).

Sediment-trace surveys during TCs have revealed the relationship between the movement of sediments and wave conditions on coral reefs in the Ryukyu Islands (Table 11.2). Typhoons 9609 (Herb) and 9612 (Kirk) approached Kudaka Island in Okinawa on July 30–August 01 and August 12, 1996, respectively. Their minimum central pressures were 925 and 955 hPa, respectively, and maximum wind speeds were 49 and 39 m/s, respectively (Digital Typhoon, http://agora.ex.nii.ac.jp/digital-typhoon/; accessed September 13 2015). The *Hs* and the *Ts* values were 7.39 m and 11.5 s, respectively, for Typhoon 9609 and 6.81 m and 9.9 s, respectively, for Typhoon 9612, as recorded at the Nakagusuku Bay station of NOWPHAS (Nagai et al. 1998), 12 km off the Kudaka Island. Onda (1999) surveyed the movement of marked boulders (long axis >1 m) on Kudaka Island between 1994 and 1996. The boulders were 0.13–3.5 $m^3$ in volume and 0.25–7.0 tons in weight, with a density of 2,010 $kg/m^3$ (Goto et al. 2009), and were transported from the shallow lagoon to the beach by Typhoon 9609 and from the grooves to the reef crest by Typhoon 9612. The horizontal movement of the boulders recorded was a distance of over 50 m during both typhoons, but no boulders were moved by other typhoons in 1995 (e.g., Typhoons 9503 (Faye), 9507 (Janis), and 9514 (Ryan); Onda 1999), and the maximum *Hs* was 5.32 m during Typhoon 9503 (Nagai et al. 1997). Therefore, Onda (1999) concluded that an *Hs* of at least 6 m is necessary to transport these boulders (<7.0 tons in weight) from the grooves to the reef crest and to move them on the reef crest. Moreover, a boulder weighing 54 tons was emplaced on the reef crest from the spurs and grooves on this island between 1993 and 2005 (Goto et al. 2009). At the period of 1993–2005, the maximum height and period of *Hs* and *Ts* during Typhoon 0423 (Tokage) in 2004 were 11.93 m and 13.6 s, respectively, recorded at the Nakagusuku Bay station of NOWPHAS (Nagai and Satomi 2006). Another example is a boulder of 94 tons in weight that was horizontally moved 2.8 m by Typhoon 9021 (Hattie) on a cliff top (15 m high) at

**Table 11.2** Relationship between movement of sediments and wave conditions on coral reefs in the Ryukyu Islands

| Site | *Hs* (m) | *Ts* (s) | Wave observation site | TC no | Sediment size | Movement | Distance of movement | Cited reference |
|---|---|---|---|---|---|---|---|---|
| Kudaka Island, Okinawa | 7.39 | 11.5 | Nakagusuku Bay of NOWPHAS | 9609 | 0.25–7.0 tons | From shallow lagoon to beach | Over 50 m (horizontal) | Onda (1999) |
| Kudaka Island, Okinawa | 6.81 | 9.9 | Nakagusuku Bay of NOWPHAS | 9612 | 0.25–7.0 tons | From reef slope to reef crest | Over 50 m (horizontal) | Onda (1999) |
| Kudaka Island, Okinawa | 11.9 | 13.6 | Nakagusuku Bay of NOWPHAS | 0423 | 54 tons | From spurs and grooves to reef crest | – | Goto et al. (2009) |
| Zanpa Cape, Okinawa Island, Okinawa | 9.24 | 14.1 | Naha | 9021 | 94 tons | On criff | 2.8 m (horizontal) | Kato et al. (1991) |
| Chinen reef, Okinawa Island, Okinawa | 10[a] | – | Off south Okinawa Island | 5115 | 100 tons | From reef slope to reef crest | 70 m (horizontal) | Kawana (2008) and Goto et al. (2011) |
| Kume Island, Okinawa | 6.48 | 12.5 | Nakagusuku Bay of NOWPHAS | 9307 | 4.1 kg | Reef slope at depths of 5 m | 2 m | Kan (1994) |
| Kume Island, Okinawa | 6.48 | 12.5 | Nakagusuku Bay of NOWPHAS | 9307 | 1.05 kg | Reef slope at depths of 17 m | – | Kan (1994) |
| Kume Island, Okinawa | 6.48 | 12.5 | Nakagusuku Bay of NOWPHAS | 9307 | Sand | Reef slope at depths of 26 m | 1.8 m | Kan (1994) |
| Shiraho reef, Ishigaki Island, Okinawa | 10 | 13 | Coastal grid point "O" at JMA | 0416 and/or 0417 | 47 tons | Reef crest | Ca. 50 m | Goto et al (2010b) |
| Shiraho reef, Ishigaki Island, Okinawa | 11 | 13 | Coastal grid point "O" at JMA | 0613 | | | | |

[a]The value is the maximum wave height

Zanpa Cape on Okinawa Island (Kato et al. 1991). Typhoon Hattie had an *Hs* of 9.24 m and a *Ts* of 14.1 s, which were recorded at the Naha station of NOWPHAS (Nagai et al. 1992), 22 km off the study site. On October 14, 1951, southern Okinawa Island was approached by Typhoon 5115 (Ruth) during high tide and experienced wave heights that exceeded 10 m (Kawana 2008). One coralline boulder (5.6 m long axis, 4.8 m short axis, 2.9 m height, and 100 tons in weight, known locally as "Yuisa-ishi") was detached from the reef slope (Kawana 2008) and redeposited on the reef crest up to 70 m from the reef edge (Goto et al. 2011). Furthermore, on August 8–9, 1993, Typhoon 9307 (Robyn) approached the Ryukyu Islands. Kan (1994) surveyed the movement of marked clasts (0.15–4.1 kg in weight) at the reef edge and reef slope on Kume Island. *Hs* and *Ts* values were recorded at the NOWPHAS Nakagusuku Bay station (125 km from Kume Island) of 6.48 m and 12.5 s, respectively, during the typhoon (Nagai et al. 1995). The largest clast (4.1 kg) was moved 2 m in a water depth of 5 m. Movement extended to a depth of 17 m for clasts (1.05 kg) and to a depth of 26 m for sand. Many boulders were also deposited on the reef crest of Shiraho reef on Ishigaki Island (Goto et al. 2010b). The largest boulder weighed 47 tons and was displaced approximately 50 m landward between 1977 and 2007, based on an analysis of aerial photographs taken in 1977 and 2007 (Goto et al. 2010b). Many TCs approached the island over the period 1977–2007, and the boulder was probably moved by waves during some of these typhoons. For example, *Hs* and the *Ts* values were 10 m and 13 s, respectively, for Typhoons 0416 (Chaba) and/or 0417 (Aere), which were recorded at coastal grid point "O" of the JMA (JMA wave analysis chart: 240000TC AUG 2004), 36 km off the study site. For Typhoon 0613 (Shanshan), the *Hs* and the *Ts* values were 11 m and 13 s, respectively, also recorded at point "O" (JMA, http://www.data.jma.go.jp/gmd/kaiyou/data/db/wave/chart/daily/pdf/jp/06/09/06091600jp.pdf; accessed December 1, 2015).

A number of studies have reported on the impact of TCs on coral communities in Japan. For example, the impact of TCs on the corals is one of the major disturbance factors on Kushimoto reef in the southernmost part of Honsyu (Nomura 2009). A coral monitoring survey at Kushimoto reef at a period of 2002–2008 showed a major decline in coral cover of *Acropora muricata*, the dominant species on the reef, from 60 % to 27 % in October 2004 after Typhoon 0423 (Tokage; Nomura 2009). The *Hs* and *Ts* values associated with Tokage were 10.22 m and 15.7 s, respectively, as recorded at the Shionomisaki station of NOWPHAS (Nagai and Satomi 2006), which is within 10 km of Kushimoto reef. Another significant example is the impact of typhoons 1102 (Songda) and 1109 (Muifa) on the mesophotic reef at Kume Island in Okinawa on May 28, 2011, and August 5–6, 2011, respectively (Fujita et al. 2012). At Nanhanari on Kume Island, arborescent *Acropora* communities dominated by *Acropora horrida* at depths of 25–35 m decreased from >90 % to between 20 % and 30 % as a consequence of the damage caused by these TCs (breakage, death, and covered by algae/cyanobacteria; Fujita et al. 2012).The *Hs* and *Ts* values were 10.22 m and 11.3 s, respectively, for Typhoon 1102 and 4.64 m and 10.0 s, respectively, for Typhoon 1109, as recorded at the Naha station of NOWPHAS (Kawaguchi et al. 2013), 83 km off the study site. In 2012, a mesophotic reef located off Okinawa Island was damaged by Typhoon 1217 (Jelawat; White et al. 2013). Large foliose *Pachyseris foliosa* at depths of 32–42 m was apparently less resistant to the TC waves than the shallower high-diversity coral communities at depths of 25–32 m, because large foliose corals are just as susceptible to TC waves as shallower branching corals (White et al. 2013).The associated *Hs* and *Ts* values were 7.1 m and 13 s, respectively, based on wave analysis data from coastal grid point "P" (JMA, http://www.data.jma.go.jp/gmd/kaiyou/data/db/wave/chart/daily/pdf/jp/12/09/12092900jp.pdf; accessed September 28, 2015), 39 km off the study site.

The potential future effects of TCs on coral reefs under conditions of global warming have been simulated using engineering and fluid dynamics models, e.g., Hongo et al. (2012) predicted mechanical damage to massive and tabular corals subject to the effects of super-extreme TCs using the CADMAS-SURF. Their results were consistent with the observed distributions of corals on Ibaruma reef, Ishigaki Island, which is prone to frequent present-day strikes by moderate to extreme TCs (wave height = 10 m, wave period = 10–15 s). For example, massive corals are found in the shallow lagoon on the reef, whereas tabular corals are distributed on the reef crest and upper reef slope. This study indicates that if a super-extreme TC (wave height = 20 m and wave period = 20 s) strikes the reef in the near future, massive corals will survive on the landward side of the shallow lagoon, but most tabular corals will be destroyed and removed from the reef.

Reconstructing the past hydrodynamic impact of TCs on coral reefs has been attempted in the Ryukyu Islands. Many boulders of coral and reef blocks have been deposited on coral reefs and shorelines around the world (Goto et al. 2010a; Richmond et al. 2011), and these deposits may be useful markers of past TCs and/or tsunamis. However, at present there is probably no way to distinguish between boulders deposited by TC-generated waves and those deposited by tsunamis (Nott 2003; Paris et al. 2011; Nakamura et al. 2014). The Ryukyu Islands are a suitable location for research into the various influences of TC waves and tsunamis on the boulders deposited around reefs (Goto et al. 2010b), thus; many numerical models, sedimentological surveys, and historical studies have been carried out on the islands (e.g., Kato and Kimura 1983; Imamura et al. 2008; Goto et al. 2010b). On Kudaka Island, many boulders were deposited at the reef crest within 275 m of the reef edge, and the size distribution of the boulders was characterized by an exponential landward fining trend (Goto et al. 2009). This pattern was interpreted to be controlled by TC waves because the height of TC waves decreases exponentially from the reef edge to the reef flat (Egashira et al. 1985). The same distribution pattern of boulders was observed on Okinawa and Tsuken Islands (Goto et al. 2011). This indicates that the transport limit of boulders by TC waves is approximately 300 m from the reef edge (Goto et al. 2011). In contrast, a bimodal distribution of boulders was observed on Ishigaki Island: boulders within about 300 m of the reef edge (group 1; Fig. 11.5) and around the shoreline (group 2; Fig. 11.5; Goto et al. 2010b). The boulders in group 2 were significantly heavier than those in group 1 (Goto et al. 2010b). The distribution and size of the boulders in group 2 cannot be explained by the action of TC-generated waves, which suggest a tsunami origin (Goto et al. 2010b). The Meiwa Tsunami is a well-documented event that affected Ishigaki Island in 1771 (Goto et al. 2010a; Araoka et al. 2013). Recently, Watanabe et al. (2016) calculated the past maximum wave velocity at the reef crest and the shoreline on Ishigaki Island, for both TCs and the Meiwa Tsunami, using the data relating to the distribution of the boulders. The maximum velocity decreased from approximately 10 m/s at the reef crest to approximately 4 m/s at the shoreline during an assumed TC (10 m wave height and 15 s wave period; Watanabe et al. 2016). In contrast, the maximum velocity associated with the Meiwa Tsunami was 13.8–15.7 m/s at the shoreline (Watanabe et al. 2016).

## 11.5 Future Perspectives

The relationship between the threshold of coral destruction (e.g., breaking branched coral, overturning massive coral, and dislodging tabular coral) and the hydrodynamic force of TCs remains unknown. In Japan, wave height and wave period data for coral reefs during TCs, based on field observations and projections by the government, universities, and research institutes, are accumulating. However, data related to wave velocity, wave-induced acceleration, and wave force

**Fig. 11.5** (**a**) Deposited reef boulders (long axis >1 m) by TC waves on the reef crest on Shiraho reef on Ishigaki Island. (**b**) Deposited reef boulder (*Porites* sp., long axis >1 m) by a tsunami on the beach at Ibaruma reef on Ishigaki Island

on reefs is limited or nonexistent. For example, tabular corals can be dislodged when the wave force exceeds the bending stress (0.2 MPa) of the attachment substratum on the Great Barrier Reef (Madin 2005; Madin and Connolly 2006). However, this threshold probably varies among species, location, and substratum conditions. The hydrodynamic impact of TCs on branching and foliaceous corals is still unknown because these corals are characterized by a complicated growth form. Moreover, ocean acidification has probably weakened the skeletons and substratum (Madin et al. 2012). To evaluate the impact of hydrodynamic forces on coral destruction, the development of in situ observation systems on reefs during TCs should be considered for a modeled coral reef in Japan. A small wave pressure device for field observation has recently been developed in Japan (Nakaza et al. 2008). It has a pressure sensor, three-directional acceleration sensors, and a self-recording system. If we wish to develop a better understanding of the relationship between the threshold of coral destruction and hydrodynamic forces associated with TCs, more water tank experiments should be conducted. This knowledge will be essential if we are to protect reef ecosystems as they come under attack from intensified TCs in the near future.

To accurately project the hydrodynamic impacts of TCs on coral reefs and coral communities in the future, the reconstruction of paleo-hydrodynamic records should be attempted for various reefs in Japan. The Geospatial Information Authority of Japan began tidal observations in the 1890s, but *Hs* and *Ts* values have been recorded since the1970s by the JMA and MLIT. Boulders deposited on reefs reflect the occurrence of past large wave events associated with TCs and/or tsunamis (Kato and Kimura 1983; Imamura et al. 2008). In particular, the weight of boulders and distance from their original position are key parameters that can be used to reconstruct past hydrodynamic conditions (Goto et al. 2009, 2010b). Recently, Watanabe et al. (2016) developed a new numerical model based on the hydraulic force, friction force, and gravitational force of boulders on the Ibaruma reef off Ishigaki Island. Their results show that the wave height and wave period that fit the observed distribution of boulders on the reef well were a maximum of 8 m and 10 s, respectively, assuming a mean tidal level. Many reef boulders have been deposited on coral reefs around Japan (e.g., Kato et al. 1991; Onda 1999; Goto et al. 2009, 2011, 2013); consequently, further research into the recurrence interval and magnitude of past TCs using such boulders should be considered and the data used to forecast the potential impact of TCs on coral reefs in the future.

More intense TCs will generate increasing wind speeds, wave heights, and wave periods, and they will inevitably devastate coastal areas around the world. Coral reefs are well known to act as natural breakwaters (Ferrario et al. 2014). However, destructive tsunami-like waves are generated by intense TCs and they caused extensive damage and casualties (Roeber and Bricker 2015). Therefore, we must recognize the need for research into the adaptation to the impact of TCs and disaster risk reduction in coastal areas. In addition to the reduction in wave energy, these natural breakwaters contribute to the protection from coastal erosion and to a reduction of inundation height. However environmental factors, such as elevated SST, ocean acidification, outbreaks of crown-of-thorns sea stars, terrestrial runoff, and coastal development, have led to the recent decrease in the abundance and diversity of coral species. These negative impacts gradually cause a decline in the ability of a reef to act as a natural breakwater and so will cause an increase in the risks posed by TCs to coral reefs. However, the natural breakwater role of coral reefs in the present-day and the near future is poorly understood. Consequently, a policy marker in response to the disaster is also poorly understood. The following questions remain to be addressed: (1) How much will wave energy, wave velocity, and inundation height change on coral reefs affected by more intense TCs in the near future? (2) Which species of coral will contribute to the natural breakwater effect in the near future? (3) Is there an appropriate technique to recover and maintain the natural breakwater role of damaged coral reefs? (4) What is the impact on society and how can our scientific knowledge be used in future policy-making? It will be difficult to answer these questions based on evidence from only a single scientific field; thus, a multidisciplinary approach that combines earth science, ecology, biology, coastal engineering, and natural disaster science should be considered.

**Acknowledgments** The author thanks Lucien F. Montaggioni and Kazuhisa Goto for their comments, which significantly improved the clarity of an earlier version of the manuscript. This research was supported by Nippon Life Insurance Foundation and by JST (Japan Science and Technology Agency)/JICA (Japan International Cooperation Agency)-SATREPS (Science and Technology Research Partnership for Sustainable Development).

## References

Araoka D, Yokoyama Y, Suzuki A, Goto K, Miyagi K, Miyazawa K, Matsuzaki H, Kawahata H (2013) Tsunami recurrence revealed by *Porites* coral boulders in the southern Ryukyu Islands, Japan. Geology 41:919–922

Balaguru K, Chang P, Saravanan R, Leung LR, Xu Z, Li M, Hsieh JS (2012) Ocean barrier layers' effect on tropical cyclone intensification. Proc Natl Acad Sci U S A 109:14343–14347

Coastal Development Institute of Technology (2001) Research and development of numerical wave flume: CADMAS-SURF. Coastal Development Institute of Technology, Tokyo. (in Japanese)

Coastal Development Institute of Technology (2010) Development and research for super roller flume for computer aided design of maritime structure 3D. Coastal Development Institute of Technology, Tokyo. (in Japanese)

Cooper TF, De'ath G, Fabricius KE, Lough JM (2008) Declining coral calcification in massive *Porites* in two nearshore regions of the northern Great Barrier Reef. Glob Chang Biol 14:529–538

De'ath G, Fabricius KE, Sweatman H, Puotinen M (2012) The 27–year decline of coral cover on the Great Barrier Reef and its causes. Proc Natl Acad Sci U S A 109:17995–17999

Done TJ (1992) Effects of tropical cyclone waves on ecological and geomorphological structures on the Great Barrier Reef. Cont Shelf Res 12:859–872

Done TJ (2011) Tropical cyclone/hurricane. In: Hopley D (ed) Encyclopedia of modern coral reefs. Springer, Dordrecht, pp 1092–1096

Dunn JG, Sammarco PW, LaFleur G (2012) Effects of phosphate on growth and skeletal density in the scleractinian coral *Acropora muricata*: a controlled experimental approach. J Exp Mar Biol Ecol 411:34–44

Egashira K, Fukuda I, Kishira Y, Nishimura T (1985) Field measurement of the wave deformation on the reef. Proc Coast Eng, JSCE 32:90–94. (in Japanese)

Fabricius KE (2005) Effects of terrestrial runoff on the ecology of corals and coral reefs: review and synthesis. Mar Pollut Bull 50:125–146

Fabricius KE, De'ath G, Puotinen ML, Done T, Cooper TF, Burgess SC (2008) Disturbance gradients on inshore and offshore coral reefs caused by a severe tropical cyclone. Limnol Oceanogr 53:690–704

Ferrario F, Beck MW, Storlazzi CD, Micheli F, Shepard CC, Airoldi L (2014) The effectiveness of coral reefs for coastal hazard risk reduction and adaptation. Nat Commun 5:3794

Fujita Y, Kimura T, Shioiri A, Naruse T (2012) Typhoon damage of large-scaled coral communities dominated by *Acropora horrida* (Dana 1846) (Scleractinia: Acroporidae) in the mesophotic zone off Kumejima Island, the Ryukyu Islands, Japan. Biol Mag Okinawa 50:61–66. (in Japanese)

Goto K, Okada K, Imamura F (2009) Characteristics and hydrodynamics of boulders transported by storm waves at Kudaka Island, Japan. Mar Geol 262:14–24

Goto K, Kawana T, Imamura F (2010a) Historical and geological evidence of boulders deposited by tsunamis, southern Ryukyu Islands, Japan. Earth-Sci Rev 102:77–99

Goto K, Miyagi K, Kawamata H, Imamura F (2010b) Discrimination of boulders deposited by tsunamis and storm waves at Ishigaki Island, Japan. Mar Geol 269:34–45

Goto K, Miyagi K, Kawana T, Takahashi J, Imamura F (2011) Emplacement and movement of boulders by known storm waves—field evidence from the Okinawa Islands, Japan. Mar Geol 283:66–78

Goto K, Miyagi K, Imamura F (2013) Localized tsunamigenic earthquakes inferred from preferential distribution of coastal boulders on the Ryukyu Islands, Japan. Geology 41:1139–1142

Hasegawa H (1990) Shoreline changes of the Hatenohama coral sand cays at the eastern end of Kume Island, the Ryukyus, Japan. Geogr Rev Jpn 63A:676–692. (in Japanese)

Hearn CJ (2011) Hydrodynamics of coral reef systems. In: Hopley D (ed) Encyclopedia of modern coral reefs. Springer, Dordrecht, pp 563–573

Hongo C, Kawamata H, Goto K (2012) Catastrophic impact of typhoon waves on coral communities in the Ryukyu Islands under global warming. J Geophys Res 117:G02029

Imamura F, Goto K, Ohkubo S (2008) A numerical model for the transport of a boulder by tsunami. J Geophys Res 113:C01008

Kan H (1994) Typhoon effects on sediment movement on reef edges and reef slopes. In: Bellwood O, Choat H, Saxena N (eds) Recent advances in marine science and technology '94. Pacon International, James Cook University of North Queensland, Queensland, pp 191–201

Kan H (2011) Ryukyu Islands. In: Hopley D (ed) Encyclopedia of modern coral reefs. Springer, Dordrecht, pp 940–945

Kan H, Hori N, Kawana T, Kaigara T, Ichikawa K (1997) The evolution of a Holocene fringing reef and island: reefal environmental sequence and sea level change in Tonaki Island, the Central Ryukyus. Atoll Res Bull 443:1–20

Kato Y, Kimura M (1983) Age and origin of so-called "Tsunami-ishi", Ishigaki Island, Okinawa Prefecture. J Geol Soc Jpn 89:471–474. (in Japanese)

Kato Y, Akamine N, O'hori K, Tamaki T, Tamura K, Matsuo K (1991) Movement of limestone-blocks by wind waves-an example by Typhoon No. 21, 1990, at Zanpa Cape, Okinawa Island, southwestern Japan. Bull Coll Sci Univ Ryukyus 51:19–33. (in Japanese)

Kawaguchi K, Inomata T, Seki K (2013) Annual report on Nationwide Ocean Wave information network for Ports and HArbourS (NOWPHAS 2011). Technical note of the Port and Harbour Research Institute, Ministry of Transport, Japan, Yokosuka (in Japanese)

Kawaguchi K, Inomata T, Seki K, Fujiki T (2015) Annual report on Nationwide Ocean Wave information network for Ports and HArbourS (NOWPHAS 2013). Technical note of the Port and Harbour Research Institute, Ministry of Transport, Japan, Yokosuka (in Japanese)

Kawana T (2008) Yuisa-Ishi on the reef crest off Chinen, Nanjo City, Okinawa Island: coralline block transported by the Ruth Typhoon in 1951. Bull Coll Educ, Univ Ryukyus 72:135–139. (in Japanese)

Kayanne H, Hongo C, Yamano H (2004) Coral reef landforms in Japan. In: The Japanese Coral Reef Society, Ministry of the Environment (ed) Coral reefs of Japan. Ministry of the Environment, Tokyo, pp 14–19

Knutson TR, McBride JL, Chan J, Emanuel K, Holland G, Landsea C, Held I, Kossin JP, Srivastava AK, Sugi M (2010) Tropical cyclones and climate change. Nat Geosci 3:157–163

Kossin JP, Emanuel KA, Vecchi GA (2014) The poleward migration of the location of tropical cyclone maximum intensity. Nature 509:349–352

Kuroyama J, Toyota T, Tsutsui H, Nakashima T (2000) Study of environmental factors around a reef slope and the effect of a typhoon in the Okinawa Sekisei Lagoon. Rep Jpn Mar Sci Technol Cent (Biol) 40:47–54. (in Japanese)

Madin JS (2005) Mechanical limitations of reef corals during hydrodynamics disturbances. Coral Reefs 24:630–635

Madin JS, Connolly SR (2006) Ecological consequences of major hydrodynamic disturbances on coral reefs. Nature 444:477–480

Madin JS, Hughes TP, Connolly SR (2012) Calcification, storm damage and population resilience of tabular corals under climate change. PLoS One 7:e46637

Manzello DP, Brandt M, Smith TB, Lirman D, Hendee JC, Nemeth RS (2007) Hurricanes benefit bleached corals. Proc Natl Acad Sci U S A 104:12035–12039

Manzello DP, Kleypas JA, Budd DA, Eakin CM, Glynn PW, Langdon C (2008) Poorly cemented coral reefs of the eastern tropical Pacific: possible insights into reef development in a high-$CO_2$ world. Proc Natl Acad Sci U S A 105:10450–10455

Maragos JE, Baines GBK, Beveridge PJ (1973) Tropical Cyclone Bebe creates a new land formation on Funafuti Atoll. Science 181:1161–1164

Montaggioni LF, Braithwaite CJR (2009) Quaternary coral reef systems. Elsevier, Oxford

Nadaoka K, Nihei Y, Kumano R, Yokobori T, Omija T, Wakaki K (2001) A field observation on hydrodynamic and thermal environments of a fringing reef at Ishigaki Island under typhoon and normal atmospheric conditions. Coral Reefs 20:387–398

Nagai T, Satomi S (2006) Annual report on Nationwide Ocean Wave information network for Ports and HArbourS (NOWPHAS 2004). Technical note of the Port and Harbour Research Institute, Ministry of Transport, Japan, Yokosuka (in Japanese)

Nagai T, Sugahara K, Hashimoto N, Asai T, Hirano T (1992) Annual report on Nationwide Ocean Wave information network for Ports and HArbourS (NOWPHAS 1990). Technical note of the Port and Harbour Research Institute, Ministry of Transport, Japan, Yokosuka (in Japanese)

Nagai T, Sugahara K, Shimizu K, Hashimoto N (1995) Annual report on Nationwide Ocean Wave information network for Ports and HArbourS (NOWPHAS 1993). Technical note of the Port and Harbour Research Institute, Ministry of Transport, Japan, Yokosuka (in Japanese)

Nagai T, Sugahara K, Watanabe H, Kawaguchi K (1997) Annual report on Nationwide Ocean Wave information network for Ports and HArbourS (NOWPHAS 1995). Technical note of the Port and Harbour Research Institute, Ministry of Transport, Japan, Yokosuka (in Japanese)

Nagai T, Watanabe H, Sugahara K, Kawaguchi K (1998) Annual report on Nationwide Ocean Wave information network for Ports and HArbourS (NOWPHAS 1996). Technical note of the Port and Harbour Research Institute, Ministry of Transport, Japan, Yokosuka (in Japanese)

Nakamura M, Arashiro Y, Shiga S (2014) Numerical simulations to account for boulder movements on Lanyu Island, Taiwan: tsunami or storm? Earth Planet Space 66:1–20

Nakaza E, Hino M (1988) Coastal disasters due to Bore-like surf beat on reef coasts. Proc Jpn Conf Coast Eng 35:202–206. (in Japanese)

Nakaza E, Tsukayama S, Takara N (1990) Waves, wave set-up and surf beat on reef coasts. Bull Fac Eng, Univ Ryukyus 40:37–48. (in Japanese)

Nakaza E, Tsukayama S, Tanaka S (1994) Waves and surf beats induced by wave group on coral reef flat. Proc Coast Eng, JSCE 41:86–90. (in Japanese)

Nakaza E, Ohtani A, Kawakami K, Kaneshi T, Takeuchi T, Takeuchi R (2008) Development of a pressure sensor for field measurements and its application to a breakwater on coral reef Proc Civ Eng Ocean 24:963–967. in Japanese

Nomura K (2009) Recent changes in coral communities in Kushimoto, the southernmost part of Honshu, Japan. J Jpn Coral Reef Soc 11:39–49. (in Japanese)

Nott J (2003) Waves, coastal boulder deposits and the importance of the pre-transport setting. Earth Planet Sci Lett 210:269–276

Onda M (1999) The distribution of large reef blocks and the effect of germorphic development in Kudaka Island, the Ryukyu Islands, Japan. Geogr Rev Jpn 72A:746–762. (in Japanese)

Oouchi K, Yoshimura J, Yoshimura H, Mizuta R, Kusunoki S, Noda A (2006) Tropical cyclone climatology in a global-warming climate as simulated in a 20 km-mesh global atmospheric model: frequency and wind intensity analyses. J Meteo Soc Jpn 84:259–276

Paris R, Naylor LA, Stephenson WJ (2011) Boulders as a signature of storms on rock coasts. Mar Geol 283:1–11

Pielke RA, Pielke RA (1997) Hurricanes: their nature and impacts on society. Wiley, Chichester

Richmond BM, Watt S, Buckley M, Jaffe BE, Gelfenbaum G, Morton RA (2011) Recent storm and tsunami coarse-clast deposit characteristics, southeast Hawai'i. Mar Geol 283:79–89

Roeber V, Bricker JD (2015) Destructive tsunami-like wave generated by surf beat over a coral reef during Typhoon Haiyan. Nat Commun 6:7854

Scoffin TP (1993) The geological effects of hurricanes on coral reefs and the interpretation of storm deposits. Coral Reefs 12:203–221

Simpson RH, Riehl H (1981) The hurricane and its impact. Louisiana State Univ Press, Baton Rouge

Sowa K, Watanabe T, Kan H, Yamano H (2014) Influence of land development on Holocene *Porites* coral calcification at Nagura Bay, Ishigaki Island, Japan. PLoS One 9:e88790

Stoddart DR (1971) Coral reefs and islands and catastrophic storms In: Steers JA (ed) Applied coastal geomorphology. MIT press Cambridge, MA, pp 155–197

Takahashi K, Katayama H, Sekimoto T, Aoki K, Kayanne H, Isobe M (2011) Influence of waves, currents and gravel supplies on the formation of coral cays. J Jpn Soc Civ Eng, Ser B2 (Coast Eng) 67:I_636–I_640. (in Japanese)

Takayama T, Kamiyama Y, Kikuchi O (1977) Wave transformation on a reef. Technical note of the Port and Harbour Research Institute, Ministry of Transport, Japan. (in Japanese)

Tamura H, Nadaoka K, Paringit EC (2007) Hydrodynamic characteristics of a fringing coral reef on the east coast of Ishigaki Island, southwest Japan. Coral Reefs 26:17–34

Tsuboki K, Yoshioka MK, Shinoda T, Kato M, Kanada S, Kitoh A (2015) Future increase of supertyphoon intensity associated with climate change. Geophys Res Lett 42:646–652

Watanabe M, Goto K, Imamura F, Hongo C (2016) Numerical identification of tsunami boulders and estimation of local tsunami size at Ibaruma reef of Ishigaki Island, Japan. Island Arc 25:316–332

White KN, Ohara T, Fujii T, Kawamura I, Mizuyama M, Montenegro J, Shikiba H, Narure T, McClelland TY, Denis V, Reimer JD (2013) Typhoon damage on a shallow mesophotic reef in Okinawa, Japan. PeerJ 1:e151

Yamamoto T, Nadaoka K, Nishimoto T, Blanco AC, Ishimaru T (2006) Dynamic process of current and water quality fluctuations in a coral reef due to episodic atmospheric disturbances. Proc Coast Eng, JSCE 53:1076–1080. (in Japanese)

Yamano H, Hongo C, Sugihara K, Yara Y, Nakao Y, Fujii M (2014) Current status of national coral database in Japan: dataset development, applications, and future directions. In: Nakano S, Yahara T, Nakashizuka T (eds) The biodiversity observation network in the Asia-Pacific region: integrative observations and assessments of Asian biodiversity Springer, Japan, pp 65–81

# Concluding Remarks: Future Perspectives on Coral Reef Studies of Japan – From Biology, Earth Science, and Conservation and Restoration

12

Akira Iguchi and Chuki Hongo

**Abstract**

In this chapter, we discuss the future direction of coral reef studies of Japan to facilitate multidisciplinary approaches for understanding various aspects of coral reefs. In particular, we emphasize that the scientific knowledge of coral reefs of Japan would be applicable to practices for conservation and restoration of coral reef ecosystems in the world.

**Keywords**

Multidisciplinary approach • Global-local impacts • Human activities • Conservation • Restoration

In previous chapters, this book has provided the information on recent important studies in coral reefs of Japan. It is based on various research fields and their significant advances to deepen coral reef science as the results of recent studies of biology, ecology, physiology, ocean chemistry, sedimentology, paleoclimatology, and coastal engineering. It would be difficult to understand coral reefs by focusing on only specific research field. Hence, multidisciplinary insights and approaches will be required to provide a better understanding of coral reefs, to which this book would significantly contribute.

As introduced in some chapters in this book, reef-building coral, which is a most important animal supporting the diversity of coral reef ecosystem, has been well studied from various biological research fields such as ecology, genetics, physiology, and so on. Recent development of high-throughput sequencing technology has been also applied to corals at earlier stage, and decoding of genomic information of corals has been well progressed (e.g., Shinzato et al. 2011). The genomic information has provided various DNA markers for understanding population dynamics and health condition of corals under some environmental conditions, which would contribute to the prediction of coral communities in Japan after disturbances such as coral bleaching, coral disease, and outbreaks of crown-of-thorns starfish (e.g., van Woesik et al. 2011; Harii et al. 2014). Hence, unifying the approaches of biological research fields based on genomic information of corals would be necessary to develop coral reef science further. In parallel, the geological data of coral reef have rapidly increased for the past several decades (Montaggioni and Braithwaite 2009). For example, sedimentological knowledge regarding the growth pattern and internal facies of coral reefs has shown a marked improvement. In this context, knowledge for its controlling factors such as sea level, sea surface temperature, wave condition, and water quality has increased, as the results of field observation and analysis of proxy for paleoclimate. Similarly, knowledge of global-/local-scale future projection for the controlling factors has increased as a result of numerical simulation (Stocker et al. 2013). These findings above imply that future reef growth pattern and its internal facies will be probably projected, but the researches are poorly documented. Generally,

A. Iguchi (✉)
Department of Bioresources Engineering, National Institute of Technology, Okinawa College, Nago, Okinawa, Japan
e-mail: iguchi.a0218@gmail.com

C. Hongo
Department of Chemistry, Biology, and Marine Science, University of the Ryukyus, Nishihara, Okinawa, Japan
e-mail: g123001@sci.u-ryukyu.ac.jp


A. Iguchi, C. Hongo (eds.), *Coral Reef Studies of Japan*, Coral Reefs of the World 13,
https://doi.org/10.1007/978-981-10-6473-9_12

the geological records provide the information on pristine reefs and its surrounding environmental conditions before the human occupation of the reefs. However, present-day coral reefs have been subject to dramatic declines as a consequence of both global and local impacts. In order to predict future coral reefs, it is necessary to combine the knowledge of pristine coral reefs based on analysis of geological records with that of present-day coral reefs based on analysis of biology, ecology, and the related research fields.

We reemphasize that coral reefs around Japan will provide important knowledge on the topics above because the area is located at the northern limit of the coral distribution, composed by mainly fringing reefs along archipelago, and easily impacted by human activities around coastal areas. Actually, a number of coral reef researches from past to present based on biological and geological studies have been well documented in Japan (e.g., Japanese Coral Reef Society and Ministry of the Environment 2004; Kayanne 2016), and a reanalysis of the accumulated data will probably contribute to understanding the future coral reefs in Japan. In order to collect paleoecological data on coral reefs, the Ryukyu Archipelago is one of the key regions in Japan because the region is characterized by a tectonically active situation, and the Pleistocene and Holocene uplift reefs have developed (e.g., Konishi et al. 1970; Koba et al. 1982). This implies that fossil corals and other marine organisms such as calcareous algae and foraminifera are well preserved. Moreover, archeological and anthropological data in Ryukyu Islands since late Pleistocene are recently accumulated (Nakagawa et al. 2010; Takamiya 2012), and the knowledge of the impact of human activities such as fisheries around coral reefs after human settlements on the region has also been accumulated as a result of analysis of anthropologic remains (e.g., Yamaguchi 2016). Consequently, the area is one of the unique regions to understand temporal change in reef condition from past to future. The knowledge regarding coral reefs around this region would be also informative for the studies of conservation (e.g., marine protected areas: MPAs, world heritages) and restoration (e.g., coral transplantation) of coral reef ecosystems in the world, especially Southeast Asia because the coral reefs around this region have been already degraded by human impacts and emergent measures have been intensively searched.

## References


Harii S, Hongo C, Ishihara M, Ide Y, Kayanne H (2014) Impacts of multiple disturbances on coral communities at Ishigaki Island, Okinawa, Japan, during a 15 year survey. Mar Ecol Prog Ser 509:171–3180

Kayanne H (2016) Coral reef science. Springer, Netherlands

Koba M, Nakata T, Takahashi T (1982) Late Holocene eustatic sea-level changes deduced from geomorphological features and their $^{14}$C dates in the Ryukyu Islands, Japan. Palaeogeogr Palaeoclimatol Palaeoecol 39:231–260

Konishi K, Schlanger SO, Omura A (1970) Neotectonic rates in the central Ryukyu Islands derived from $^{230}$Th coral ages. Mar Geol 9:225–240

Montaggioni LF, Braithwaite CJR (2009) Quaternary coral reef systems. Elsevier, Oxford

Nakagawa R, Doi N, Nishioka Y, Nunami S, Yamauchi H, Fujita M, Yamazaki S, Yamamoto M, Katagiri C, Mukai H, Matsuzaki H, Gakuhari T, Takigami M, Yoneda M (2010) Pleistocene human remains from Shiraho-Saonetabaru Cave on Ishigaki Island, Okinawa, Japan, and their radiocarbon dating. Anthropol Sci 118:173–183

Shinzato C, Shoguchi E, Kawashima T, Hamada M, Hisata K et al (2011) Using the *Acropora digitifera* genome to understand coral responses to environmental change. Nature 476:320–323

Stocker TF, Qin D, Plattner G-K, Tignor M, Allen SK, Boschung J, Nauels A, Xia Y, Bex V, Midgley PM (2013) Climate change 2013: the physical science basis. Cambridge University Press, Cambridge/New York

Takamiya H (2012) Prehistoric Ryukyu culture: biogeographic perspective. Quat Res 51:239–245. (in Japanese)

The Japanese Coral Reef Society and Ministry of the Environment (2004) Coral reefs of Japan. Ministry of the Environment, Tokyo

van Woesik R, Sakai K, Ganase A, Loya Y (2011) Revisiting the winners and the losers a decade after coral bleaching. Mar Ecol Prog Ser 434:67–76

Yamaguchi T (2016) A review of coral studies of the Ryukyu Island arc to reconstruct its long-term landscape history. In: Kayanne H (ed) Coral reef science. Springer, Dordrecht, pp 55–63

# Erratum to: Distribution Expansion and Historical Population Outbreak Patterns of Crown-of-Thorns Starfish, *Acanthaster planci* sensu lato, in Japan from 1912 to 2015

Nina Yasuda

**Erratum to:**
**Chapter 9 in: A. Iguchi, C. Hongo (eds.), *Coral Reef Studies of Japan*, Coral Reefs of the World 13, https://doi.org/10.1007/978-981-10-6473-9_9**

The original version of this chapter was inadvertently published with incorrect value of Latitude and Longitude of Okinawa E and Okinawa W region in the table 9.3. The correct values are updated in the current version.

The updated original online version for this chapter can be found at https://doi.org/10.1007/978-981-10-6473-9_9


A. Iguchi, C. Hongo (eds.), *Coral Reef Studies of Japan*, Coral Reefs of the World 13,
https://doi.org/10.1007/978-981-10-6473-9_13

# Index




A. Iguchi, C. Hongo (eds.), *Coral Reef Studies of Japan*, Coral Reefs of the World 13,
https://doi.org/10.1007/978-981-10-6473-9

MIX
Papier aus verantwortungsvollen Quellen
Paper from responsible sources
FSC® C105338

FSC
www.fsc.org

If you have any concerns about our products,
you can contact us on
**ProductSafety@springernature.com**

In case Publisher is established outside the EU,
the EU authorized representative is:
**Springer Nature Customer Service Center GmbH**
**Europaplatz 3, 69115 Heidelberg, Germany**

Printed by Libri Plureos GmbH
in Hamburg, Germany